# PERSONAL KNOWLEDGE

# PERSONAL KNOWLEDGE

Towards a Post-Critical
Philosophy

by

MICHAEL POLANYI

THE UNIVERSITY OF CHICAGO PRESS

THE UNIVERSITY OF CHICAGO PRESS, CHICAGO 60637
Routledge & Kegan Paul Ltd, London EC4V 5EL

81 80 79 78 77      10 9 8 7 6

ISBN: 0-226-67287-5 (clothbound), 0-226-67288-3 (paperback)
Library of Congress Catalog Card Number: 58-5162

MICHAEL POLANYI, Fellow of the Royal Society and member of the American Academy of Arts and Sciences, has been Gifford lecturer at the University of Aberdeen, professor of physical chemistry and of social studies at the University of Manchester, and Fellow of Merton College, Oxford. His published works include *Science, Faith, and Society* (1946), *The Logic of Liberty* (1951), *The Study of Man* (1958), *The Tacit Dimension* (1966), and *Knowing and Being* (1969).

# PREFACE

THIS is primarily an enquiry into the nature and justification of scientific knowledge. But my reconsideration of scientific knowledge leads on to a wide range of questions outside science.

I start by rejecting the ideal of scientific detachment. In the exact sciences, this false ideal is perhaps harmless, for it is in fact disregarded there by scientists. But we shall see that it exercises a destructive influence in biology, psychology and sociology, and falsifies our whole outlook far beyond the domain of science. I want to establish an alternative ideal of knowledge, quite generally.

Hence the wide scope of this book and hence also the coining of the new term I have used for my title: Personal Knowledge. The two words may seem to contradict each other: for true knowledge is deemed impersonal, universally established, objective. But the seeming contradiction is resolved by modifying the conception of knowing.

I have used the findings of Gestalt psychology as my first clues to this conceptual reform. Scientists have run away from the philosophic implications of gestalt; I want to countenance them uncompromisingly. I regard knowing as an active comprehension of the things known, an action that requires skill. Skilful knowing and doing is performed by subordinating a set of particulars, as clues or tools, to the shaping of a skilful achievement, whether practical or theoretical. We may then be said to become 'subsidiarily aware' of these particulars within our 'focal awareness' of the coherent entity that we achieve. Clues and tools are things used as such and not observed in themselves. They are made to function as extensions of our bodily equipment and this involves a certain change of our own being. Acts of comprehension are to this extent irreversible, and also non-critical. For we cannot possess any fixed framework within which the re-shaping of our hitherto fixed framework could be critically tested.

Such is the *personal participation* of the knower in all acts of understanding. But this does not make our understanding *subjective*. Comprehension is neither an arbitrary act nor a passive experience, but a responsible act claiming universal validity. Such knowing is indeed *objective* in the sense of establishing contact with a hidden reality; a contact that is defined as the condition for anticipating an indeterminate range of yet unknown (and perhaps yet inconceivable) true implications. It seems reasonable to describe this fusion of the personal and the objective as Personal Knowledge.

Personal knowledge is an intellectual commitment, and as such inherently hazardous. Only affirmations that could be false can be said to convey objective knowledge of this kind. All affirmations published in this book are my own personal commitments; they claim this, and no more than this, for themselves.

Throughout this book I have tried to make this situation apparent. I have shown that into every act of knowing there enters a passionate contribution of the person knowing what is being known, and that this coefficient is no mere imperfection but a vital component of his knowledge. And around this central fact I have tried to construct a system of correlative beliefs which I can sincerely hold, and to which I can see no acceptable alternatives. But ultimately, it is my own allegiance that upholds these convictions, and it is on such warrant alone that they can lay claim to the reader's attention.

*Manchester*                                                                               M. P.
   *August 1957*

# ACKNOWLEDGMENTS

THIS book is based on my Gifford Lectures 1951–2, delivered in the University of Aberdeen. I wish to thank the University for this opportunity to develop my thoughts. Since subsequent work has not essentially changed my views, large parts of the lectures could be retained unchanged; other parts have been reconsidered, some cut out and others amplified.

Manchester University has made it possible for me to accept the invitation of Aberdeen and to spend nine years almost exclusively on the preparation of this book. The generosity of Senate and Council in allowing me to exchange my Chair of Physical Chemistry for a Professorial appointment without lecturing duties, has placed me deeply in their debt. I want to thank particularly Sir John S. B. Stopford, then Vice-Chancellor, and Lord Simon of Wythenshawe, then Chairman of the Council.

Many of my colleagues at the University have helped me in my enquiries; I have never ceased to admire their patience. May I thank them here once more. I recall also with gratitude the weeks spent on two occasions with the Committee on Social Thought in Chicago, where I lectured on these subjects.

This work owes much to Dr. Marjorie Grene. The moment we first talked about it in Chicago in 1950 she seemed to have guessed my whole purpose, and ever since she has never ceased to help its pursuit. Setting aside her own work as a philosopher, she has devoted herself for years to the service of the present enquiry. Our discussions have catalysed its progress at every stage and there is hardly a page that has not benefited from her criticism. She has a share in anything that I may have achieved here. Dr. J. H. Oldham, Mr. Irving Kristol, Miss Elizabeth Sewell and Professor Edward Shils have read the whole manuscript; Mr. W. Haas, Dr. W. Mays, Professor M. S. Bartlett and Dr. C. Lejewski have read parts of it. They have all suggested improvements, for which I thank them. Miss Olive Davies has carried the burden of secretarial work connected with this book for ten years. Her skill and hard work have given me invaluable assistance. Expenses of books, travel and assistance in the service of this enquiry were covered by grants received from the Rockefeller Foundation, the Volker Fund and the Congress for Cultural Freedom.

## Acknowledgments

Finally, I want to express my admiration for a person who unhesitatingly shared with me the risks of this unusual enterprise and sustained year after year the stresses radiating from me as the centre of this unaccustomed activity; I mean my wife.

I have published the following papers in the period of 1952-8 on the subject of this book. The corresponding pages of the book are given in brackets.

'The Hypothesis of Cybernetics', *The British Journal for the Philosophy of Science*, **2**, (1951-2). (Chapter 8, pp. 261-3.)

'Stability of Beliefs', *The British Journal for the Philosophy of Science*, November, 1952. (Chapter 9, pp. 286-94.)

'Skills and Connoisseurship', *Atti del Congresso di Metodologia*, Torino, December 17-20th, 1952. (Chapter 4, pp. 49-57.)

'On the Introduction of Science into Moral Subjects', *The Cambridge Journal*, No. 4, January, 1954. (Survey of one aspect of the argument.)

'Words, Conceptions and Science', *The Twentieth Century*, September, 1955. (Chapter 5, passim.)

'From Copernicus to Einstein', *Encounter*, September, 1955. (Chapter 1, pp. 3-18.)

'Pure and Applied Science and their appropriate forms of Organization', *Dialectica*, **10**, No. 3, 1956. (Chapter 6, pp. 174-84.)

'Passion and Controversy in Science', *The Lancet*, June 16th, 1956. (Chapter 6, pp. 134-60.)

'The Magic of Marxism', *Encounter*, December, 1956. (Chapter 7, pp. 226-48.)

'Scientific Outlook: its Sickness and Cure', *Science*, **125**, March 15th, 1957. (A brief survey of the main argument.)

'Beauty, Elegance and Reality in Science', *Symposium on Observation and Interpretation*, Bristol, April 1st, 1957. (Survey of Chapters 5 and 6.)

'Problem Solving', *The British Journal for the Philosophy of Science*, August, 1957. (Chapter 5, pp. 120-31.)

'On Biassed Coins and Related Problems', *Zs. f. Phys. Chem.*, 1958. (Chapter 3, pp. 37-40; Chapter 13, pp. 390-402.)

# CONTENTS

## Contents

### PART TWO: THE TACIT COMPONENT

# Contents

## PART THREE: THE JUSTIFICATION OF PERSONAL KNOWLEDGE

# Contents

## PART FOUR: KNOWING AND BEING

# PART ONE
# THE ART OF KNOWING

# I

# OBJECTIVITY

## 1. THE LESSON OF THE COPERNICAN REVOLUTION

IN the Ptolemaic system, as in the cosmogony of the Bible, man was assigned a central position in the universe, from which position he was ousted by Copernicus. Ever since, writers eager to drive the lesson home have urged us, resolutely and repeatedly, to abandon all sentimental egoism, and to see ourselves objectively in the true perspective of time and space. What precisely does this mean? In a full 'main feature' film, recapitulating faithfully the complete history of the universe, the rise of human beings from the first beginnings of man to the achievements of the twentieth century would flash by in a single second. Alternatively, if we decided to examine the universe objectively in the sense of paying equal attention to portions of equal mass, this would result in a lifelong preoccupation with interstellar dust, relieved only at brief intervals by a survey of incandescent masses of hydrogen—not in a thousand million lifetimes would the turn come to give man even a second's notice. It goes without saying that no one—scientists included—looks at the universe this way, whatever lip-service is given to 'objectivity'. Nor should this surprise us. For, as human beings, we must inevitably see the universe from a centre lying within ourselves and speak about it in terms of a human language shaped by the exigencies of human intercourse. Any attempt rigorously to eliminate our human perspective from our picture of the world must lead to absurdity.

What is the true lesson of the Copernican revolution? Why did Copernicus exchange his actual terrestrial station for an imaginary solar standpoint? The only justification for this lay in the greater intellectual satisfaction he derived from the celestial panorama as seen from the sun instead of the earth. Copernicus gave preference to man's delight in abstract theory, at the price of rejecting the evidence of our senses, which present us with the irresistible fact of the sun, the moon, and the stars rising daily in the east to travel across the sky towards their setting in the

3

west. In a literal sense, therefore, the new Copernican system was as anthropocentric as the Ptolemaic view, the difference being merely that it preferred to satisfy a different human affection.

It becomes legitimate to regard the Copernican system as more object-ive than the Ptolemaic only if we accept this very shift in the nature of intellectual satisfaction as the criterion of greater objectivity. This would imply that, of two forms of knowledge, we should consider as more objective that which relies to a greater measure on theory rather than on more immediate sensory experience. So that, the theory being placed like a screen between our senses and the things of which our senses otherwise would have gained a more immediate impression, we would rely increasingly on theoretical guidance for the interpretation of our experience, and would correspondingly reduce the status of our raw impressions to that of dubious and possibly misleading appearances.

It seems to me that we have sound reasons for thus considering theoretical knowledge as more objective than immediate experience.

(*a*) A theory is something other than myself. It may be set out on paper as a system of rules, and it is the more truly a theory the more completely it can be put down in such terms. Mathematical theory reaches the highest perfection in this respect. But even a geographical map fully embodies in itself a set of strict rules for finding one's way through a region of otherwise uncharted experience. Indeed, all theory may be regarded as a kind of map extended over space and time. It seems obvious that a map can be correct or mistaken, so that to the extent to which I have relied on my map I shall attribute to it any mistakes that I made by doing so. A theory on which I rely is therefore objective knowledge in so far as it is not I, but the theory, which is proved right or wrong when I use such knowledge.

(*b*) A theory, moreover, cannot be led astray by my personal illusions. To find my way by a map I must perform the conscious act of map-reading and I may be deluded in the process, but *the map* cannot be deluded and remains right or wrong in itself, impersonally. Consequently, a theory on which I rely as part of my knowledge remains unaffected by any fluctuations occurring within myself. It has a rigid formal structure, on whose steadfastness I can depend whatever mood or desire may possess me.

(*c*) Since the formal affirmations of a theory are unaffected by the state of the person accepting it, theories may be constructed without regard to one's normal approach to experience. This is a third reason why the Copernican system, being more theoretical than the Ptolemaic, is also more objective. Since its picture of the solar system disregards our terrestrial location, it equally commends itself to the inhabitants of Earth, Mars, Venus, or Neptune, provided they share our intellectual values.

Thus, when we claim greater objectivity for the Copernican theory, we do imply that its excellence is, not a matter of personal taste on our part, but an inherent quality deserving universal acceptance by rational creatures. We abandon the cruder anthropocentrism of our senses—but

4

only in favour of a more ambitious anthropocentrism of our reason. In doing so, we claim the capacity to formulate ideas which command respect in their own right, by their very rationality, and which have in this sense an objective standing.

Actually, the theory that the planets move round the sun was to speak for itself in a fashion that went far beyond asserting its own inherent rationality. It was to speak to Kepler (sixty-six years after the death of Copernicus) and inspire his discovery of the elliptic path of planets and of their constant angular surface velocity; and to inspire again, ten years later, his discovery of the Third Law of planetary motion, relating orbital distances to orbital periods. And another sixty-eight years later, Newton was to announce to the world that these laws were but an expression of the underlying fact of general gravitation. The intellectual satisfaction which the heliocentric system originally provided, and which gained acceptance for it, proved to be the token of a deeper significance unknown to its originator. Unknown but not entirely unsuspected; for those who wholeheartedly embraced the Copernican system at an early stage committed themselves thereby to the expectation of an indefinite range of possible future confirmations of the theory, and this expectation was essential to their belief in the superior rationality and objective validity of the system.

One may say, indeed, quite generally, that a theory which we acclaim as rational in itself is thereby accredited with prophetic powers. We accept it in the hope of making contact with reality; so that, being really true, our theory may yet show forth its truth through future centuries in ways undreamed of by its authors. Some of the greatest scientific discoveries of our age have been rightly described as the amazing confirmations of accepted scientific theories. In this wholly indeterminate scope of its true implications lies the deepest sense in which objectivity is attributed to a scientific theory.

Here, then, are the true characteristics of objectivity as exemplified by the Copernican theory. Objectivity does not demand that we estimate man's significance in the universe by the minute size of his body, by the brevity of his past history or his probable future career. It does not require that we see ourselves as a mere grain of sand in a million Saharas. It inspires us, on the contrary, with the hope of overcoming the appalling disabilities of our bodily existence, even to the point of conceiving a rational idea of the universe which can authoritatively speak for itself. It is not a counsel of self-effacement, but the very reverse—a call to the Pygmalion in the mind of man.

This is not, however, what we are taught today. To say that the discovery of objective truth in science consists in the apprehension of a rationality which commands our respect and arouses our contemplative admiration; that such discovery, while using the experience of our senses as clues, transcends this experience by embracing the vision of a reality beyond the impressions of our senses, a vision which speaks for itself in guiding us to

an ever deeper understanding of reality—such an account of scientific procedure would be generally shrugged aside as out-dated Platonism: a piece
of mystery-mongering unworthy of an enlightened age. Yet it is precisely
on this conception of objectivity that I wish to insist in this introductory
chapter. I want to recall how scientific theory came to be reduced in the
modern mind to the rank of a convenient contrivance, a device for recording events and computing their future course, and I wish to suggest then
that twentieth-century physics, and Einstein's discovery of relativity in
particular, which are usually regarded as the fruits and illustrations of this
positivistic conception of science, demonstrate on the contrary the power of
science to make contact with reality in nature by recognizing what is
rational in nature.

## 2. The Growth of Mechanism

The story is in three parts, of which the first begins long before Copernicus, though it leads straight up to him. It starts with Pythagoras, who
lived a century before Socrates. Even so, Pythagoras was a late-comer in
science, for the scientific movement was started almost a generation
earlier on rather different lines by the Ionian school of Thales. Pythagoras
and his followers did not, like the Ionians, try to describe the universe in
terms of certain material elements (fire, air, water, etc.) but interpreted it
exclusively in terms of numbers. They took numbers to be the ultimate
substance, as well as the form, of things and processes. When sounding an
octave they believed they could hear the simple numerical ratio of 1:2 in
the harmonious chiming of the sounds from two wires whose lengths had
the ratio 1:2. Acoustics made the perfection of simple numerical relations audible to their ear. They turned their eyes towards the heavens and
saw the perfect circle of the sun and moon; they watched the diurnal
rotation of the firmament and, studying the planets, saw them governed
by a complex system of steady circular motions; and they apprehended
these celestial perfections in the way one listens to a pure musical interval.
They listened to the music of the spheres in a state of mystic communion.

The revival of astronomical theory by Copernicus after two millennia was
a conscious return to the Pythagorean tradition. While studying law in
Bologna, he worked with the professor of astronomy, Novara, a leading
Platonist, who taught that the universe was to be conceived in terms of
simple mathematical relationships. Then, on his return to Cracow, with the
thought of a heliocentric system in his mind, he made a further study of the
philosophers and traced his new conception of the universe back to writers
of antiquity standing in the Pythagorean tradition.

After Copernicus, Kepler continued wholeheartedly the Pythagorean
quest for harmonious numbers and geometrical excellence. In the volume
containing the first statement of his Third Law, we can see him speculating
intensely on the way the sun, which is the centre of the cosmos and there-

fore somehow *nous* (Reason) itself, apprehends the celestial music performed by the planets: 'Of what sort vision is in the sun, what are its eyes, or what other impulse it has . . . even without eyes . . . for judging the harmonies of the (celestial) motions,' it would be 'for those inhabiting the earth, not easy to conjecture'—yet one may at least dream, 'lulled by the changing harmony of the band of planets', that 'in the sun there dwells an intellect simple, intellectual fire or mind, whatever it may be, the fountain of all harmony'.[1] He even went so far as to write down the tune of each planet in musical notation.

To Kepler astronomic discovery was ecstatic communion, as he voiced it in a famous passage of the same work:

> What I prophesied two-and-twenty years ago, as soon as I discovered the five solids among the heavenly orbits—what I firmly believed long before I had seen Ptolemy's Harmonics—what I had promised my friends in the title of this fifth book, which I named before I was sure of my discovery— what sixteen years ago I urged to be sought—that for which I have devoted the best part of my life to astronomical contemplations, for which I joined Tycho Brahe . . . at last I have brought it to light, and recognized its truth beyond all my hopes. . . . So now since eighteen months ago the dawn, three months ago the proper light of day, and indeed a very few days ago the pure Sun itself of the most marvellous contemplation has shone forth—nothing holds me; I will indulge my sacred fury; I will taunt mankind with the candid confession that I have stolen the golden vases of the Egyptians, in order to build of them a tabernacle to my God, far indeed from the bounds of Egypt. If you forgive me, I shall rejoice; if you are angry, I shall bear it; the die is cast, the book is written, whether to be read now or by posterity I care not; it may wait a hundred years for its reader, if God himself has waited six thousand years for a man to contemplate His work.[2]

What Kepler claimed here about the Platonic bodies was nonsense, and his exclamation about God's having waited for him for thousands of years was a literary fancy; yet his outburst conveys a true idea of the scientific method and of the nature of science; an idea which has since been disfigured by the sustained attempt to remodel it in the likeness of a mistaken ideal of objectivity.

Passing from Kepler to Galileo, we see the transition to a dynamics in which for the first time numbers enter as measured quantities into mathematical formulae. But with Galileo this usage applies only to terrestrial events, while in respect to heavenly motions he still holds the Pythagorean view that the book of nature is written in geometrical characters.[3] In the *Two Great Systems of the World* (1632), he argues in the Pythagorean tradition from the principle that the parts of the world are perfectly ordered.[4] He still believes that the motion of the heavenly bodies

---

[1] J. Kepler, *Harmonices Mundi*, Book V, ch. 10.
[2] *ibid.*, Prooemium to Book V.
[3] *Il Saggiatore* (*Opere*, **6**, p. 232), quoted by H. Weyl, *Philosophy of Mathematics and Natural Science*, Princeton (1949), p. 112.
[4] *Opere*, **1**, Florence (1842), p. 24.

—in fact all natural motion as such—must be circular. Rectilinear motion implies change of place, and this can occur only from disorder to order: that is, either in the transition from primeval chaos to the right disposition of the parts of the world, or in violent motion, i.e. in the endeavour of a body artificially moved to return to its 'natural' place. Once world order is established, all bodies are 'naturally' at rest or in circular motion. Galileo's observations of inertial motion along a plane terrestrial surface were interpreted by him as circular motions around the centre of the earth.

Thus the first century after the death of Copernicus was inspired by Pythagorean intimations. Their last great manifestation was perhaps Descartes's universal mathematics: his hope of establishing scientific theories by the apprehension of clear and distinct ideas, which as such were necessarily true.

But a different line of approach was already advancing gradually, stemming from the other line of Greek thought which lacked the mysticism of Pythagoras, and which recorded observations of all kinds of things, however imperfect. This school, derived from the Ionian philosophers, culminated in Democritus, a contemporary of Socrates, who first taught men to think in materialistic terms. He laid down the principle: 'By convention coloured, by convention sweet, by convention bitter; in reality only atoms and the void.' [1] With this Galileo himself agreed; the mechanical properties of things alone were primary (to borrow Locke's terminology), their other properties were derivative, or secondary. Eventually it was to appear that the primary qualities of such a universe could be brought under intellectual control by applying Newtonian mechanics to the motions of matter, while its secondary qualities could be derived from this underlying primary reality. Thus emerged the mechanistic conception of the world which prevailed virtually unchanged till the end of the last century. This too was a theoretical and objective view, in the sense of replacing the evidence of our senses by a formal space-time map that predicted the motions of the material particles which were supposed to underlie all external experience. In this sense the mechanistic world-view was fully objective. Yet there is a definite change from the Pythagorean to the Ionian conception of theoretical knowledge. Numbers and geometrical forms are no longer assumed to be inherent as such in Nature. Theory no longer reveals perfection; it no longer contemplates the harmonies of Creation. In Newtonian mechanics the formulae governing the mechanical substratum of the universe were differential equations, containing no numerical rules and exhibiting no geometrical symmetry. Henceforth 'pure' mathematics, formerly the key to nature's mysteries, became strictly separated from the *application* of mathematics to the formulation of empirical laws. Geometry became the science of empty space; and analysis, affiliated since

[1] H. Diels, *Die Fragmente der Vorsokratiker* (6th edn.), Berlin (1952), **2**, p. 97 (Democritus A 49).

Descartes to geometry, seceded with it into the region beyond experience. Mathematics represented all rational thinking which appeared necessarily true; while reality was summed up in the events of the world which were seen as contingent—that is, merely such as happened to be the case.

The separation of reason and experience was pressed further by the discovery of non-Euclidean geometry. Mathematics was thereafter denied the capacity of stating anything beyond sets of tautologies formulated within a conventional framework of notations. Physical theories were correspondingly also subjected to a further reduction of status. Towards the end of the nineteenth century a new positivist philosophy arose, denying to the scientific theories of physics any claim to inherent rationality, a claim which it condemned as metaphysical and mystical. The earliest, most energetic and influential development of this idea was due to Ernst Mach, who by his book, *Die Mechanik*, published in 1883, founded the Vienna school of positivism. Scientific theory, according to Mach, is merely a convenient summary of experience. Its purpose is to save time and trouble in recording observations. It is the most economical adaptation of thought to facts, and just as external to the facts as a map, a timetable, or a telephone directory; indeed, this conception of scientific theory would include a time-table or a telephone directory among scientific theories.

Accordingly, scientific theory is denied all persuasive power that is intrinsic to itself, as theory. It must not go beyond experience by affirming anything that cannot be tested by experience; and above all, scientists must be prepared immediately to drop a theory the moment an observation turns up which conflicts with it. In so far as a theory cannot be tested by experience—or appears not capable of being so tested—it ought to be revised so that its predictions are restricted to observable magnitudes.

This view, which can be traced back to Locke and Hume, and which in its massive modern absurdity has almost entirely dominated twentieth-century thinking on science, seems to be the inevitable consequence of separating, in principle, mathematical knowledge from empirical knowledge. I shall now proceed to the story of relativity, which is supposed to have brilliantly confirmed this view of science, and shall show why in my opinion it has supplied on the contrary some striking evidence for its refutation.

## 3. RELATIVITY

The story of relativity is a complicated one, owing to the currency of a number of historical fictions. The chief of these can be found in every text-book of physics. It tells you that relativity was conceived by Einstein in 1905 in order to account for the negative result of the Michelson-Morley experiment, carried out in Cleveland eighteen years earlier, in 1887. Michelson and Morley are alleged to have found that the speed of light measured by a terrestrial observer was the same in whatever direction the signal was sent out. That was surprising, for one would have expected that the observer would catch up to some extent with signals sent out in the direction in which the earth was moving, so that the speed would appear slower in this direction, while the observer would move away from the signal sent out in the opposite direction, so that the speed would then appear faster. The situation is easily understood if we imagine the extreme case that we are moving in the direction of the signal exactly at the speed of light. Light would appear to remain in a fixed position, its speed being zero, while of course at the same time a signal sent out in the opposite direction would move away from us at twice the speed of light.

The experiment is supposed to have shown no trace of such an effect due to terrestrial motion, and so—the textbook story goes on—Einstein undertook to account for this by a new conception of space and time, according to which we could expect invariably to observe the same value for the speed of light, whether we are at rest or in motion. So Newtonian space, which is 'necessarily at rest without reference to any external object', and the corresponding distinction between bodies in absolute motion and bodies at absolute rest, were abandoned and a framework set up in which only the relative motion of bodies could be expressed.

But the historical facts are different. Einstein had speculated already as a schoolboy, at the age of sixteen, on the curious consequences that would occur if an observer pursued and kept pace with a light signal sent out by him. His autobiography reveals that he discovered relativity

> after ten years' reflection . . . from a paradox upon which I had already hit at the age of sixteen: If I pursue a beam of light with the velocity $c$ (velocity of light in a vacuum), I should observe such a beam of light as a spatially oscillatory electromagnetic field at rest. However, there seems to be no such thing, whether on the basis of experience or according to Maxwell's equations. From the very beginning it appeared to me intuitively clear that, judged from the standpoint of such an observer, everything would have to happen according to the same laws as for an observer who, relative to the earth, was at rest.[1]

There is no mention here of the Michelson-Morley experiment. Its findings were, on the basis of pure speculation, rationally intuited by Einstein before he had ever heard about it. To make sure of this, I addressed an enquiry to the late Professor Einstein, who confirmed the fact that 'the Michelson-Morley experiment had a negligible effect on the discovery of relativity'.[2]

[1] *Albert Einstein: Philosopher-Scientist*, Evanston, 1949, p. 53.
[2] This statement was approved for publication by Einstein early in 1954. Dr. N. Balazs, who was working with Einstein in Princeton in Summer 1953, introduced my questions to him and reported his replies. The result of his first interview with Einstein was described by Mr. Balazs in a letter of July 8th, 1953, as follows:
'Today I discussed with Einstein the basic ideas which have led to the foundation of the special theory of relativity.
The result is about the following:
There were basically two problems whose contemplation was of fundamental im-

Actually, Einstein's original paper announcing the Special Theory of Relativity (1905) gave little grounds for the current misconception concerning the origins of his discovery. It opens with a long paragraph referring to the anomalies in the electrodynamics of moving media, mentioning in particular the lack of symmetry in its treatment, on the one hand, of a wire with current flowing through it moving relative to a magnet at rest, and on the other of a magnet moving relative to the same electric current at rest. It then goes on to say that 'similar examples, as well as the unsuccessful attempts to observe the relative motion of the earth in respect to the medium of light lead to the conjecture that, as in mechanics, so also in electrodynamics, absolute rest is not observable. . . .'[1] The usual textbook account of relativity as a theoretical response to the Michelson-Morley experiment is an invention. It is the product of a philosophical prejudice. When Einstein discovered rationality in nature, unaided by any observation that had not been available for at least fifty years before, our positivistic textbooks promptly covered up the scandal by an appropriately embellished account of his discovery.

There is an aspect of this story that is even more curious. For the programme which Einstein carried out was largely prefigured by he very positivist conception of science which his own achievement so flagrantly refuted. It was formulated explicitly by Ernst Mach, who, as we have seen, had first advanced the conception of science as a timetable or telephone directory. He had extensively criticized Newton's definition of space and absolute rest on the grounds that it said nothing that could be tested by experience. He condemned this as dogmatic, since it went beyond experience, and as *meaningless*, since it pointed to nothing that could conceivably be tested by experience.[2] Mach urged that Newtonian dynamics should be reformulated so as to avoid referring to any movement of bodies except as the relative motion of bodies with respect to each other, and Einstein acknowledged the 'profound influence' which Mach's book exercised on him as a boy and subsequently on his discovery of relativity.[3]

portance. (1) The problem he is referring to in his autobiographical sketch about the impressions of an observer moving with the velocity of light and viewing a lightwave; (2) the lack of symmetry of action between phi current elements and phi magnets. (In the pre-relativistic electrodynamics of moving media it made a lot of difference whether you move a wire with a current relative to a magnet, or the magnet relative to the wire.) (1) suggested to him that the velocity of light must play a privileged role; (2) seemed strange since, among other reasons, he felt that the situation is to be determined by the relative velocities which are the same. I hope I do not misrepresent him.

The Michelson-Morley experiment had no role in the foundation of the theory. He got acquainted with it while reading Lorentz's paper about the theory of this experiment (he of course does not remember exactly when, though prior to his papers), but it had no further influence on Einstein's considerations and the theory of relativity was not founded to explain its outcome at all.'

[1] Albert Einstein, 'Zur Elektrodynamik bewegter Körper'; *Annalen der Physik* (4), **17** (1905), p. 891.

[2] E. Mach, *Die Mechanik in ihrer Entwicklung*, 2nd edn., Leipzig (1889), pp. 213–14.

[3] *Albert Einstein: Philosopher-Scientist*, p. 21.

Yet if Mach had been right in saying that Newton's conception of space as absolute rest was meaningless—because it said nothing that could be proven true or false—then Einstein's rejection of Newtonian space could have made no difference to what we hold to be true or false. It could not have led to the discovery of any new facts. Actually, Mach was quite wrong: he forgot about the propagation of light and did not realize that in this connection Newton's conception of space was far from untestable. Einstein, who realized this, showed that the Newtonian conception of space was not *meaningless* but *false*.

Mach's great merit lay in possessing an intimation of a mechanical universe in which Newton's assumption of a single point at absolute rest was eliminated. His was a super-Copernican vision, totally at variance with our habitual experience. For every object we perceive is set off by us instinctively against a background which is taken to be at rest. To set aside this urge of our senses, which Newton had embodied in his axiom of an 'absolute space' said to be 'inscrutable and immovable', was a tremendous step towards a theory grounded in reason and transcending the senses. Its power lay precisely in that appeal to rationality which Mach wished to eliminate from the foundations of science. No wonder therefore that he advanced it on false grounds, attacking Newton for making an empty statement and overlooking the fact that—far from being empty—the statement was false. Thus Mach prefigured the great theoretic vision of Einstein, sensing its inherent rationality, even while trying to exorcise the very capacity of the human mind by which he gained this insight.

But there yet remains an almost ludicrous part of the story to be told. The Michelson-Morley experiment of 1887, which Einstein mentions in support of his theory and which the textbooks have since falsely enshrined as the crucial evidence which compelled him to formulate it, actually did not give the result required by relativity! It admittedly substantiated its authors' claim that the relative motion of the earth and the 'ether' did not exceed a quarter of the earth's orbital velocity. But the actually observed effect was not negligible; or has, at any rate, not been proved negligible up to this day. The presence of a positive effect in the observations of Michelson and Morley was pointed out first by W. M. Hicks in 1902[1] and was later evaluated by D. C. Miller as corresponding to an 'ether-drift' of eight to nine kilometres per second. Moreover, an effect of the same magnitude was reproduced by D. C. Miller and his collaborators in a long series of experiments extending from 1902 to 1926, in which they repeated the Michelson-Morley experiment with new, more accurate apparatus, many thousands of times.

The layman, taught to revere scientists for their absolute respect for the observed facts, and for the judiciously detached and purely provisional manner in which they hold scientific theories (always ready to abandon a theory at the sight of any contradictory evidence), might well have

[1] W. M. Hicks, *Phil. Mag.*, 6th ser., **3** (1902), pp. 9–42.

thought that, at Miller's announcement of this overwhelming evidence of a 'positive effect' in his presidential address to the American Physical Society on December 29th, 1925, his audience would have instantly abandoned the theory of relativity. Or, at the very least, that scientists— wont to look down from the pinnacle of their intellectual humility upon the rest of dogmatic mankind—might suspend judgment in this matter until Miller's results could be accounted for without impairing the theory of relativity. But no: by that time they had so well closed their minds to any suggestion which threatened the new rationality achieved by Einstein's world-picture, that it was almost impossible for them to think again in different terms. Little attention was paid to the experiments, the evidence being set aside in the hope that it would one day turn out to be wrong.[1]

The experience of D. C. Miller demonstrates quite plainly the hollow- ness of the assertion that science is simply based on experiments which anybody can repeat at will. It shows that any critical verification of a scientific statement requires the same powers for recognizing rationality in nature as does the process of scientific discovery, even though it exercises these at a lower level. When philosophers analyse the verification of scientific laws, they invariably choose as specimens such laws as are not in doubt, and thus inevitably overlook the intervention of these powers. They are describing the practical demonstration of scientific law, and not its critical verification. As a result we are given an account of the scientific method which, having left out the process of discovery on the grounds that

[1] In his Presidential Address to Section A of the British Association, Cambridge, 1938, C. G. Darwin says of D. C. Miller's experiments: 'We cannot see any reason to think that this work would be inferior to Michelson's, as he had at his disposal not only all the experience of Michelson's work, but also the very great technical development of the intervening period, but in fact he failed to verify the exact vanishing of the aether drift. What happened? Nobody doubted relativity. There must therefore be some unknown source of error which had upset Miller's work.'—I can confirm from my own experience that this was the attitude of contemporary physicists all during that period. Only Soviet physicists, who objected to relativity for ideological reasons, felt that Miller's experiments casted a doubt on the theory. I owe this information to Mme. T. Ehrenfest, who was a professor of physics in Soviet Russia at the time.

The true position was explicitly stated by J. L. Synge, *Scientific Proc. Royal Dublin Society*, **26**, N.S. (1952), pp. 45–54. The special theory is accepted on other grounds than the experiments of Michelson and Morley. Among these are the observations by G. Joos, *Ann. d. Physik*, **7** (1930), p. 385; R. J. Kennedy, *Proc. Nat. Acad. Science*, **12** (1926), p. 621; K. K. Illingworth, *Phys. Rev.*, **30** (1927), p. 692; Michelson, Pease and Pearson, *J. Opt. Soc. Amer.*, **18** (1929), 181, which have shown the absence of ether- drift by other methods than the Michelson interferometer. Hence Synge rejects the explanation given by D. C. Miller for his experiments and accepts 'the theorist's description' of the Michelson-Morley experiment which 'is to be found in any book on relativity'.

Synge thinks that Miller's results are to be explained by the fact that the interferometer is not carried in a uniform straight motion, but in a circle, by the rotating earth. More recently, some of Miller's original data sheets have been analysed by R. S. Shankland, S. W. McCuskey, F. C. Leone and G. Kuerti in *Rev. Modern Phys.*, **27** (1955), p. 167, who conclude that the apparent ether drift was simulated by statistical fluctuations and temperature effects.

it follows no definite method.[1] overlooks the process of verification as well, by referring only to examples where no real verification takes place.

At the time that Miller announced his results, relativity had yet made few predictions that could be confirmed by experiment. Its empirical support lay mainly in a number of already known observations. The account which the new theory gave of these known phenomena was considered rational, since it derived them from one single convincingly rational principle. It was the same as when Newton's comprehensive account of Kepler's Three Laws, of the moon's period and of terrestrial gravitation—in terms of a general theory of universal gravitation—was immediately given a position of surpassing authority, even before any predictions had been deduced from it. It was this inherent rational excellence of relativity which moved Max Born, despite the strong empirical emphasis of his accounts of science, to salute as early as 1920 'the grandeur, the boldness, and the directness of the thought' of relativity, which made the world-picture of science 'more beautiful and grander'.[2]

Since then, the passing years have brought wide and precise confirmation of at least one formula of relativity; probably the only formula ever sent sprawling across the cover of *Time* magazine. The reduction of mass ($m$) by the loss of energy ($e$) accompanying nuclear transformation has been repeatedly shown to confirm the relation $e = mc^2$, where $c$ is the velocity of light. But such verifications of relativity are but confirmations of the original judgment of Einstein and his followers, who committed themselves to the theory long before these verifications. And they are an even more remarkable justification of the earlier strivings of Ernst Mach for a more rational foundation of mechanics, setting out a programme for relativity at a time when no avenues could yet be seen towards this objective.

The beauty and power inherent in the rationality of contemporary physics is, as I have said, of a novel kind. When classical physics superseded the Pythagorean tradition, mathematical theory was reduced to a mere instrument for computing the mechanical motions which were supposed to underlie all natural phenomena. Geometry also stood outside nature, claiming to offer an *a priori* analysis of Euclidean space, which was regarded as the scene of all natural phenomena but not thought to be involved in them. Relativity, and subsequently quantum mechanics and modern physics generally, have moved back towards a mathematical conception of reality. Essential features of the theory of relativity were anticipated as

[1] Take the following two statements: 'The philosopher of science is not much interested in the thought processes which lead to discovery . . .' (H. Reichenbach in *Einstein: Philosopher-Scientist*, Evanston (1949), p. 289); or 'The gist of the scientific method is . . . verification and proof, not discovery' (H. Mehlberg in *Science and Freedom*, London (1955), p. 127). Actually, philosophers deal extensively with induction as a method of scientific discovery; but when they occasionally realize that this is not how discoveries are made, they dispose of the facts to which their theory fails to apply by relegating them to psychology.

[2] Max Born, *Einstein's Theory of Relativity*, translated by H. L. Brose, London (1924), p. 289.

mathematical problems by Riemann in his development of non-Euclidean geometry; while its further elaboration relied on the powers of the hitherto purely speculative tensor calculus, which by a fortunate accident Einstein got to know from a mathematician in Zürich. Similarly, Max Born happened to find the matrix calculus ready to hand for the development of Heisenberg's quantum mechanics, which could otherwise never have reached concrete conclusions. These examples could be multiplied. By them, modern physics has demonstrated the power of the human mind to discover and exhibit a rationality which governs nature, before ever approaching the field of experience in which previously discovered mathematical harmonies were to be revealed as empirical facts.

Thus relativity has restored, up to a point, the blend of geometry and physics which Pythagorean thought had first naïvely taken for granted. We now realize that Euclidean geometry, which until the advent of general relativity was taken to represent experience correctly, referred only to comparatively superficial aspects of physical reality. It gave an idealization of the metric relations of rigid bodies and elaborated these exhaustively, while ignoring entirely the masses of the bodies and the forces acting on them. The opportunity to expand geometry so as to include the laws of dynamics was offered by its generalization into many-dimensional and non-Euclidean space, and this was accomplished by work in pure mathematics, before any empirical investigation of these results could even be imagined. Minkowski took the first step in 1908 by presenting a geometry which expressed the special theory of relativity, and which included classical dynamics as a limiting case. The laws of physical dynamics now appeared as geometrical theorems of a four-dimensional non-Euclidean space. Subsequent investigation by Einstein led, by a further generalization of this type of geometry, to the general theory of relativity, its postulates being so chosen as to produce invariant expressions with regard to all frames of reference assumed to be physically equivalent. As a result of these postulates, the trajectories of masses follow geodetics, and light is propagated along zero lines. When the laws of physics thus appear as particular instances of geometrical theorems, we may infer that the confidence placed in physical theory owes much to its possessing the same kind of excellence from which pure geometry and pure mathematics in general derive their interest, and for the sake of which they are cultivated.

## 4. Objectivity and Modern Physics

We cannot truly account for our acceptance of such theories without endorsing our acknowledgement of a beauty that exhilarates and a profundity that entrances us. Yet the prevailing conception of science, based on the disjunction of subjectivity and objectivity, seeks—and must seek at all costs—to eliminate from science such passionate, personal, human appraisals of theories, or at least to minimize their function to that of a negligible by-play. For modern man has set up as the ideal of knowledge the conception of natural science as a set of statements which is 'objective' in the sense that its substance is entirely determined by observation, even while its presentation may be shaped by convention. This conception, stemming from a craving rooted in the very depths of our culture, would be shattered if the intuition of rationality in nature had to be acknowledged as a justifiable and indeed essential part of scientific theory. That is why scientific theory is represented as a mere economical description of facts; or as embodying a conventional policy for drawing empirical inferences; or as a working hypothesis, suited to man's practical convenience—interpretations that all deliberately overlook the rational core of science.

That is why, also, if the existence of this rational core yet reasserts itself, its offensiveness is covered up by a set of euphemisms, a kind of decent understatement like that used in Victorian times when legs were called limbs—a bowdlerization which we may observe, for example, in the attempts to replace 'rationality' by 'simplicity'. It is legitimate, of course, to regard simplicity *as a mark* of rationality, and to pay tribute to any theory as a triumph of simplicity. But great theories are rarely simple in the ordinary sense of the term. Both quantum mechanics and relativity are very difficult to understand; it takes only a few minutes to memorize the facts accounted for by relativity, but years of study may not suffice to master the theory and see these facts in its context. Hermann Weyl lets the cat out of the bag by saying: 'the required simplicity is not necessarily the obvious one but we must let nature train us to recognize the true inner simplicity.'[1] In other words, simplicity in science can be made equivalent to rationality only if 'simplicity' is used in a special sense known solely by scientists. We understand the meaning of the term 'simple' only by recalling the meaning of the term 'rational' or 'reasonable' or 'such that we ought to assent to it', which the term 'simple' was supposed to replace. The term 'simplicity' functions then merely as a disguise for another meaning than its own. It is used for smuggling an essential quality into our appreciation of a scientific theory, which a mistaken conception of objectivity forbids us openly to acknowledge.

What has just been said of 'simplicity' applies equally to 'symmetry' and 'economy'. They are contributing elements in the excellence of a theory, but can account for its merit only if the meanings of these terms are stretched far beyond their usual scope, so as to include the much deeper qualities which make the scientists rejoice in a vision like that of relativity. They must stand for those peculiar intellectual harmonies which reveal, more profoundly and permanently than any sense-experience, the presence of objective truth.

I shall call this practice a pseudo-substitution. It is used to play down man's real and indispensable intellectual powers for the sake of maintain-

[1] H. Weyl, *op. cit.*, p. 155.

ing an 'objectivist' framework which in fact cannot account for them. It works by defining scientific merit in terms of its relatively trivial features, and making these function then in the same way as the true terms which they are supposed to replace.

Other areas of science will illustrate even more effectively these indispensable intellectual powers, and their passionate participation in the act of knowing. It is to these powers and to this participation that I am referring in the title of this book as 'Personal Knowledge'. We shall find Personal Knowledge manifested in the appreciation of probability and of order in the exact sciences, and see it at work even more extensively in the way the descriptive sciences rely on skills and connoisseurship. At all these points the act of knowing includes an appraisal; and this personal co-efficient, which shapes all factual knowledge, bridges in doing so the disjunction between subjectivity and objectivity. It implies the claim that man can transcend his own subjectivity by striving passionately to fulfil his personal obligations to universal standards.

# 2

# PROBABILITY

## 1. Programme

THE purpose of this book is to show that complete objectivity as usually attributed to the exact sciences is a delusion and is in fact a false ideal. But I shall not try to repudiate strict objectivity as an ideal without offering a substitute, which I believe to be more worthy of intelligent allegiance; this I have called 'personal knowledge'. In this First Part, entitled 'The Art of Knowing', I hope sufficiently to foreshadow the perspective which the conception of personal knowledge will open up, to justify my persistence—which otherwise may appear merely captious —in rattling all the skeletons in the cupboard of the current scientific outlook. This apology is necessary, for every system of thought has of course some loose ends tucked away out of sight, and the system that I am trying to build round the conception of personal knowledge will also leave many questions in abeyance. Yet it is a fact that time and again men have become exasperated with the loose ends of current thought and have changed over to another system, heedless of similar deficiencies within that new system. There is no other way in philosophy than this; and this is my reason for continuing now my re-valuation of science.

## 2. Unambiguous Statements

The avowed purpose of the exact sciences is to establish complete intellectual control over experience in terms of precise rules which can be formally set out and empirically tested. Could that ideal be fully achieved, all truth and all error could henceforth be ascribed to an exact theory of the universe, while we who accept this theory would be relieved of any occasion for exercising our personal judgment: we should only have to follow the rules faithfully. Classical mechanics approaches this ideal so closely that it is often thought to have achieved it. But this leaves out of account the element of personal judgment involved in applying the formulae of mechanics to the facts of experience. Take for example a single planet circling round the sun. Newtonian mechanics supplies us with an exact formula by the aid of which we can compute the configuration of this two-body system for the most distant future or for the remotest past, provided only that we are given one single set of data, describing the system at one single moment of time. Supposing we observed the motion of the planet from the earth, it would suffice to know its longitude ($l_0$) and elevation ($e_0$) at a time ($t_0$), in order to compute any pair of longitudes ($l$) and elevation ($e$) for any other time ($t$). Such an operation would be quite impersonal and could indeed be done by a machine, automatically, so that it does look as if it predicted certain facts of experience from other anterior facts of experience quite impersonally. But this would overlook the fact that the numbers giving longitudes, elevations, and times which enter into the formulae of celestial mechanics are not facts of experience. The facts are readings on the instruments of a particular observatory: readings from which we derive the data on which we base our computation and by which we check the results of such computations. The derivation of data and checking of data that bridge the gap between our instrument readings and the magnitudes figuring in our formulae can never be fully automatic. For any correlation between a measured number introduced into an exact theory and the corresponding instrument readings, rests on an estimate of observational errors which cannot be definitively prescribed by rule. This indeterminacy is due in the first place to the statistical fluctuations of observational errors, to which I shall yet return. In consequence of such random errors we can only proceed from the probable values of initial data to probable values of predicted magnitudes, and since no strict relationship exists between these two sets of figures, the process remains to this extent indeterminate. Apart from these fluctuations we have always the possibility of systematic errors. Even the most strictly mechanized procedure leaves something to personal skill in the exercise of which an individual bias may enter.

We should remember always the famous case of the Astronomer Royal, Nicholas Maskeleyne, who dismissed his assistant Kinnebrook for persistently recording the passage of stars more than half a second later than he, his superior.[1] Maskelyne did not realize that an equally watchful observer may register systematically different times by the method employed by him; it was only Bessel's realization of this possibility which 20 years later resolved the discrepancy and belatedly justified Kinnebrook.

[1] Maskeleyne wrote in the Greenwich *Astronomical Observations* for July 31st, 1795: 'I think it necessary to mention that my assistant, Mr. David Kinnebrook . . . began from the beginning of August last, to set them [the transits] down half a second of time later than he should do, according to my observations; in January of the succeeding year, 1796, he increased his error to $\frac{8}{10}$ of a second. As he had unfortunately continued a considerable time in this error before I noticed it, and did not seem to me likely ever to get over it, and return to a right method of observing, therefore, though with great reluctance . . . I parted with him.' (Quoted by R. L. Duncombe, 'Personal Equation in Astronomy', *Pop. Astron.*, **53** (1945), 2–13, 63–76, 110–21, p. 3.)

Experimental psychology, of which Bessel thus laid the foundation, has since taught us universally to expect such individual variations in perceptive faculties. We must always assume, therefore, that some trace of a hidden personal bias may systematically affect the result of a series of readings.[1]

Such residual indeterminacies, governed by no definite rules, can usually be disposed of according to routine practice. Yet even so, this process always sets aside conceivable doubts regarding the application of any definite set of rules, and without this no scientific work could ever be accomplished and no scientific statement could be asserted. We have here an essential personal participation of the scientist even in the most exact operations of science.

There is an even wider area of personal judgment in every verification of a scientific theory. Contrary to current opinion, it is not the case that a proven discrepancy between theoretical predictions and observed data suffices in itself to invalidate a theory. Such discrepancies may often be classed as anomalies. The perturbations of the planetary motions that were observed during 60 years preceding the discovery of Neptune, and which could not be explained by the mutual interaction of the planets, were rightly set aside at the time as anomalies by most astronomers, in the hope that something might eventually turn up to account for them without impairing—or at least not essentially impairing—Newtonian gravitation. Speaking more generally, we may say that there are always some conceivable scruples which scientists customarily set aside in the process of verifying an exact theory. Such acts of personal judgment form an essential part of science.[2]

## 3. PROBABILITY STATEMENTS

Yet the theories of classical physics differ from all other chapters in the domain of science by the fact that events are conceivable which would strictly falsify them. It is conceivable, for example, that a sun with a planet

---

[1] I could substantiate this by quoting the great Princeton astronomer H. N. Russell on the 'extremely troublesome errors' varying from observer to observer, which affect the use of the modern transitmicrometer (H. N. Russell, R. S. Dugan and J. Q. Stewart, *Revision of C. A. Young's Manual of Astronomy I. The Solar System*. Boston (1945), p. 63). But we may take instead a more homely illustration, even though it may be slightly off my point. The award of the winner's place in a horse race in England used to be a highly skilled performance entrusted to the stewards of the Jockey Club, until the advent of the photo-finish camera which seemed to render the decision altogether obvious. However, some years ago the late A. M. Turing showed me the print of a photo finish where one horse's nose is seen a fraction of an inch ahead of another's, but the second horse's nose extends forward by six inches or so well ahead of that of its rival by virtue of the projection of a thick thread of saliva. Since such a situation was not foreseen by the rules, the case had to be referred to the stewards and the award made on the grounds of their personal judgment. Turing gave me this as an example for the ultimate vagueness of even the most objective methods of observation, in confirmation of my views in this matter.

[2] See Part Two, ch. 6 and Part Two, ch. 9.

circling around it might be so far removed from other celestial bodies as to render negligible any perturbations caused by these, and that we should know this to be the case. Assuming further, for the sake of argument, that we could observe the position of the planet exactly at successive moments of time, the formulae of mechanics would acquire the power of making quite impersonal predictions, which would be strictly falsified by the fact that the planet failed to turn up in any predicted position at a predicted time. A finite deviation, however slight, would entail a complete refutation of the theory.

By such assumptions as the above we may succeed in restoring at least fictionally the conception of impersonal knowledge in classical mechanics. But the pretences of any such claim become altogether transparent if we pass on to statements of probability. Probability statements can never be strictly contradicted by experience, even if we assume that all external perturbations and all observational errors are entirely eliminated. The only difficulty in demonstrating this fact is that it is so obvious; for nobody will be ready to believe that the matter is so simple, when so many volumes have been written around it without clearly saying so.

Let me illustrate the point by the example of a hydrogen atom as described by quantum mechanics. It presents us with a map which assigns to every point of infinite space a number which is a function $f(r)$ of its distance $r$ from the nucleus. This number denotes the probability of finding the electron of the hydrogen atom at this particular point and likewise at any other point having the same distance $r$ from the nucleus. The simple reason why this statement cannot be contradicted by any conceivable event lies in the fact that it admits that the electron may be found or not be found at the designated place on the specified occasion. There is a story of a dog-owner who prided himself on the perfect training of his pet. Whenever he called: 'Here! will you come or not!' the dog invariably either came or not. That is exactly how electrons behave when controlled by probability.

Statements of this kind are essentially ambiguous and may therefore seem to say nothing. However, if there is, as I believe that there is, some meaning in assigning a numerical value to the probability of our finding an electron at a certain place on a particular occasion, such an assignment must imply some restriction on this ambiguity; and if no strictly objective restriction can be derived from the assignment of this probability, we may expect to find in it instead some guidance to our personal participation in the event to which the probability statement refers.

It is indeed easy to recognize in principle our participation in chance events if we relax for a moment our objectivist sophistication and revert to ordinary usage. We commonly describe certain events as remarkable coincidences; we have our stories of memorable pieces of luck or ill-luck. These are appraisals of events governed by chance. We make such appraisals both before and after such an event has taken place, and if their probability is expressed numerically, this number guides and largely expresses our appraisal. If I accept the probability statement that the chances of throwing three double sixes in immediate succession are one in 46656, I shall entertain a correspondingly small expectation of doing so; while if it should still happen, I shall be surprised, to a degree corresponding to the reciprocal of this numerical probability. Such is my participation in the event to which a probability statement refers, and this I regard as the proper meaning of its probability.

This is not to ascribe a subjective meaning to the probability of an event —either in the laws of quantum mechanics or in the statement that the chances of throwing a double six are 1/36. I ascribe universal validity to my appraisals of probability, in spite of the fact that they make no predictions which could be contradicted by any conceivable events. I shall mention in the next chapter a wide range of universally valid appraisals within the exact sciences which are all essentially incapable of being contradicted by any conceivable event.

There is, of course, an important sense in which a probability statement can be controverted (though not contradicted) by the events. If the expectations based on a statement of probability are repeatedly disappointed and the ensuing events appear to have been correspondingly improbable in the light of the anterior statement of their probability, we shall begin to suspect the correctness of this statement. The process of deciding that a certain statistical statement is untenable has indeed been systematically developed by Sir Ronald Fisher in his famous treatise, *The Design of Experiments*.

I shall give a brief outline of Fisher's standard example for the application of this procedure, dealing with Charles Darwin's experiments on the influence of cross-fertilization in contrast to self-fertilization on the height of plants.[1] 15 plants of each kind were measured and 15 pairs formed at random from which 15 differences in height (measured in eighths of an inch) were obtained. The differences being denoted by $X_1, X_2, X_3 \ldots$, their mean is $\bar{X}$. The value of $\bar{X}$ shows that on the average the cross-fertilized plants were 20·93 eighths of an inch taller than the self-fertilized plants. The heart of the question is then whether this difference is significant or due to mere chance. To decide this, we shall have to compare the magnitude of this difference with the range of accidental variations which appear to occur in our sample, and $\bar{X}$ will be acknowledged as significant only if it sufficiently exceeds the range of such variations. Technically, we describe this range by a magnitude called the standard deviation $\sigma$ which is computed by forming the sum of the square of the deviations from the mean, dividing this (in the case of 15 observations) first by 14 and then by 15, and taking the square root of the result, so that

$$\sigma = \sqrt{\frac{\Sigma(X-\bar{X})^2}{14 \times 15}}$$

[1] R. A. Fisher, *The Design of Experiments*, London, 1935, Part III (pp. 30 ff.).

In our case $\sigma$ comes out at 9·746 eighth inches. Thus it is immediately apparent that $\bar{X}$ is larger than the standard deviation of the individual heights. But the question still remains whether it is sufficiently larger than $\sigma$ to be incapable of being accounted for by chance variations of heights.

*To answer this question is to bring a statement of probability to the test of experience.* Let us watch how Fisher proceeds to do this. He forms the ratio $\bar{X}/\sigma = t$, which comes out to 2·148; then he consults a table which tells what the probability is for $t$ to have any particular value when based on 14 independent discrepancies, and finds that $t = 2\cdot148$ is reached or exceeded by chance in exactly 5 per cent of such random trials.

This tells us that on the hypothesis (which Fisher calls the null-hypo-thesis) that the differences in the heights of self-fertilized and cross-fertilized plants are purely accidental, the probability for our actually observed sample having occurred was less than 5 per cent. Such a statement entitles us to be surprised by the observed result to the same extent as we would be if we drew a black ball from a sack supposed to contain only 5 black balls in a hundred otherwise undistinguishable balls. Now suppose that we had ourselves placed the balls, 95 per cent of them white and 5 per cent of them black, into the sack, and then having shaken them up, we drew out a black ball. We should be very surprised, yet remain unshaken in our belief that the bag contained the balls we had put into it. Not so, however, for our null-hypothesis. In this case Sir Ronald Fisher suggests (and I am prepared to follow him) that we must abandon the assumption that cross-fertilization as against self-fertilization has no effect on the height of plants, since the probability of Darwin's results, being less than 5 per cent, renders the assumption untenable.

Indeed, we may accept Fisher's recommendation of a standard procedure for disproving a null-hypothesis, based on the exclusion of probabilities that are less than 5 per cent. But it is already apparent that this procedure can apply only to hypothetical assumptions that we consider to be of a likelihood comparable to that of the ineffectiveness of cross-fertilization as against self-fertilization, and not to assumptions of such a high degree of likelihood as we would hold in respect of the continued presence in a bag of the black and white balls which we placed in it.

Of course, a series of actual results of sufficiently low probability might shake our initial assumptions even when they are most firmly held. Thus the card-guessing experiments of Rhine in the U.S. and of Soal in England have rendered untenable to these observers and their followers the null-hypothesis that in these experiments the card to be guessed had no effect on the guessing of it. But in these cases the probabilities of the observed re-sults as evaluated on the basis of the null-hypothesis had to fall very far below 5 per cent in order to shake one's belief in it. There is of course no finite limit to the confidence we may reasonably place in a null-hypothesis, nor can there be therefore any definite lower limit either to the probability of events which we may assume to have occurred on the basis of some null-hypothesis. Hence it is clear that a probability statement cannot be strictly contradicted by any event, however improbable this event may appear in its light. The contradiction must be established by a personal act of appraisal which rejects certain possibilities as being too improbable to be entertained as true.

## 4. PROBABILITY OF PROPOSITIONS

The concept of a personal knowledge concerning matters of chance will come out more clearly if we examine by contrast some current attempts to avoid facing up to the fact of our holding a personal knowledge of this kind. Thus we may deny that probability statements imply any reference to objects and suggest that they are concerned only with propositions. This interpretation of probability has indeed widely prevailed in the modern theory of probability since J. M. Keynes first proposed it in his *Treatise on Probability*, published in 1921.

Taking for our example Darwin's investigation of the effect of cross-fertilization as against self-fertilization on the height of plants, we would consider in this light that its result is a proposition $H$, 'cross fertilization enhances growth', rendered probable on the evidence that can be summed up in the proposition $E$, 'the mean of the 15 observed differences is $2\cdot148$ times larger than the calculated standard error of the 15 observed differences'. Thus we would establish a probability relation $P(H/E)$ between two propositions, which is a belief not about events but about a relation between propositions. Some authors describe a result of this kind as conveying a certain degree of belief in $H$ based on the evidence $E$, and symbolize it accordingly by $P_B(H/E)$.[1]

But this analysis does not correspond to actual practice or indeed to any acceptable practice. Darwin's intention was to establish the effect of cross-fertilization on plant growth and not the relation of a proposition asserting such an effect to a proposition about observed heights of plants. When Rhine undertook to investigate the chances of card guessing he wanted to find out whether extra-sensory perception exists, and not whether there obtains a relation between the assertion of its existence and the proportion of guesses recorded. What both these investigators (as interpreted by Fisher) established is *a probability statement H*, namely the contradictory of the null-hypothesis—which is in each case the law of nature that they alleged in conclusion. But such a result is something quite different from *the probability of a statement H*, or the particular degree of belief in $H$ which would correspond to the observed evidence $E$.

This difference between *a probability statement* on the one hand, and

---

[1] See H. Jeffreys, *Theory of Probability*, Oxford, 1939; I. J. Good, *Probability and the Weighing of Evidence*, London, 1950. Jeffreys uses $P(H/E)$, Good $P_B(H/E)$. Keynes in the *Treatise* used the expression $a/h$, where $h$ stands for the evidence and $a$ for the proposition inferred from it.

*the probability of a statement*, or the degree of belief in a statement on the other, may seem elusive, but is actually quite obvious. Take the throw of a die. I say that the probability of a six to be thrown is 1/6; that is a 'probability statement *H*'. There are six such probability statements referring to the throw, such as 'the probability of a one to be thrown is 1/6', 'the probability of a two to be thrown is 1/6, . . .' etc., all six of which I jointly hold to be true. If, on the other hand, we are to make statements *H* about the throw that are *not* probability statements, they must be of the form 'a six will be thrown', 'a five will be thrown', 'a four will be thrown', etc. These six contradictory statements are supposed to become mutually compatible and severally acceptable, by being held not with certainty but with a degree of probability or belief to which we ascribe the number 1/6. But obviously nobody can believe that a die will fall with each of its six sides uppermost at the same time, and no reduction of the degree of this belief will make it acceptable. Nor is it true to say—as a matter of psychology— that we believe that a die will always fall with a six on top but are rather uncertain of this, while at the same time we believe that it will always fall with a five on top but are rather uncertain of this too, and so on. It is absurd to describe our state of mind in these terms; and any attempt to do so can only be prompted by a desperate desire to avoid saying that the chance of throwing a six is 1/6, which would make an ambiguous and yet significant statement about an external event. I conclude, therefore, that in so far as we arrive at probability statements on the lines of the statistical method illustrated by Darwin's or Rhine's investigations, or as made every day about the toss of a coin, these are *statements about probable events* and *not probable statements about events*.

This logical argument will gain in scope if we relate it to psychological observations made on the expectations induced in animals and men by exposing them to a variable series of events. Experiments by Humphreys have shown that persons will acquire the habit of blinking when a light is shown, both if the showing is invariably followed by a puff of air blown into the eye or if the puff is administered only on frequent occasions, at random. But the expectations involved in the two habits were shown to be different when the administration of the puff of air was eventually discontinued. Subjects trained according to the first method rapidly lost the habit of blinking, while those trained according to the second persisted in it through a larger number of tests. A vivid illustration of this effect can be given in terms of a statistical guessing experiment in which a signal light was followed by a second light, either invariably or in 50 per cent of the cases, at random. On the completion of their training the subjects of the first experiment were guessing the occurrence of the second light correctly with 100 per cent frequency, while those of the second experiment were guessing at chance level, i.e. about 50 per cent right. The curves show that after the showing of the second light was definitely discontinued, subjects of group (1) soon ceased to expect that it would turn up again while those of group (2) at first increased their percentage of positive guesses and then comparatively slowly ceased to expect it altogether.[1]

The expectations induced in group (1) appear similar to those affirmed by classical physics. Arising from a confrontation of the subject with an unambiguous correlation of sign and event, these expectations are sharply disappointed the moment the correlation is discontinued and they are quickly abandoned in consequence. By contrast, the expectations induced in group (2) appear similar to those of quantum mechanics or any other probability statement such as refers, e.g., to the spin of a coin. They are not easily disappointed by any turn of events, though they are gradually weakened and eventually extinguished altogether, when they can be upheld only by considering the events which have actually occurred as having been extremely improbable.

We can relate these psychological observations to our logical analysis of empirical inference by endorsing the process which they describe as a rational mode of behaviour on the part of the subjects. Having acknowledged that the observed subjects were forming justifiable expectations and were abandoning them later on reasonable grounds, we may try to enlarge this acknowledgment by analysing their performance in further detail.

We will note then, in the first place, that both kinds of expectations were held by the subjects with varying degrees of confidence at various stages of their experience, and that their confidence was finally reduced to zero by a series of consistent disappointments. We note that the fiduciary element contained both in an unambiguous affirmation and in the affirmation of a probability may vary from a sense of unshakable certitude down to a mere lingering trace of suspicion. I shall acknowledge it as reasonable to make either kind of affirmation, and to entertain the corresponding expectations the more confidently, the more consistently they are borne out by experience. I shall acknowledge it as reasonable also to allow our confidence to ebb away and gradually to vanish altogether if experience continues to conflict with these affirmations, or if it can be reconciled with them only on the assumption that the events that have occurred were exceedingly improbable. In case we are testing a numerical probability law, we may assess how improbable it is that a particular series of observations should be compatible with that law. We may then follow R. A. Fisher in trying to set a specific limit to the improbability which we are prepared to countenance before abandoning the law in question. But since no such rule can be firmly upheld, it merely expresses a personal judgment that is subject to similar variations of confidence as the original probability statement, the validity of which it was intended to test.

This raises an important question; namely whether these varying degrees

[1] L. G. Humphreys, 'The Effect of random alternation of reinforcement on the acquisition and extinction of conditioned eyelid reactions', *J. exp. Psychol.*, 25 (1939), pp. 141–58. 'Acquisition and extinction of verbal expectations in a situation analogous to conditioning', *J. exp. Psychol.*, 25 (1939), pp. 294–301. Reported in E. R. Hilgard, *Theories of Learning*, New York, 1948, pp. 373–5.

of confidence might be themselves expressed as probability statements, the intensity of our confidence being equated to the improbability of the evidence having come to hand accidentally and not attributed to the correctness of the affirmations for which it appears to speak. In order to meet this suggestion, widely current in various forms throughout modern literature on probability since Keynes' treatise of 1921, I have to digress on the nature of affirmations in general.

## 5. THE NATURE OF ASSERTIONS

A sincere allegation is an act that takes place in speaking or in writing down certain symbols. Its agent is the speaking or writing person. Like all intelligent actions, such assertions have a passionate quality attached to them. They express conviction to those to whom they are addressed. We have on record the outcries of dizzy exultation to which Kepler gave vent at the dawning of discovery, as well as those of others at the false dawn of supposed discoveries. We know the violence with which great pioneers like Pasteur have upheld their claims against their critics and can hear the same angry impatience expressed today by fanatical cranks like Lysenko. A doctor deciding on a serious diagnosis in a difficult case or a juryman bringing in a fatal verdict in dubious circumstances will feel the weight of a heavy personal responsibility. In routine observations, unobstructed by opposition and unworried by doubts, these passions are dormant but not absent; no sincere assertion of fact is essentially unaccompanied by feelings of intellectual satisfaction or of a persuasive desire and a sense of personal responsibility. Therefore, in a strict usage the same symbol should never represent the act of sincerely asserting something and the content of what is asserted.

For the symbolic distinction between the two, Frege (1893) has introduced the 'signpost' symbol $\vdash$. This is prefixed to a statement $p$ if $\vdash . p$ is to signify the actual assertion of $p$, while the bare symbol $p$ must henceforth be used only as part of a sentence, whether asserted or unasserted. Written down by itself the signpost symbol $\vdash$ conveys as little meaning as would a solitary question mark or exclamation mark, which are its nearest analogues among existing symbols. This incompleteness of the symbol has an important and perhaps not so readily acceptable correlate. It suggests that a declaratory sentence is by itself also an incomplete symbol. If language is to denote speech it must reflect the fact that we never say anything that has not a definite impassioned quality. It should be clear from the modality of a sentence whether it is a question, a command, an invective, a complaint or an allegation of fact. Since an unasserted declaratory sentence could not stand for an allegation of fact, its modality would be unspecified and could therefore denote no spoken sentence. There are words like 'however', 'altogether' or 'into', and clauses like 'if I were king', which, though not meaningless, can have definite significance only as part

of a sentence. Similarly, I suggest, a sentence itself has only vague significance until supplemented by the symbol defining its modality. In the case
of sentences intended to convey a factual communication this requirement
is fulfilled by a prefixed assertion sign. An unasserted sentence is no better
than an unsigned cheque; just paper and ink without power or meaning.

But our signpost symbol is yet incompletely defined. It is clear that I can
make use of the sign ⊢ to put on paper an allegation of my own; but it
has not been explained how this sign is to function between different
persons and between successive periods in the same person's life. If the
sign is to signify the passionate act of sincerely pronouncing the asserted
sentence—while there are many people in the world and innumerable
moments in any single person's life—the symbol ⊢ . $p$ must be supplemented,
so that it may tell us whose allegation it represents and at what time the
person in question had alleged $p$. In the case of an assertion made on
paper, we may take all this to be expressed by the act that the symbol
⊢ . $p$ is written down by a particular person at a particular moment. This
is how Whitehead and Russell define the use of the sign in their introduction to *Principia Mathematica*. They say that if an asserted sentence is
printed in a book and the assertion turns out to be false the author will
be blamed. Unfortunately the translation of the sign into words, which
Whitehead and Russell suggest, tends to obscure their correct interpretation of it. They translate, for example, '⊢ . $p$ implies $q$' into the words 'it
is asserted that $p$ implies $q$'. But the phrase 'it is asserted' suggests an
impersonal happening of assertions: 'it is asserted', as 'it is raining' or
'it happens'. The value of the assertion sign is lost if we allow ourselves to
revert in our verbal translation of it to the muddle of a declaratory sentence
which asserts itself or is impersonally asserted by nobody in particular.

To avoid this, I may read the ⊢ sign in a book by Whitehead and Russell
as 'W. and R. assert . . .'; from which, after accepting their conclusions,
I may proceed to 'I assert . . .' But on closer scrutiny I shall reject any
wording which mentions assertion. For the significance of my writing
down '⊢ . $p$' is not that I make an assertion but that I commit myself to
it; it is not the act of *my uttering* a sentence $p$ that I express by '⊢ . $p$' but
the fact that *I believe* what the sentence $p$ says. The correct reading of
'⊢ . $p$' written down by me in good faith is therefore 'I believe $p$', or some
other words expressing the same fiduciary act.

It follows, further, that we cannot use the assertion sign as a prefix to
'I believe $p$'. For '⊢ . $p$' and its verbal equivalent 'I believe $p$' stand for a
present fiduciary act of my own, and an act cannot be asserted. A declaratory sentence can be asserted, because it is an incomplete symbol, of
indeterminate modality; while a question, a command, an invective, or
any other sentence of fixed intention can no more be asserted than could
my act of hewing wood or of drinking tea. It would be as meaningless to
prefix the words 'I believe', or the assertion sign which denotes these
words, to such sentences of fully qualified modality as it would be to any

inarticulate act.[1] It follows that the words 'I believe' in the assertion 'I believe *p*' must not be taken to form a declaratory sentence, and indeed no sentence at all. They are more in the nature of an exclamation like 'By Jove!' or thumping on the table; they seal a commitment, a vouching or asseveration. Like the signpost symbol which it transposes into words, the phrase 'I believe' acquires meaning only in conjunction with the clause which follows it. The symbol and the phrase convey in their respective terms the personal endorsement of the sentence prefixed by them.

As a result of this enquiry into the act of affirmation we must deny the possibility of casting the fiduciary element of an affirmation in the form of a probability statement. A bare probability statement is impersonal. This is equally true of a sentence like 'the probability of throwing a double six is 1/36' and of a formula like $P(H/E)$ stating that the probability of an hypothesis $H$ on the evidence $E$ has the value $P$. Being impersonal, these are all incomplete symbols, requiring to be accompanied by the utterance of a personal commitment in order that they may become the content of an assertion. But the act by which I set my seal to any statement—be it an unambiguous statement or a statement of probability—is a personal act of my own. It cannot, therefore, be expressed by any symbol which would have the same meaning if uttered by somebody else: that is, by any impersonal clause, such as an unasserted probability statement.[2]

However, we must allow for the fact that a personal act can be *partly formalized*. By reflecting on the way we are performing it we may seek to establish rules for our own guidance in this act. But such formalization is likely to go too far unless it acknowledges in advance *that it must remain within a framework of personal judgment*. All attempts to formulate the process of inductive inference go astray precisely in this respect. The interpretation of our growing confidence in an empirical proposition in view of the accumulating evidence in its favour, derived from a calculus of probability such as that proposed by Keynes and his followers, falls into this fallacious category. This type of theory points out that any hypothesis $H$ having a finite initial probability will, if it happens to be true, be confirmed by subsequent evidence until its degree of probability approaches complete certainty. Assuming that the universe is such that hypotheses $H$ of finite initial probability do somehow present themselves to our minds,

[1] I am assuming here that the hewing is done purposefully and not by accident or in a hypnotic trance and that similarly I am drinking my tea and liking it. The mechanical performances of hewing and drinking could quite well be prefixed by an assertion sign or an equivalent exclamation to signify the zest and appetite impelling the act.

[2] It is of course possible to make two different uses of the word 'probable', one in the manner which includes it in probability statements and another which would take the place of 'I believe . . .' as a reading of the assertion sign. We would then have to avoid any impersonal wording such as 'it is probable . . .', which would still lack a personal prefix if it is to express an affirmation, and would have to use instead words like 'I consider it probable . . .'. Such a phrase, if understood as synonymous with 'I believe with moderate assurance . . .', would effectively express the affirmation of a sentence or formula following by it. It would not need to be prefixed nor allow itself to be prefixed by an assertion sign.

it concludes that by testing all hypotheses thus presenting themselves to us we shall eventually come to believe all those which are true with a degree of probability approaching certainty.

To this it must be objected, first, that this condition would not be sufficient in practice. In order that the method should work in practice, the frequency of *H* being true must be not merely finite but of quite appreciable magnitude. Life is too short to allow us to go on testing millions of false *H*'s in order to hit on a true one. It is of the essence of the scientific method to select for verification hypotheses having a *high* chance of being true. To select good questions for investigation is the mark of scientific talent, and any theory of inductive inference in which this talent plays no part is a *Hamlet* without the prince. The same holds for the process of verification. Things are not labelled 'evidence' in nature, but are evidence only to the extent to which they are accepted as such by us as observers. This is true even for the most exact sciences. The Cambridge astronomer Challis who undertook to verify the hypothesis of Leverrier and Adams concerning the existence of a new planet, sighted the undiscovered planet four times during the summer of 1846, and once even noticed that it appeared to have a disc, but these facts made no impression on him, for he distrusted altogether the hypothesis which he was testing.[1] Challis acted mistakenly; but the example of D. C. Miller has shown that it may be equally wrong to go on investigating facts that seem to contradict a theory which is sufficiently well established on different grounds. Indeed, no scientist can forgo selecting his evidence in the light of heuristic expectations. And besides, we shall see that he may well be unable to tell on what evidence *E* his belief in a hypothesis *H* is founded. It is a travesty of the scientific method to conceive of it as a process which depends on the speed of accumulating evidence presenting itself automatically in respect to hypotheses selected at random.[2]

## 6. MAXIMS

Yet the foregoing considerations should not make us reject the calculus of probability as irrelevant to the elucidation of the process of scientific discovery. It has its place there when regarded as a partial formalization of a personal act, which is to be interpreted within the context of this personal act. The selection and testing of scientific hypotheses are personal acts, but like other such acts they are subject to rules and the probability scheme may be accepted as a set of such rules. In the chapter on skills (Part One, Ch. 4) I shall have more to say about the curious nature of *rules*

[1] See W. M. Smart, 'John Couch Adams and the Discovery of Neptune', *Nature*, **158** (1946), pp. 648–52.

[2] Keynes' principle of Limited Variability does not effectively restrict this choice; since any explicit hypothesis already presupposes this principle by the use of denotative terms, it cannot operate in the choice between one explicit hypothesis and another.

*of art*, which I should like to call maxims. Maxims are rules, the correct application of which is part of the art which they govern. The true maxims of golfing or of poetry increase our insight into golfing or poetry and may even give valuable guidance to golfers and poets; but these maxims would instantly condemn themselves to absurdity if they tried to replace the golfer's skill or the poet's art. Maxims cannot be understood, still less applied by anyone not already possessing a good practical knowledge of the art. They derive their interest from our appreciation of the art and cannot themselves either replace or establish that appreciation. Another person may use my scientific maxims for the guidance of his inductive inference and yet come to quite different conclusions. It is owing to this manifest ambiguity that maxims can function only—as I have said— within a framework of personal judgment. Once we have accepted our commitment to personal knowledge, we can also face up to the fact that there exist rules which are useful only within the operation of our personal knowing, and can realize also how useful they can be as part of such acts. The probability schemes of Keynes and his followers, purporting to represent the scientific process, may be granted some value of this kind.

## 7. Grading of Confidence

I have argued that my confident utterance of a hypothesis $H$ cannot be expressed by the impersonal symbol $P(H/E)$. Accordingly, my commitment to an empirical inference $H$ based on the evidence $E$ would always have to be asserted in the form $\vdash . H/E$, where the assertion sign would embody the degree of confidence I place in $H$ on the grounds of $E$.

But we must not disregard the fact that an inference $H$ may be reached with a numerically ascertainable degree of confidence. We may test a 'normal population'—i.e. an aggregate assumed to show purely random variations of a certain measured quantity—by observing a certain number of samples from it. We may evaluate the samples, for example, with a view to establishing the spread, or standard deviation, ($\sigma$) of the measured quantity within the population. We thus obtain a series of upper limits for the value of $\sigma$, each of which is ascertained with a different degree of probability. Starting from the truism that $\sigma < \infty$ can be asserted with absolute certainty, we can see that our assertions will gradually lose in confidence, as the asserted upper limit decreases. A useful compromise might then be found by asserting an upper limit which can be asserted with a reasonably high degree of confidence, say with a probability of 95 per cent.[1]

In this case all three elements of the symbol $P(H/E)$ can be substantiated. We have a specific evidence $E$ on which we base the inference $H$ asserting

---

[1] This example, based in principle on R. A. Fisher, 'Inverse Probability', *Cambridge Phil. Soc. Proc.*, **26** (1929-30), p. 528, was provided to the author by Professor M. S. Bartlett of Manchester University.

a numerical upper limit for $\sigma$, and we assert concurrently that the probability of this inference being right is 0·95. It is legitimate to denote this probability then by $P(H/E)$, which would function as an incomplete symbol and have to be prefixed by the assertion sign in order to convey a confident utterance of our own. We would write $\vdash . P(H/E)$.

This notation, $\vdash . P(H/E)$, could be generalized further by applying it to all processes of inference based on admittedly incomplete or even erroneous evidence. We could then use $\vdash . H/E$ and $\vdash . P(H/E)$ for uttering two different statements. The first alleges an inference $H$, based on $E$; while the second credits someone with having carried out a process of inference believed by him to be based on $E$. This distinction can be made more definite only later, within a framework which reconciles the universal intent of our own assertions with the divergences between equally strong convictions of different people, or of the same people at different times.

The conclusions of this chapter bear a resemblance to the dual theory of probability, advocated for example by Carnap.[1] But its relation to these antecedents is rather complex, for it acknowledges a greater variety of elements and also a certain number of combinations between them. Whether a statement is unambiguous ($p_u$) or statistical ($p_s$), it can be uttered with different grades of confidence. These modalities are expressed by prefixing the assertion sign, so as to write: $\vdash . p_u$ or $\vdash . p_s$. This sign brings in the second kind of probability, which may be further specifiable numerically, as in the case of judging a population from a sample. This situation can be represented by the symbol $P(H/E)$, which must then be completed to read $\vdash . P(H/E)$. Alternatively, we may designate in the same terms a belief $H$—whether unambiguous or statistical—that is, or was, held on some specific evidence $E$ by another person, or by ourselves at another time. Such a belief may again be considered then with varying grades of approval, and this establishes the bearing of the symbol $P(H/E)$ on the whole range of beliefs, all the way from what are approved as rational beliefs to beliefs as compulsive conditions observed psychologically.[2]

[1] R. Carnap, *Logical Foundations of Probability*, Chicago and London, 1950.
[2] See p. 373 below.

# 3

# ORDER

## 1. CHANCE AND ORDER

IN the last chapter I discussed how science teaches us to decide that a particular set of events has occurred accidentally, rather than because certain laws of nature, which these events seem to confirm, are in fact valid. I now want to urge that any such decision is based on two different but mutually correlated appraisals. When I say that an event is governed by chance, I deny that it is governed by order. Any numerical assessment of the probability that a certain event has occurred by chance can be made only with a view to the alternative possibility of its being governed by a particular pattern of orderliness.

It may help to bring out my point better and at the same time to extend its generality if I introduce a fresh example of the kind of statistical judgment I have in mind here. At the border between England and Wales you pass a small town called Abergele. Its railway station has a beautifully kept garden in which, sprawling across the lawn, you are faced with the inscription, set out in small white pebbles: 'Welcome to Wales by British Railways.' No one will fail to recognize this as an orderly pattern, deliberately contrived by a thoughtful station-master. And we could refute anyone who doubted this by computing as follows the odds against the arrangement of the pebbles having come about by mere chance. Suppose that the pebbles had originally all belonged to the garden and would, if left to chance, be found in any part of this area with equal probability; we could compare the large number of arrangements open to the pebbles, if distributed at random all over the garden, with the incomparably smaller number of arrangements in which they would spell out the inscription 'Welcome to Wales by British Railways'. The ratio of the latter small number over the former very large number would represent the fantastically small chance of the pebbles having arranged themselves in the form of the inscription merely by accident; and this would crushingly refute any supposition of this having been the case.

33

But suppose that some years later, the thoughtful station-master having died, the pebbles became scattered all over the station garden of Abergele, and that on returning to the place we were to seek out the previously eloquent stones and map out on a sheet of paper exactly their present position. Might we not get into serious difficulty if we were now asked once more: what is the chance of the pebbles having arranged themselves in this particular manner by mere accident? The previous computation—dividing by the number of all possible configurations of the pebbles within the garden the narrowly restricted number of configurations which represent their present arrangement shown by our map—would again yield a fantastically small value for the probability of this particular arrangement. Yet obviously we are *not* prepared to say that this arrangement has not come about by chance.

Now why this sudden change in our methods of inference? Actually, there is no change: we have merely stumbled on a tacit assumption of our argument which we ought to make explicit now. We have assumed from the start that the arrangement of the pebbles which formed an intelligible set of words appropriate to the occasion represented a distinctive pattern. It was only in view of this orderliness that the question could be asked at all whether the orderliness was accidental or not. When the pebbles are scattered irregularly over the whole available area they possess no pattern and therefore the question whether the orderly pattern is accidental or not cannot arise.

Another example puts this in a nutshell. It would be rational for someone returning from a visit to an exhibition to relate the strange coincidence that he happened to be the 500,000th visitor. He may even have been offered, as such, a complimentary gift by the management, as was the case at the Festival of Britain in 1951. But no one would claim it as a strange coincidence that he was the 573,522nd visitor, although the chances for that are even less than those of being the 500,000th. The difference is obviously that 500,000 is a round number while 537,522 is not. The significance of round numbers can be seen in the commemoration of centenaries, bicentenaries, etc. Everyone knew that it was disingenuous on the part of the Soviets to convene an international meeting in the summer of 1945 to celebrate the 225th anniversary of the foundation of their academy, for 225 is not a round number.

My story branches out at this point in a variety of directions which can only be outlined at this stage. One point of interest is that we can see now why it is a mistake to say generally (as has been done) that the probability of a past event to have occurred by chance in precisely the way that it did is vanishingly small. You may justly speak of the improbability of particular past events if you recognize in them a distinctive pattern, for example the fulfilment of a horoscope—*and if at the same time* you deny the reality of this pattern *and* assert instead that the events occurred at random within a wide range of possible alternatives, and were therefore much more likely to have taken a different course. The appearance of the alleged astrological pattern must then be regarded as the specious result of a very unlikely accidental coincidence.[1]

This bears on the theory that the different living species have come into existence by accidental mutations. This can be affirmed only if, first you accredit the distinctive pattern of living beings as exhibiting a peculiar orderliness which you trust yourself to appraise, and second you accept at the same time the belief that evolution has taken place by a vastly improbable coincidence of random events combining to an orderly shape of a highly distinctive character. However, if we are to identify—as I am about to suggest—the presence of significant order with the operation of an ordering principle, no highly significant order can ever be said to be solely due to an accidental collocation of atoms, and we must conclude therefore that the assumption of an accidental formation of the living species is a logical muddle. It appears to be a piece of equivocation, unconsciously prompted by the urge to avoid facing the problem set to us by the fact that the universe has given birth to these curious beings, including people like ourselves. To say that this result was achieved by natural selection is entirely beside the point. Natural selection tells us only why the unfit failed to survive and not why any living beings, either fit or unfit, ever came into existence. As a solution for our problem it is logically on a par with the method of catching a lion by catching two and letting one escape. I shall fully elaborate this argument in Part Four, ch. 13.

But let me pause here to point out that I have fulfilled in the course of the previous survey the promise that I would generalize the correlation between probability and order. I have done so by introducing some

---

[1] A small probability of all chance events having happened exactly as they did, arises from fictitiously attributing significance to the precise way in which the events happened. An important instance of this fallacy has gained currency under the authority of Sir Ronald Fisher (see his 'Retrospect of Criticisms of Natural Selection' in Huxley, Hardy and Ford, *Evolution as a Process*, London, 1954, pp. 91-2), who used it in defence of the theory of natural selection against the objection that the probability of the evolutionary process having occurred by random mutations is prohibitively small. He argues that by the same token the chances of any man having descended through a thousand ancestral generations could be rejected as much too improbable, since the probability of one ancestral individual having a descendant in the thousandth succeeding generation can be shown to be infinitesimal. However, nothing distinctive is known about the thousandth ancestor of a man, and the question as to the probability of his having descended from any particular ancestor is therefore without foundation. The degree of chance involved in the chain of procreation, which produced any particular individual, is the probability of *any* member of the thousandth ancestral generation producing a descendant in our own day. And this probability is—in the light of Sir Ronald Fisher's own calculations—quite enough. I have illustrated the principles involved here by the example of the pebbles in the station garden at Abergele. One may reasonably ask what the probability was for the pebbles having arranged themselves by chance to form an English sentence, but it is unreasonable to ask what the chances were for them to become scattered in a particular way; for when they are scattered at random they do *not* form a pattern. The theory of natural selection claims to explain the formation of certain significant patterns and not the formation of a particular random collocation of atoms.

instances of a new kind of order, not based on natural law—as were the cases mentioned in my first two chapters—but produced by human artifice, like the inscription 'Welcome to Wales by British Railways' at the railway station in Abergele. This expansion of the concept of orderly patterns was helpful to my purpose, which I shall now declare more fully.

I wish to suggest that the conception of events governed by chance implies a reference to orderly patterns which such events can simulate only by coincidence. To test the probability of such coincidences and hence the permissibility of assuming that they have taken place, is the method of Sir Ronald Fisher for establishing *a contrario* the reality of an orderly pattern. On these grounds I suggest, quite generally, that the appraisal of order is an act of personal knowledge, exactly as is the assessment of probability to which it is allied. This is, of course, quite evident when the ordered pattern is contrived by ourselves; such cases may help us therefore to recognize the principle asserted here and to see that it holds quite generally.

This line of thought may seem in danger of defeating itself. If *all* knowledge can be shown to be personal, it may appear that this does no more than attach new labels to our customary concepts. This is avoided, however, by the fact that the degree of our personal participation varies greatly within our various acts of knowing. We can normally distinguish in everything we know some relatively objective facts supporting a supervening personal fact. For example, we may regard the throw of three consecutive double sixes as an objective fact, and our evaluation of this event as striking a coincidence as the stating of a supervening personal fact. Similarly, the location of the pebbles in the station garden of Abergele is an objective fact as compared with the personal fact that the pebbles form a sentence in the English language. I have already acted on this policy in my previous chapter when contrasting the relative objectivity of classical dynamics with the more massively personal knowledge of quantum mechanics and of probability statements in general.

The modern theory of communications throws this whole matter into sharp relief. Suppose we get twenty consecutive signals transmitted over a line: twenty dots or dashes which we shall write down as twenty noughts and crosses:

$$X O X X X O O X O X X O O O X O X O X X.$$

We may take it that this sequence of noughts and crosses is an objective fact. But it may also be a personal fact, and here there are two alternatives: it may be a coded *message* or it may arise from random disturbances which are merely a *noise*. Communication theory tells us that if the sequence is a message, the maximum amount of communication that can be packed into the sequence numbers 948,576, or technically expressed, 20 binary units. The figure 20 measures, as it were, the amount of distinctiveness that can be imparted to a sequence of 20 choice between two altern-

36

atives. Much more distinctiveness could, of course, be imparted into our sequence if we had ordinary numerals running from 0 to 9 at our disposal. Twenty such digits would carry a message of the magnitude $10^{20}$, amounting to about 66 binary units.

If, alternatively, our sequence of binary signals had been obtained as a result of random disturbances, this noise would also be measured on this scale and its numerical value would be $2^{20}$ or 20 binary units. This number is called the amount of equivocation caused by such a noise in any messages transmitted through the same channel.

It is a curious fact that modern communication theory, which has been used by cyberneticists to build around it a fully mechanized model of mental processes, turns out to be based on a clear recognition of personal acts of intelligent appreciation, for the distinctiveness of which it provides for the first time a quantitative measure. I shall deal with this subject more fully later.

Meanwhile I shall carry forward the conclusion that the distinctiveness of an orderly pattern—whether deliberately contrived or found inherent in nature—is revealed by its improbability, and that as such it cannot be strictly contradicted by experience. This, however, is not to say that orderly patterns are subjective. My recognition of a pattern *may* be subjective, but only in the sense that it is mistaken. The shapes of the constellations are subjective patterns, for they are due to accidental collocations; and the alleged confirmations of horoscopes recorded by astrologers are likewise subjective. But, as we have seen in the chapter on Objectivity, man has the power to establish real patterns in nature, the reality of which is manifested by the fact that their future implications extend indefinitely beyond the experience which they were originally known to control. The appraisal of such order is made with universal intent and conveys indeed a claim to an unlimited range of as yet unspecifiable true intimations.

## 2. RANDOMNESS AND SIGNIFICANT PATTERN

But the conceptions introduced so far are yet insufficiently rooted in their subject matter. We must correct this now by shifting our attention to the nature of randomness and of significant patterns. We can sum up the conclusions of the last chapter in these terms, as follows. Statements of probability can be made about random systems and about significantly ordered systems in so far as these are affected by interaction with random systems. Though intimations of significant order may be rendered uncertain by random disturbances, such heuristic surmises remain essentially different from the act of guessing the outcome of a random event. We can reformulate in this sense also the lessons reached so far on the subject of Chance and Order in this chapter. Randomness alone can never produce a significant pattern, for it consists in the absence of any such pattern; and we must not treat the configuration of a random event as a significant pattern, whether by attributing to it fictitiously a distinctiveness that it does not possess, as in the case of the scattered pebbles, or by granting it erroneously a specious significance, such as the fulfilment of a horoscope.[1]

Probability statements are therefore always based on an anterior knowledge of randomness. But how can we tell that certain aggregates are disposed at random, or that certain events are occurring at random? My answer to this question will have to be postponed until much later. But I shall anticipate it here by asserting my belief that random systems exist and can be recognized as such, though it is logically impossible to give any precise definition of randomness.

Indeed, I shall suggest that the contrast between identifiable objects and their accidental surroundings which underlies all acts of visual perception can be expressed in these terms. When the eye divides the field of vision into 'figure' and 'background', it prepares to see the figure retain its identity while moving forward, backward or sideways against a background which, by contrast, is essentially at rest and retains its background character even while undergoing an indefinite variety of changes. No feature of the background may be linked in an orderly manner to the figure. Hence all relations of the background features to the figure must be random, and this will be best safeguarded if the background is random in itself. Similarly, a process unambiguously determined by an ordering principle, such as the motion of the planets round the sun, can be said to constitute a closed system of events only to the extent to which its relations to other objects and events are found to be purely random. Any entity—whether an object or determinate process—will be the more clearly set off against its background, the more amply its internal particulars show steadiness and regularity—combined with an amply confirmed absence of any co-variance between these particulars and those of the background.[2]

We may even grade the intensity of coherent existence on this scale. Owing to its more significant internal structure a human being is a more substantial entity than a pebble. The difference can be appreciated by comparing the sciences of anatomy and physiology with the range of interest offered by the structure of a particular type of pebble. Every kind of human knowing, ranging from perception to scientific observation, includes an appreciation both of order contrasted to randomness and of the degree of this order. We have seen that information theory ascribes in fact a numerical value to the degree of order present in an ordered system forming a message.

A solid object bombarded by the random elements of the medium form-

[1] The conception of a significant pattern used in this chapter excludes the orderly distribution of averages; the reasons for assigning these random features to an essentially different class will be explained in chapter 13.

[2] In designing an experiment we must try to discriminate against irrelevant features by making sure that they vary at random. In agricultural experiments sites may be assigned by the tossing of a coin. (R. A. Fisher, *The Design of Experiments*, London, 1935, p. 48.)

ing its background will itself be set into random motion. The Brownian movement of microscopic particles caused by the thermal motion of the surrounding molecules exemplifies this principle. The calculus of probability applies pre-eminently to the Brownian movement of symmetrical solids. A perfectly unbiassed die, resting on one of its six sides, will occasionally be tumbled over by an exceptionally violent Brownian shock. *We can then say that the chances of the die resting on any particular side are equal.* The randomness of the impacts to which the die is subject transposes the orderliness of its cubic symmetry into the identical frequency of its six alternative stable positions.[1] Such dynamic interaction between order and randomness is a necessary and sufficient condition for the applicability of probability statements to mechanical systems. We shall see later that it is also an ultimate condition, not reducible to any more fundamental terms.[2]

An ordering principle can be *extrinsic*, as in the case of a message or any other artefact, or *intrinsic*, as shown in the ordered coherence of a solid body and in any stable configuration, whether static or dynamic. I shall now describe three imaginary experiments which reveal the characteristic behaviour of both kinds of ordered systems under random impacts.

1. Take a large number of perfect dice resting on a plane surface and all showing the same face—say a one—on top. The orderliness of these dice is purely extrinsic. Prolonged Brownian motion will destroy this orderliness and ultimately produce a state of maximum disorder, in which all faces will show on top with nearly the same frequency.

2. Take a similar set of dice showing the one on top, but let them be biassed in favour of showing a six on top. This shall be contrived by weighting the upper half of the dice so that when they are turned round to show the six on top their potential energy shall decrease by $\triangle E$. Prolonged Brownian motion acting at low temperatures, where $\triangle E \gg kT$, ($k =$ Boltzman's constant, $T =$ absolute temperature) will cause a rearrangement in the sense that *most* dice will show a six on top. This is a stable pattern due to an intrinsic (dynamic) ordering principle.

3. Having produced the dynamically stable pattern we increase the temperature so that $kT \gg \triangle E$. Prolonged Brownian motion will destroy this pattern once more and produce instead the same kind of random aggregate as in experiment (1), with all faces showing on top with nearly the same frequency.

Experiment 2 shows that random impulses may release the operation

[1] Thus we arrive first at the definition of alternative probabilities and secondly at an identification of these with relative frequencies. All attempts at deriving, on the contrary, alternative probabilities from relative frequencies have turned out to be logically unsound, for the statements of frequencies are themselves probability statements. This objection would be met if frequencies could be defined in unambiguous terms, but that is self-contradictory (see Part Four, ch. 13). For a more detailed presentation of the following argument see my paper 'On Biassed Coins and Related Subjects' in *Zs. f. Phys. Chem.* (1958).

[2] See Part Four, ch. 13, p. 15.

of forces which tend to produce a stable pattern. Where such a dynamic ordering principle is lacking, as in Experiment 1, the existing order is destroyed in the long run even by the weakest random impulses. But random impulses of a sufficient strength as applied in Experiment 3 will destroy likewise any dynamically stable order, even though this order came into existence originally by the impact of random impulses of lesser intensity.[1]

Communication theory has computed the blurring of a message by background noises. This illustrates Experiment 1, i.e. the purely destructive effect of random impulses on a meaningful artefact. Experiment 2 can be illustrated by the annealing of a piece of cold-worked metal. The atomic pattern, shattered by hammering or rolling, spontaneously re-crystallizes under the influence of moderate heating. But heating to higher temperatures once more disorganizes the crystalline pattern; when its temperature is raised beyond its melting point, the metal fuses and finally evaporates. Thus Experiment 3 follows on Experiment 2.

This model represents in principle the whole sweep of statistical thermo-dynamics and kinetics and generalizes at the same time the laws of thermal motion to any random impacts.[2] It also extends the range of ordering principles and thereby includes information theory. By a further generalization we shall take in later also the principles ordering the growth and functioning, as well as the reproduction and evolution, of living beings, which will substantiate the critique of natural selection at which I have hinted already.

For the moment, it is enough to recognize here that, in affirming these fundamental laws of nature, we accredit our capacity for knowing random-ness from order in nature and that this distinction cannot be based on considerations of numerical probabilities, since the calculus of probabilities presupposes, on the contrary, our capacity to understand and recognize randomness in nature.

### 3. The Law of Chemical Proportions

The comparison of pebbles with living beings has shown us that our appreciation of order includes an appraisal of the degree of order. I shall illustrate this now within the exact sciences by our knowledge of the

---

[1] Experiment 3 shows that the effect of a bias tends to disappear at higher tempera-tures, and will vanish correspondingly also under the effect of more violent shaking. Note that the energy $E_t$ required for causing a die to tumble over should always be large compared with $\triangle E$, so that even at the higher temperatures $kT$ may be kept well below $E_t$. We should have $E_t \gg kT \gg \triangle E$; and this should apply correspondingly also to the conditions of 'more violent shaking'. Otherwise the die will keep rolling all the time.

[2] The principal scope of thermodynamics lies in variable combinations between the operations of ordering principles according to Experiment 2 and of the counteracting randomizing effect of thermal motion as exemplified in Experiment 3. But these com-binations may be disregarded for our present purpose.

chemical composition of compounds and afterwards, even more emphatically, by our appreciation of the symmetry of crystals.

Everyone knows of the law of simple chemical proportions and can understand a simple chemical formula. If the composition of chloroform is denoted by $CHCl_3$, this means that it consists of one part of carbon measured in units of 12 grams, one part of hydrogen measured in units of about 1 gram and three parts of chlorine measured in units of 35·5 grams. These units of weight, which differ from element to element, are called the atomic weights of the elements. Once these units are adopted, every combination of carbon, hydrogen and chlorine can be written down in similar simple forms, like $CH_3Cl$ for methylchloride, $CH_2Cl_2$ for methylene dichloride, and so on.

This seems quite straightforward, yet this theory makes a claim which relies in a peculiar way on acts of personal appraisal—far more heavily so than classical mechanics, which can be verified with a minimum of personal participation on the part of the observer. The chemical formulae which I have quoted assert that the composition of the compounds in question is represented (when measured in appropriate units) by such ratios as 1 : 1 : 3; or 1 : 3 : 1; or 1 : 2 : 2. To ascertain a simple proportion of integers from a measurement of weights demands that we go beyond the method of ascertaining measured quantities from sets of instrument readings, as we do for the verification of predictions in classical dynamics. We must take the further step of identifying arithmetical proportions of measured quantities with integer fractions. The transition from sets of instrument readings to numbers accepted as measured can be formalized up to a point by the assumption of random errors to account for the spread of instrument readings; but there is no formal rule for ascertaining the integer fractions corresponding to any particular proportion of measured numbers.

The step from measured data to integer relations is rendered indeterminate by the inevitably implied demand that the integers be small. We may consider it obvious that if the ratio of the measured proportions of carbon to hydrogen in a sample of chloroform and a sample of methylene dichloride comes out at 0·504 with a probable error of $\pm 0·04$, we should take it that the ratio is to be represented by the integer fraction 1/2; but this is only so because we readily assume that the ratio must be simple, i.e. made up of *small* integers. Much closer approximations would be offered, of course, by taking the ratios of larger integers. By choosing from these we could always achieve a perfect fit, as we would by taking 1008 : 2000 to represent the measured ratio 0·504.

It is indeed meaningless to speak of establishing a correspondence between measured quantities and integers unless the condition is included that the integers should be small and their fractions simple. In accepting as significant a law of nature like that of simple chemical proportions, we claim that we can evaluate observed magnitudes in terms of simple integer fractions.

Note the word 'simple'! To the extent to which the attribute of simplicity is vague, the demands which a law of simple proportions makes on experience are indeterminate. If future observations of chemical proportions could be represented only by larger integers than those found apposite to previously analysed compounds, we might feel increasingly disappointed in the theory and be eventually altogether discouraged from relying on it. But the process would resemble more closely the gradual relinquishing of a supposed statistical law which has repeatedly failed to find corroboration, than the rejection of an unambiguous theory which has met with a series of conflicting observations.

It is true that the chemical analysis of a substance with a high molecular weight may lead to proportions described by large integers. The end-group of a long chain of carbon atoms may be formed by some element $X$, so that the proportion of $X$ to carbon and hydrogen as well (measured in atomic weight units) may be 1 : 1000 or even higher. When a chemical analysis is interpreted in these terms we no longer rely on the law of simple proportions, but on the atomic theory which has come to replace it as the conceptual framework of chemistry. Atoms can be counted, and their counting would necessarily lead to integer proportions of chemical compounds. Proportions obtained by counting are *observed integer fractions* which need not be simple. Indeed, if we could count the number of sodium and chlorine particles in a crystal of rock salt, we would find a slight excess of one or the other kind of particles and the proportion of the two would be something like 1,000,000,000 : 1,000,000,001. We may say quite generally that chemical proportions which cannot be expressed by small integers may nevertheless be interpreted as integer proportions if this appears justified by more direct evidence concerning the atomic structure of the analysed substances.

But we must remember that the laws of simple chemical proportions were established, or at least strongly urged, before the atomic theory was invoked for their explanation. By the time John Dalton's atomic theory was taking shape, the German Richter had stated this law for the combination of acids and bases, and the Frenchman Proust was about to overcome the opposition of his compatriot Berthollet in an attempt to extend these laws to certain metal compounds. It would seem that by 1808, well before Dalton's ideas became known in France, Proust had convincingly established this claim for copper carbonate, the two oxides of tin and the sulphides of iron. Dalton's discovery of the atomic theory was itself based on the evidence of simple chemical proportions and thus confirmed the intimations of reality contained in the appraisal of this orderly pattern. He is quoted as saying that 'the doctrine of definite proportions appears mysterious unless we adopt the atomic hypothesis'. It appears, he said, like the mystical ratios of Kepler which Newton so happily elucidated.[1]

[1] *Enc. Brit.*, 11th edn.; Article 'Atom' by F. H. Neville.

As time went on the significance of this orderly pattern was even more richly revealed. Dalton's atom proved a mere shadowy prefiguration of its successor, the atom of Rutherford and Bohr. Once more it was proved—and this time on a vast scale—that a scientific theory, when it conforms to reality, gets hold of a truth that is far deeper than its author's understanding of it.[1]

The difficulty of establishing an integer character of a magnitude from any measurements of it may be illustrated by an example in which this process is still sharply controversial. Eddington had deduced that the reciprocal of the 'fine structure constant' which in the usual symbols has the formula $\frac{hc}{2\pi e^2}$ is equal to the integer 137. When Eddington's assertion was originally made, the value computed from observation was 137·307, with a probable error $\pm$ 0·048, which seemed to contradict his assertion. However, the accepted experimental value has changed in the last 20 years and is now 137·009.[2] Yet this close agreement between theory and observation is considered fortuitous by the overwhelming majority of physicists. It is merely a source of annoyance to them.

## 4. CRYSTALLOGRAPHY

I shall now turn to my last, but in many ways most telling example of the theoretical appraisal of order in the exact sciences. This is the story of crystallography and its application to experience.

From earliest times men were fascinated by stones of distinctive shapes. Regularity is one of the distinctive characteristics which pleases the eye and stimulates the imagination. Stones, bounded on many sides by plane surfaces which met in straight edges, attracted attention, particularly if they were also beautifully coloured like rubies, sapphires or emeralds. This first attraction held the intimation of a still hidden and greater significance, which the primitive mind expressed by ascribing magical powers to gems. Later, it stimulated the scientific study of crystals, which established and elaborated in formal terms all systems of appraisals that are inherent in any intelligent appreciation of crystals.

The·system sets up first an ideal of shapeliness, by which it classifies solid bodies into such as tend to fulfil this ideal and others in which no such shapeliness is apparent. The first are crystals, the second the shapeless (or amorphous) non-crystals, like glass. Next, each individual crystal is taken to represent an ideal of regularity, all actual deviations from which are regarded as imperfections. This ideal shape is found by assuming that the approximately plane surfaces of crystals are geometrical planes which

---

[1] Mendel's observation of simple integer relations between the numbers of individuals with alternative inheritable characters (1866) was similarly confirmed by the genic structure of chromosomes about half a century later.

[2] Sir E. Whittaker, *Eddington's Principle in the Philosophy of Science*, Cambridge, 1951, p. 23.

extend to the straight edges in which such planes must meet, thus bounding the crystal on all sides. This formalization defines a polyhedron which is taken to be the theoretical shape of a crystal specimen. It embodies only such aspects of the specimen as are deemed regular and in respect to these it is required to fit the facts of experience; but otherwise, however widely the crystal specimen deviates from the theory, this will be put down as a shortcoming of the crystal and not of the theory.

To each crystal specimen there is thus assigned a different ideal polyhedron, and crystallographic theory proceeds next to discover a principle characterizing the regularity of these polyhedra. This principle is found residing in the symmetry of crystals. 'Symmetry' is a word with connotations almost as wide as 'order'. We may use it, applied to objects, to distinguish an unsymmetrical face from another having perfect symmetry. A scalene triangle is unsymmetrical, an isosceles triangle is symmetrical, but an equilateral triangle is more highly symmetrical than the isosceles triangle. Symmetry is used here as a standard to which observed objects may approximate and which itself may be said to possess different degrees of its own kind of perfection.

This kind of symmetry implies the possibility of transforming one part of a figure or body into another part by applying to it a prescribed operation, such as mirroring. By mirroring a right hand I can transpose it into a left hand, and hence a body with two hands is symmetrical. The fact that an equilateral triangle is more symmetrical than an isosceles may be expressed by pointing out that it has three planes of symmetry instead of one. Alternatively, we may introduce a new symmetry operation by observing that the equilateral triangle can be brought to coincide with itself by rotating it by 120° around a vertical axis passing through its centre. We may readily think of symmetry operations for other regular figures and, in extension of the same principle, also for regular polyhedra. The example of the equilateral triangle shows that the presence of three planes of symmetry, crossing each other along one line and forming angles of 120° with each other, will turn their crossing line into a threefold axis of symmetry. The geometry of regular solids explores such relationships between coexisting elementary symmetries and determines the possibilities for combining such symmetries in one and the same polyhedron. The principle of crystal symmetry was discovered by assuming that crystals contained only six elementary symmetries (mirroring, inversion, twofold, threefold, fourfold and sixfold rotations). From this it was concluded that the 32 possible combinations of these six elementary symmetries represented all distinct kinds of crystal symmetry.

The only sharp distinction laid down by this theory is that between the 32 classes of symmetry. They are distinct forms of a certain kind of order. As the ideal polyhedron of a crystal specimen exhaustively represents the regularity of a crystal specimen, so the class of symmetry into which the polyhedron falls exhaustively represents the regularity of the polyhedron.

And just as the same polyhedron could fit innumerable specimens disfigured by different flaws, the same class of symmetry can be embodied in innumerable polyhedra constituted by an indefinite series of surfaces having an infinite range of relative extension.

Each class of symmetry is a distinctive standard of perfect order to which observed specimens approximate, but these standards themselves possess different degrees of their own form of perfection. The 32 classes of symmetry can be arranged roughly in a line of descending symmetries, from the highest cubic to the lowest triclinic class. The variation down this series is extensive and only the higher classes possess sufficient beauty to make their specimens valued as precious stones.

We have here, in brief, the exhaustive formalization of our appreciation of regularity in crystals, including that of the existence of distinctive kinds of such regularities and of the different grades of regularity represented by each kind. I shall postpone a further analysis of the relation of this formalism to experience until I have supplemented it by an account of the hidden structural pattern of which it is today regarded as the overt manifestation.

The atomic theory of crystals defining this hidden structure, which was prophetically mooted in the nineteenth and triumphantly vindicated early in the twentieth century, has unified and greatly extended the system of order enframed in the 32 classes of symmetry. In this theory the significance of the planes and edges exhibited by a crystal is further reduced. These distinctive features are now regarded as merely indicating the presence of an underlying atomic orderliness, from which the 32 classes of symmetry can be rigorously derived.

The principle of atomic orderliness is an extension of the conception of symmetry. If an operation which brings one part of a figure into coincidence with another part of it is defined as constituting a symmetry, a repetitive pattern like that of a wall paper may be regarded as symmetrical, in view of the fact that its parallel displacement brings it to coincide with itself—except for the edges, which we may disregard if the sheet is very large compared with the spacing of the pattern. Such regular rhythms can be readily conceived in one, two, three or more dimensions. The structural theory of crystals assumes that they are built as regularly repetitive three-dimensional arrays of atoms.

Such arrays, when taken to extend in all directions to infinity, can be readily seen to possess symmetries of the kind observed in crystals, and it can be proved that they can possess only those six elementary symmetries which are found in crystals. Owing to certain alternative possibilities of regular atomic structure not affecting the symmetry of the crystal as observed macroscopically, the underlying three-dimensional atomic patterns can have 230 distinctive rhythms; though these are manifested in only 32 distinctive principles of crystalline regularity.

We may now turn to the question, on what principles our acceptance of crystallographic theory rests.

The theories of the 32 classes of symmetry and of the 230 repetitive patterns called 'space groups' are geometrical statements. As such they speak in terms defined only by the fact that they satisfy the axioms of the theory. The spatial pictures by which we keep in mind their meaning are merely a possible model which embodies this meaning. However, geometry even in this form says nothing definite about experience. Its acceptance rests primarily on our validation of its consistency, ingenuity and profundity. But it does bear potentially on experience, for there is always a possibility that experience may present us with models for a geometrical theory. Such experience may be contrived, consisting in an artificial model. An instance of this is the banking firm described by Cohen and Nagel, having seven partners who form seven managing committees, so that each partner is chairman of one committee and every partner serves on three and only three committees. The constitution of these committees can be shown to embody the seven axioms of a geometry, so that all the theorems of this geometry apply to the relations between the banking firm, its partners and the various committees.[1]

Alternatively, the interpretation of a geometry may be found in the natural order of things. Our conceptual imagination, like its artistic counterpart, draws inspiration from contacts with experience. And like the works of imaginative art, the constructions of mathematics will tend therefore to disclose those hidden principles of the experienced world of which some scattered traces had first stimulated the imaginative process by which these constructions were conceived.

When experienced orderliness is taken to be an embodiment of geometry, it may become possible to test its correspondence to experience. The observation of relativistic phenomena has served as an experimental test for deciding whether the material universe was an instance of Riemann's geometry formulated in space-time by Einstein's rules, when combined with the assumption of trajectories being geodetics.

Take again our 32 classes of symmetry and the 230 space groups. The 32 classes define groups of polyhedra and the 230 space groups define indefinitely extended patterns of points in space. These geometrical constructions were originally initiated by a contemplation of crystals and speculations about their atomic structure; hence they will tend to refer to these matters of experience, and it is in the following up of this reference by observation that any empirical grounds for our acceptance of crystallographic theory must be found.

For the sake of brevity I shall limit my discussion mainly to the theory of space groups. Supposing that the deduction of the 230 groups is, on its own premises, correct; then experience can only teach us whether or not there are in the world instances of atomic structure which embody these premises. There may exist an infinite range of bodies which do not embody

[1] M. R. Cohen and E. Nagel, *An Introduction to Logic and Scientific Method*, London and New York, 1936, pp. 133-9.

them, among them even some (like disorderly solid solutions) which form externally well-shaped crystals; yet this would reveal no internal inconsistency and therefore cause no embarrassment to the theory. Therefore, no conceivable event could falsify this theory. I have already hinted that the relation of crystallographic theory to experience is similar in this respect to that between alternative geometries and the actually experienced universe. But an obvious difference between the two relations of theory and experience lies in the fact that there exists only one single material universe which can serve as an instance of one among many possible geometries, while there exist a great many crystals, each of which is an instance of one out of 230 possible space groups, comprising together one unitary theory. The relation of theory to experience is in this respect more akin to that between a classificatory system, such as that used by zoologists or botanists, and the specimens classified by them. But in view of the fact that the classification is based in the present case on an antecedent geometrical theory of order, the relation between theory and experience is perhaps even more akin to that established by a work of art which makes us see experience in its own light.

A classification is significant if it tells us a great deal about an object once this is identified as belonging to one of its classes. Such a system may be said to classify objects according to their distinctive nature. The distinctiveness of the 230 space groups, like that of the 32 classes of crystal symmetry, rests purely on our appreciation of order; they embody in terms of specific symmetries the claim to universality which we necessarily attach to our personal conceptions of order. Yet this system was supremely vindicated, as was the geometrical theory of crystals in general, by its classificatory functions. It has controlled the collection, description and structural analysis of an immense number of crystalline specimens and has been richly corroborated by the physical and chemical characteristics which are found to distinguish these specimens. It has proved itself a natural classificatory principle.

Here stands revealed a system of knowledge of immense value for the understanding of experience, to which the conception of falsifiability seems altogether inapplicable. Facts which are not described by the theory create no difficulty for the theory, for it regards them as irrelevant to itself. Such a theory functions as a comprehensive idiom which consolidates that experience to which it is apposite and leaves unheeded whatever is not comprehended by it.

The application of crystallographic theory to experience is open to the hazards of empirical refutation only in the same sense as a marching song played by the band at the head of a marching column. If it is not found apposite it will not be popular. Crystallographic theory may in this sense be said to transcend the experience to which it applies. But transcendence which renders an empirical theory irrefutable by experience is of course present in every form of idealization. The theory of ideal gases cannot be

47

disproved by observed deviations from it, so long as they are of the kind which we are supposed to disregard. Such idealizations do in fact express an element of the same contemplative appreciation of which the *a priori* construction and acceptance of a complete system of symmetries is a fully constituted example. We can be legitimately attracted by the concept of ideal gases only in as much as we believe in our capacity for appreciating a kind of fundamental orderliness in nature which underlies some of its less orderly appearances. But in the theory of crystal symmetries idealization goes beyond this. For the standards of excellence which are developed by this system possess a much higher degree of intrinsic significance than the formula $pv = RT$ may claim for itself. It is not merely a scientific idealization but the formalization of an aesthetic ideal, closely akin to that deeper and never rigidly definable sensibility by which the domains of art and art-criticism are governed. That is why this theory teaches us to appreciate certain things, regardless of whether we may find any of their kind in nature, and allows us also to criticize these things when we find them, to the extent to which they fall short of the standards which the theory sets for nature.

We see emerging here a substantial alternative to the usual disjunction of objective and subjective statements, as well as to the disjunction between analytic and synthetic statements. By accrediting our capacity to make valid appraisals of universal bearing within the exact natural sciences, we may yet avoid the sterility and confusion imposed by these traditional categories.

# 4

# SKILLS

## 1. THE PRACTICE OF SKILLS

THE exact sciences are a set of formulae which have a bearing on experience. We have seen that in accrediting this bearing, we must rely to varying degrees on our powers of personal knowing. I shall now try to elucidate the structure of such personal acts further, by analysing the forces engaged in them. Science is operated by the skill of the scientist and it is through the exercise of his skill that he shapes his scientific knowledge. We may grasp, therefore, the nature of the scientist's personal participation by examining the structure of skills.

I shall take as my clue for this investigation the well-known fact *that the aim of a skilful performance is achieved by the observance of a set of rules which are not known as such to the person following them.* For example, the decisive factor by which the swimmer keeps himself afloat is the manner by which he regulates his respiration; he keeps his buoyancy at an increased level by refraining from emptying his lungs when breathing out and by inflating them more than usual when breathing in: yet this is not generally known to swimmers. A well-known scientist, who in his youth had to support himself by giving swimming lessons, told me how puzzled he was when he tried to discover what made him swim; whatever he tried to do in the water, he always kept afloat.

Again, from my interrogations of physicists, engineers and bicycle manufacturers, I have come to the conclusion that the principle by which the cyclist keeps his balance is not generally known. The rule observed by the cyclist is this. When he starts falling to the right he turns the handlebars to the right, so that the course of the bicycle is deflected along a curve towards the right. This results in a centrifugal force pushing the cyclist to the left and offsets the gravitational force dragging him down to the right. This manœuvre presently throws the cyclist out of balance to the left, which he counteracts by turning the handlebars to the left; and so he continues to keep himself in balance by winding along a series of appropriate curvatures. A simple analysis shows that for a given angle of unbalance the curvature of each winding is inversely proportional to the square of the speed at which the cyclist is proceeding.

But does this tell us exactly how to ride a bicycle? No. You obviously cannot adjust the curvature of your bicycle's path in proportion to the ratio of your unbalance over the square of your speed; and if you could you would fall off the machine, for there are a number of other factors to be taken into account in practice which are left out in the formulation of this rule. Rules of art can be useful, but they do not determine the practice of an art; they are maxims, which can serve as a guide to an art only if they can be integrated into the practical knowledge of the art. They cannot replace this knowledge.

## 2. Destructive Analysis

The fact that skills cannot be fully accounted for in terms of their particulars may lead to serious difficulties in judging whether or not a skilful performance is genuine. The extensive controversy on the 'touch' of pianists may serve as an example. Musicians regard it as a glaringly obvious fact that the sounding of a note on the piano can be done in different ways, depending on the 'touch' of the pianist. To acquire the right touch is the endeavour of every learner, and the mature artist counts its possession among his chief accomplishments. A pianist's touch is prized alike by the public and by his pupils: it has a great value in money. Yet when the process of sounding a note on the piano is analysed, it appears difficult to account for the existence of 'touch'. When a key is depressed, a hammer is set in motion which hits a string. The hammer is pushed by the depressed key only for a short distance and is thereby flung into free motion, which is eventually stopped by the chord. Therefore, it is argued, the effect of the hammer on the chord is fully determined by the speed of the hammer in free motion at the moment when it hits the chord. As this speed varies, the note of the chord will sound more or less loudly. This may be accompanied by changes in colour, etc., owing to concurrent changes in the composition of overtones, but it should make no difference in what manner the hammer acquired any particular speed. Accordingly, there could be no difference as between tyro and virtuoso in the tone of the notes which they strike on a given piano; one of the most valued qualities of the pianist's performance would be utterly discredited. Such is indeed the conclusion you find in standard textbooks like Jeans' *Science and Music* (1937) and A. Wood's *Physics of Music* (1944). Yet this result relies erroneously on an incomplete analysis of the pianist's skill. This has been demonstrated (to my satisfaction) by J. Baron and J. Hollo, who called attention to the noise that the depression of a key makes when all chords

50

are removed from a piano.[1] This noise can be varied while the speed imparted to the hammer remains unaltered. The noise mingles with the note
sounded by the hammer on the chord and modifies its quality, and this
seems to account in principle for the pianist's capacity to control the tone
of the piano by the art of his touch.

This example should stand for many others which teach the same lesson;
namely that to deny the feasibility of something that is alleged to have been
done or the possibility of an event that is supposed to have been observed,
merely because we cannot understand in terms of our hitherto accepted
framework how it could have been done or could have happened, may
often result in explaining away quite genuine practices or experiences.
Yet this method of criticism is indispensable, and without its constant
exercise no scientist or technician could keep a steady course among
the many spurious observations which he has to set aside unexplained
every day.

Destructive analysis remains also an indispensable weapon against
superstition and specious practices. Take for example homeopathy. In this
case the efficacity of an alleged art, still widely practised today, can be
wholly refuted, in my opinion, by a mere analysis of its claims. Medicinal
substances used homeopathically can be shown, on the evidence of homeo-
pathic prescriptions, to be diluted to concentrations as low as, or below
that, in which they are present in ordinary food and drinking water; it
seems impossible that an additional spoonful of them administered in a
similar dilution would be medically effective.

A desperate situation may arise if a new skill, the efficacy of which is
open to doubt, is given a false interpretation by its discoverers. This is
illustrated by the tragic failures of the pioneers of hypnotism during the
century from Mesmer to Braid. The critics of Mesmer and later of Elliotson found it easy to demonstrate that the manipulations which these men
said they were performing were in themselves ineffectual. Elliotson had
expounded a whole system of laws governing the alleged transmission of
animal magnetism. He claimed that the magnetism of a glass of water,
the drinking of which caused cataleptic trance, could be graded by dipping
one finger into it, or two fingers, or the whole hand. Another 'law' declared
that mucous surfaces of the subject, like those of the tongue or the eyeball,
were capable of receiving a greater mesmeric stimulus than the skin. Later
Elliotson announced that gold and nickel were more sensitive to mesmeric
influences than base metals like lead. All this was nonsense and was easily
proved to be nonsense. And since the assumption had not yet dawned upon
anyone that hypnotic suggestion was the effective agent of Mesmerism, the
conclusion seemed inevitable that Elliotson's subjects were impostors,

[1] J. Baron and J. Hollo, *Zeitschr. fur Sinnesphysiologie*, **66** (1935), p. 23. A renewed
presentation of this view has been recently prepared for publication in *Journ. Accoust.
Soc., Amer.* by Dr. J. Baron. The manuscript, which I have seen, mentions that O. R.
Ortmann (*Physical Basis of Piano Touch and Tone*, 1925) has to some extent anticipated
the conclusions of Baron and Hollo.

who were either deluding him or colluding with him.[1] In vain did Elliotson bitterly appeal: 'I have given details of 76 painless operations in the name of common sense and humanity, what more is wanted?'[2] Not until the concept of hypnosis was established as a framework for the facts, could these facts be eventually admitted to be true. Indeed, whenever truth and error are amalgamated in a coherent system of conceptions, the destructive analysis of the system can lead to correct conclusions only when supplemented by new discoveries. But there exists no rule for making fresh discoveries or inventing truer conceptions, and hence there can be no rule, either, for avoiding the uncertainty of destructive analysis.

A process similar to that of the critique of Mesmerism, but without its obvious miscarriages, has been continuously fostered during the past decades by technical research laboratories. Great industries, like the tanneries, the potteries or steel mills, like the breweries and the whole range of textile manufactures, as well as agriculture in its numberless branches, have realized in these days that they were carrying on their activities in the manner of an art without any clear knowledge of the constituent detailed operations. When modern scientific research was applied to these traditional industries it was faced in the first place with the task of discovering what actually was going on there and how it was that it produced the goods. This situation was penetratingly recognized from the start as early as 1920 by W. L. Balls for the scientific study of cotton spinning.[3] The hitherto accepted practice of spinning Balls described as 'a thing in itself, scarcely related to physical knowledge at all', so that 'most of the initial decade's work on the part of the scientist will have to be spent merely in defining what the spinner knows'. This prediction was confirmed to me by Dr. F. C. Toy, then Director of the Shirley Institute, the world's leading cotton research laboratory.[4] The attempt to analyse scientifically the established industrial arts has everywhere led to similar results. Indeed even in the modern industries the indefinable knowledge is still an essential part of technology. I have myself watched in Hungary a new, imported machine for blowing electric lamp bulbs, the exact counterpart of which was operating successfully in Germany, failing for a whole year to produce a single flawless bulb.

---

[1] Harley Williams, *Doctors Differ*, London, 1946, pp. 51–60. The tests which destroyed Eiliotson's claims and exposed him to ridicule and suspicion were conducted by Thomas Wakley, founder of the *Lancet*. The experiments were in fact a striking demonstration of hypnotic suggestion.

[2] *ibid.*, p. 76.

[3] 'The Nature, Scope and Difficulties of Industrial Research with particular reference to the Cotton Industry', by W. Lawrence Balls, presented to the Tenth International Cotton Congress at Zurich, June 9th–11th, 1920.

[4] In a letter dated March 13th, 1951, Dr. Toy wrote to me: 'There is no question whatever that in our early years by far our most important work was to discover the scientific bases of the technical processes used in the industry, and not at that time attempt to improve on them by *ad hoc* methods.'

## 3. TRADITION

An art which cannot be specified in detail cannot be transmitted by prescription, since no prescription for it exists. It can be passed on only by example from master to apprentice. This restricts the range of diffusion to that of personal contacts, and we find accordingly that craftsmanship tends to survive in closely circumscribed local traditions. Indeed, the diffusion of crafts from one country to another can often be traced to the migration of groups of craftsmen, as that of the Huguenots driven from France by the repeal of the Edict of Nantes under Louis XIV. Again, while *the articulate contents of science* are successfully taught all over the world in hundreds of new universities, *the unspecifiable art of scientific research* has not yet penetrated to many of these. The regions of Europe in which the scientific method first originated 400 years ago are scientifically still more fruitful today, in spite of their impoverishment, than several overseas areas where much more money is available for scientific research. Without the opportunity offered to young scientists to serve an apprenticeship in Europe, and without the migration of European scientists to the new countries, research centres overseas could hardly ever have made much headway.

It follows that an art which has fallen into disuse for the period of a generation is altogether lost. There are hundreds of examples of this to which the process of mechanization is continuously adding new ones. These losses are usually irretrievable. It is pathetic to watch the endless efforts—equipped with microscopy and chemistry, with mathematics and electronics—to reproduce a single violin of the kind the half-literate Stradivarius turned out as a matter of routine more than 200 years ago.

To learn by example is to submit to authority. You follow your master because you trust his manner of doing things even when you cannot analyse and account in detail for its effectiveness. By watching the master and emulating his efforts in the presence of his example, the apprentice unconsciously picks up the rules of the art, including those which are not explicitly known to the master himself. These hidden rules can be assimilated only by a person who surrenders himself to that extent uncritically to the imitation of another. A society which wants to preserve a fund of personal knowledge must submit to tradition.

In effect, to the extent to which our intelligence falls short of the ideal of precise formalization, we act and see by the light of unspecifiable knowledge and must acknowledge that we accept the verdict of our personal appraisal, be it at first hand by relying on our own judgment, or at second hand by submitting to the authority of a personal example as carrier of a tradition.

The subject of traditionalism cannot be pursued at length here; but some peculiarities of traditional procedure are of immediate interest for the understanding of personal knowledge. They are to be found in the practice of the Common Law, which is the most important system of strictly reasoned traditional activities. Common Law is founded on precedent. In deciding a case today the Courts will follow the example of other courts which have decided similar cases in the past, for in these actions they see embodied the rules of the law. This procedure recognizes the principle of all traditionalism that practical wisdom is more truly embodied in action than expressed in rules of action. Accordingly, the Common Law allows for the possibility that a judge may interpret his own action mistakenly. The judicial maxim which sometimes goes by the name of the 'doctrine of the dictum' lays it down that a precedent is constituted by the decision of a court, irrespective of its interpretation implied in any *obiter dicta* of the judge who made the decision. The judge's action is considered more authentic than what he said he was doing.[1]

In the course of the seventeenth and eighteenth centuries British public life developed a political art and a political doctrine. The art which embodied the exercise of public liberties was naturally unspecifiable, the doctrines of political liberty were maxims of this art which could be properly understood only by those skilled in the art. But the doctrines of political freedom spread from England in the eighteenth century to France and thence throughout the world, while the unspecifiable art of exercising public liberty, being communicable only by tradition, was not transmitted with it. When the French Revolutionaries acted on this doctrine, which was meaningless without a knowledge of its application in practice, Burke opposed them by a traditionalist conception of a free society.

### 4. Connoisseurship

What has been said of skills applies equally to connoisseurship. The medical diagnostician's skill is as much an art of doing as it is an art of knowing. The skill of testing and tasting is continuous with the more actively muscular skills, like swimming or riding a bicycle.

Connoisseurship, like skill, can be communicated only by example, not by precept. To become an expert wine-taster, to acquire a knowledge innumerable different blends of tea or to be trained as a medical diagnostician, you must go through a long course of experience under the guidance of a master. Unless a doctor can recognize certain symptoms, e.g. accentuation of the second sound of the pulmonary artery, there is no use in his reading the description of syndromes of which this symptom forms part. He must personally know that symptom and he can learn this only by repeatedly being given cases for auscultation in which the symptom is authoritatively known to be present, side by side with other cases in

---

[1] Arthur Goodhart, *Essays in Jurisprudence and the Common Law,* Cambridge, 1931, p. 25, writes: 'The principle of a case is not found in the reasons given in the opinion. The principle is not found in the rule of law set forth in the opinion.' T. B. Smith, in *The Doctrines of Judicial Precedent in Scots Law,* Edinburgh, 1952, shows that this doctrine does not hold equally in Scotland.

which it is authoritatively known to be absent, until he has fully realized the difference between them and can demonstrate his knowledge practically to the satisfaction of an expert.

Wherever connoisseurship is found operating within science or technology we may assume that it persists only because it has not been possible to replace it by a measurable grading. For a measurement has the advantage of greater objectivity, as shown by the fact that measurements give consistent results in the hands of different observers all over the world, while such objectivity is rarely achieved in the case of physiognomic appreciations.[1] The large amount of time spent by students of chemistry, biology and medicine in their practical courses shows how greatly these sciences rely on the transmission of skills and connoisseurship from master to apprentice. It offers an impressive demonstration of the extent to which the art of knowing has remained unspecifiable at the very heart of science.

## 5. Two Kinds of Awareness

What I have said of the unspecifiability of skills is closely related to the findings of Gestalt psychology. Yet my evaluation of this material is so different from that of Gestalt theory, that I shall prefer not to refer here to this theory, even though I shall continue to draw on its domain and pursue some arguments on lines closely parallel to that of its teachings. This should be borne in mind for the following analysis of the often discussed situation in which we find ourselves when using a tool, for example when driving in a nail by the strokes of a hammer.

When we use a hammer to drive in a nail, we attend to both nail and hammer, *but in a different way*. We *watch* the effect of our strokes on the nail and try to wield the hammer so as to hit the nail most effectively. When we bring down the hammer we do not feel that its handle has struck our palm but that its head has struck the nail. Yet in a sense we are certainly alert to the feelings in our palm and the fingers that hold the hammer. They guide us in handling it effectively, and the degree of attention that we give to the nail is given to the same extent but in a different way to these feelings. The difference may be stated by saying that the latter are not, like the nail, objects of our attention, but instruments of it. They are not watched in themselves; we watch something else while keeping intensely aware of them. I have a *subsidiary awareness* of the feeling in the palm of my hand which is merged into my *focal awareness* of my driving in the nail.

We may think of the hammer replaced by a probe, used for exploring the interior of a hidden cavity. Think how a blind man feels his way by the use of a stick, which involves transposing the shocks transmitted to his

---

[1] For an account of the competition between connoisseurship and grading by measurement in the process of cotton-classing see M. Polanyi, 'Skills and Connoisseurship', *Atti del Congresso di Metodologia*, Turin, 1952, pp. 381–95.

hand and the muscles holding the stick into an awareness of the things touched by the point of the stick. We have here the transition from 'knowing *how*' to 'knowing *what*' and can see how closely similar is the structure of the two.

Subsidiary awareness and focal awareness are mutually exclusive. If a pianist shifts his attention from the piece he is playing to the observation of what he is doing with his fingers while playing it, he gets confused and may have to stop.[1] This happens generally if we switch our focal attention to particulars of which we had previously been aware only in their subsidiary role.

The kind of clumsiness which is due to the fact that focal attention is directed to the subsidiary elements of an action is commonly known as self-consciousness. A serious and sometimes incurable form of it is 'stage-fright', which seems to consist in the anxious riveting of one's attention to the next word—or note or gesture—that one has to find or remember. This destroys one's sense of the context which alone can smoothly evoke the proper sequence of words, notes, or gestures. Stage fright is eliminated and fluency recovered if we succeed in casting our mind forward and let it operate with a clear view to the comprehensive activity in which we are primarily interested.

Here again the particulars of a skill appear to be unspecifiable, but this time not in the sense of our being ignorant of them. For in this case we can ascertain the details of our performance quite well, and its unspecifiability consists in the fact that the performance is paralysed if we focus our attention on these details. We may describe such a performance as *logically unspecifiable*, for we can show that in a sense the specification of the particulars would logically contradict what is implied in the performance or context in question.

Take for example the identification of a thing as a tool. It implies that a useful purpose can be achieved by handling the thing as an instrument for that purpose. I cannot identify the thing as a tool if I do not know what it is for—or if knowing its supposed purpose, I believe it to be useless for that purpose. Let me denote by $p$ the affirmations which are implied in qualifying a thing as a tool. If I know or at least hypothetically entertain $p$, the thing is a tool to me; if not, it is something else. It may be an animal, like Alice's croquet hammer which walked away because it was a flamingo. But in most cases, if I come across a tool of which I do not know the use, it will merely strike me as a peculiarly shaped object. To regard it merely as such is to imply that I do not believe and do not even hypothetically entertain $p$; which of course denies that I believe or at least hypothetically entertain $p$. And since $p$ asserts something very uncommon, my not believing $p$ virtually amounts to my asserting not-$p$.

An extension of this scheme may allow us to apply it also to the classic theme of Gestalt psychology, which is that the particulars of a pattern

[1] Comp. e.g. Henri Wallon, *De l'acte à la pensée*, Paris, 1942, p. 223.

or a tune must be apprehended jointly, for if you observe the particulars separately they form no pattern or tune. It may be argued that my attending to the pattern or tune as a whole implies its being appreciated as a pattern or a tune, and this would be contradicted by switching my focal attention to the single notes of the tune or the fragments of the pattern. But it is perhaps more appropriate to formulate the contradiction in this case in more general terms, by saying that our attention can hold only one focus at a time and that it would hence be self-contradictory to be both subsidiarily and focally aware of the same particulars at the same time.

This scheme can be easily reformulated and expanded in terms of *meaning*. If we discredit the usefulness of a tool, its meaning as a tool is gone. All particulars become meaningless if we lose sight of the pattern which they jointly constitute.

The most pregnant carriers of meaning are of course the words of a language, and it is interesting to recall that when we use words in speech or writing we are aware of them only in a subsidiary manner. This fact, which is usually described as the *transparency* of language, may be illustrated by a homely episode from my own experience. My correspondence arrives at my breakfast table in various languages, but my son understands only English. Having just finished reading a letter I may wish to pass it on to him, but must check myself and look again to see in what language it was written. I am vividly aware of the meaning conveyed by the letter, yet know nothing whatever of its words. I have attended to them closely but only for what they mean and not for what they are as objects. If my understanding of the text were halting, or its expressions or its spelling were faulty, its words would arrest my attention. They would become slightly opaque and prevent my thought from passing through them unhindered to the things they signify.

## 6. WHOLES AND MEANINGS

Gestalt psychology has described the transformation of an object into a tool and the accompanying transposition of feeling, as for example from the palm to the tip of a probe, as instances of the absorption of a part in a whole. I have covered the same ground in somewhat modified terms in order to bring out the logical structure in which a person commits himself to certain beliefs and appreciations, and accepts certain meanings by deliberately merging his awareness of certain particulars into a focal awareness of a whole. This logical structure is not apparent in the automatic perception of visual and auditory wholes from which Gestalt psychology has derived its prevailing generalizations.

But it is illuminating to recast our analysis now in terms of parts and wholes. When focusing on a whole, we are subsidiarily aware of its parts, while there is no difference in the intensity of the two kinds of awareness. For example, the more sharply we scrutinize a physiognomy, the more keenly are we alert to its particulars. Also when something is seen as subsidiary to a whole, this implies that it participates in sustaining the whole, and we may now regard this function as its *meaning*, within the whole.

Indeed, we now see coming into view two kinds of wholes and two kinds of meaning. The more clear-cut cases of meaning are those in which one thing (e.g. a word) means another thing (e.g. an object). In this case the corresponding wholes are perhaps not obvious, but we may legitimately follow Tolman in amalgamating sign and object into one whole.[1] Other kinds of things, like a physiognomy, a tune or a pattern, are manifestly wholes but this time their meaning is somewhat problematic, for though they are clearly not meaningless, they mean something only in themselves. The distinction between two kinds of awareness allows us readily to acknowledge these two kinds of wholes and two kinds of meaning. Remembering the various uses of a stick, for pointing, for exploring or for hitting, we can easily see that anything that functions effectively within an accredited context has a meaning in that context and that any such context will itself be appreciated as meaningful. We may describe the kind of meaning which a context possesses in itself as *existential*, to distinguish it especially from *denotative* or, more generally, *representative* meaning. In this sense pure mathematics has an existential meaning, while a mathematical theory in physics has a denotative meaning. The meaning of music is mainly existential, that of a portrait more or less representative, and so on. All kinds of order, whether contrived or natural, have existential meaning; but contrived order usually also conveys a message.

## 7. TOOLS AND FRAMEWORKS

As a next step I shall try to strengthen and widen the distinction between subsidiary awareness and focal awareness by identifying it with another commonly known and universally accepted distinction, namely that which we feel between parts of our own body and things that are external to it. We usually take it so much for granted that our hands and feet are members of our body and not external objects, that this assumption is brought home to us only in case they happen to be disturbed by disease. There are certain psychotic patients who do not feel part of their body as belonging to them. They have all the normal sensations transmitted to them from their limbs on both sides, but they do not identify themselves with all the limbs from which these messages originate; they feel some of them, e.g. the right arm and right leg, as external objects. When stepping out of a bath it may happen that they forget to dry these unadopted limbs.[2]

[1] I am referring to Tolman's Sign-Gestalt Theory in his *Purposive Behavior in Animals and Men*, New York, 1932.

[2] W. Russell Brain, *Mind, Perception and Science*, Oxford, 1951, p. 35. For other variants of 'depersonalization' see e.g. Henderson and Gillespie, *A Textbook of Psychiatry*, Oxford Medical Publications, 7th Edn., 1951, p. 127.

Our appreciation of the externality of objects lying outside our body, in contrast to parts of our own body, relies on our subsidiary awareness of processes within our body. Externality is clearly defined only if we can examine an external object deliberately, localizing it clearly in space outside. But when I look at something, I rely for my localization of it in space on a slight difference between the two images thrown on my retina, on the accommodation of the eyes, on the convergence of their axis and the effort of muscular contraction controlling the eye motion, supplemented by impulses received from the labyrinth, which vary according to the position of my head in space. Of all these I become aware only in terms of my localization of the object I am gazing at; and in this sense I may be said to be subsidiarily aware of them.

Our subsidiary awareness of tools and probes can be regarded now as the act of making them form a part of our own body. The way we use a hammer or a blind man uses his stick, shows in fact that in both cases we shift outwards the points at which we make contact with the things that we observe as objects outside ourselves. While we rely on a tool or a probe, these are not handled as external objects. We may test the tool for its effectiveness or the probe for its suitability, e.g. in discovering the hidden details of a cavity, but the tool and the probe can never lie in the field of these operations; they remain necessarily on our side of it, forming part of ourselves, the operating persons. We pour ourselves out into them and assimilate them as parts of our own existence. We accept them existentially by dwelling in them.

## 8. COMMITMENT

We are faced here with the general principle by which our beliefs are anchored in ourselves. Hammers and probes can be replaced by intellectual tools; think of any interpretative framework and particularly of the formalism of the exact sciences. I am not speaking of the specific assertions which fill the textbooks, but of the suppositions which underlie the method by which these assertions are arrived at. We assimilate most of these pre-suppositions by learning to speak of things in a certain language, in which there are names for various kinds of objects, names by which objects can be classified, making such distinctions as between past and present, living and dead, healthy and sick, and thousands of others. Our language includes the numerals and the elements of geometry, and it refers in these terms to laws of nature whence we can pass on to the roots of these laws in scientific observations and experiments.

The curious thing is that we have no clear knowledge of what our pre-suppositions are and when we try to formulate them they appear quite unconvincing. I have illustrated already in my chapter on probability how ambiguous and question-begging are all statements of the scientific method. I suggest now that the supposed pre-suppositions of science are so futile because the actual foundations of our scientific beliefs cannot be asserted at all. When we accept a certain set of pre-suppositions and use them as our interpretative framework, we may be said to dwell in them as we do in our own body. Their uncritical acceptance for the time being consists in a process of assimilation by which we identify ourselves with them. They are not asserted and cannot be asserted, for assertion can be made only *within* a framework with which we have identified ourselves for the time being; as they are themselves our ultimate framework, they are essentially inarticulable.[1]

It is by his assimilation of the framework of science that the scientist makes sense of experience. This making sense of experience is a skilful act which impresses the personal participation of the scientist on the resultant knowledge. It includes the skill of carrying out correctly the measurements which verify scientific predictions or the observations by which scientific classifications are applied. And it includes also connoisseurship, by which the scientist appreciates a mathematical theory in the abstract—such as the theory of space groups was until 1912—and equally, the appositeness of such a theory to the appraisal of observed specimens, for which the theory of space groups has served since the discovery of the diffraction of X-Rays by crystals in 1912.

The tracing of personal knowledge to its roots in the subsidiary awareness of our body as merged in our focal awareness of external objects, reveals not only the logical structure of personal knowledge but also its dynamic sources. I have analysed previously the beliefs which are implied in using an object as a tool. In the new scheme which I have just drawn up of the process by which an external thing is given a meaning by being made to form an extension of ourselves, these beliefs are transposed into more active intentions which draw on our whole person. In this sense I should say that an object is transformed into a tool by a purposive effort envisaging an operational field in respect of which the object guided by our efforts shall function as an extension of our body. My reliance on it for some end makes an object into a tool, even though it may not achieve that end. The burning of a man's nail pairings for the purpose of bewitching him is an instrumental action based on a mistaken integration of supposed means to supposed ends. Similarly, to pronounce a magic formula, to utter a curse or give a blessing, are verbal actions into which the speaker, confident in their efficacy, pours meaning. Conversely, where the ends are achieved by means which are not intended to produce that result, these means have no instrumental character. If a rat accidentally depresses a lever which releases a food pellet it has not used it as a tool; only after the rat has learned to use it for that purpose does the lever become its tool. Buytendijk has described (as others have done in less detail before him) the radical

[1] The subject of the Premises of Science will be dealt with at length in Part Two, ch. 6, sec. 6 (pp. 160-71).

change in the behaviour of a rat when it has learned to run a maze.[1] The animal ceases to explore the details of the walls and corners on its way and attends to these now merely as signposts. It seems to have lost its focal awareness of them and developed instead a subsidiary awareness of them which now forms part of the pursuit of its purpose.

I have said that a tool is only one example of the merging of a thing in a whole (or a gestalt) in which it is assigned a subsidiary function and a meaning in respect to something that has our focal attention. I generalized this structural analysis to include the recognition of signs as indications of subsequent events and the process of establishing symbols for things which they shall signify. We may apply to these cases also what has just been said about a tool. Like the tool, the sign or the symbol can be conceived as such only in the eyes of a person who *relies on them* to achieve or to signify something. *This reliance is a personal commitment which is involved in all acts of intelligence by which we integrate some things subsidiarily to the centre of our focal attention.* Every act of personal assimilation by which we make a thing form an extension of ourselves through our subsidiary awareness of it, is a commitment of ourselves; a manner of disposing of ourselves.

But the context of purpose and commitment, as found inherent in the personal contribution of the knower to his knowledge, yet lacks dynamic character. The pouring out of ourselves into the particulars given by experience so as to make sense of them for some purpose or in some other coherent context, is not achieved effortlessly. Take the way we acquire the use of a tool or a probe. If, as seeing men, we are blindfolded, we cannot find our way about with a stick as skilfully as a blind man does who has practised it for a long time. We can feel that the stick hits something from time to time but cannot correlate these events. We can learn to do this only by an intelligent effort at constructing a coherent perception of the things hit by the stick. We then gradually cease to feel a series of jerks in our fingers as such—as we still do in our first clumsy trials—but experience them as the presence of obstacles of certain hardness and shape, placed at a certain distance, at the point of our stick. We may say, more generally, that by the effort by which I concentrate on my chosen plane of operation I succeed in absorbing all the elements of the situation of which I might otherwise be aware in themselves, so that I become aware of them now in terms of the operational results achieved through their use.

When the new interpretation of the shocks in our fingers is achieved in terms of the objects touched by the stick, we may be said to carry out unconsciously the process of interpreting the shocks. And again, in practical terms, as we learn to handle a hammer, a tennis racket or a motor car in terms of the situation which we are striving to master, we become unconscious of the actions by which we achieve this result. This lapse into

[1] F. J. J. Buytendijk, 'Zielgerichtetes Verhalten der Ratten in einer Freien Situation', *Archives Neerlandaises de Physiologie*, **15** (1930), p. 405.

unconsciousness is accompanied by a newly acquired consciousness of the experiences in question, on the operational plane. It is misleading, therefore, to describe this as the mere result of repetition; it is a structural change achieved by a repeated mental effort aiming at the instrumentalization of certain things and actions in the service of some purpose.

## 9. UNSPECIFIABILITY

We can now answer the problem of unspecifiability with which I started on this examination of skills. If a set of particulars which have subsided into our subsidiary awareness lapses altogether from our consciousness, we may end up by forgetting about them altogether and may lose sight of them beyond recall. In this sense they may have become unspecifiable. However, this seems only a minor reason for unspecifiability, which is accounted for essentially by a somewhat different, if closely related process.

A mental effort has a heuristic effect: it tends to incorporate any available elements of the situation which are helpful for its purpose. Köhler has described this for the case of a practical effort, made by an ape in the presence of an object which may serve as a tool. The animal's insight, he says, reorganizes its field of vision so that the useful object meets his eye as a tool. We may add that this will hold not only of objects which are made use of as tools, but also of the performer's own muscular actions which may subserve his purpose. If these actions are experienced only subsidiarily, in terms of an achievement to which they contribute, its performance may select from them those which the performer finds helpful, without ever knowing these as they would appear to him when considered in themselves. This is the usual process of unconscious trial and error by which we *feel our way* to success and may continue to improve on our success without specifiably knowing how we do it—for we never meet the causes of our success as identifiable things which can be described in terms of classes of which such things are members. This is how you invent a method of swimming without knowing that it consists in regulating your breath in a particular manner, or discover the principle of cycling without realizing that it consists in the adjustment of your momentary direction and velocity, so as to counteract continuously your momentary accidental unbalance. Hence the practical discovery of a wide range of not consciously known rules of skill and connoisseurship which comprise important technical processes that can rarely be completely specified, and even then only as a result of extensive scientific research.

The unspecifiability of the process by which we thus feel our way forward accounts for the possession by humanity of an immense mental domain, not only of knowledge but of manners, of laws and of the many different arts which man knows how to use, comply with, enjoy or live by, without specifiably knowing their contents. Each single step in acquiring this domain was due to an effort which went beyond the hitherto assured capacity of some person making it, and by his subsequent realization and maintenance of his success. It relied on an act of groping which originally passed the understanding of its agent and of which he has ever since remained only subsidiarily aware, as part of a complex achievement.

All these curious properties and implications of personal knowledge go back to what I have previously described as its logical unspecifiability; that is to the disorganizing effect caused by switching our attention to the parts of a whole. We can now appreciate this effect too in dynamic terms.

Since we originally gained control over the parts in question in terms of their contribution to a reasonable result, they have never been known and were still less willed in themselves, and therefore to transpose a significant whole into the terms of its constituent elements is to transpose it into terms deprived of any purpose or meaning. Such dismemberment leaves us with the bare, relatively objective facts, which had formed the clues for a supervening personal fact. It is a destructive analysis of personal knowledge in terms of the underlying relatively objective knowledge.

I have described the effort which we put into acquiring the art of knowing as the attempt to assimilate certain particulars as extensions of our body, so that by becoming imbued with our subsidiary awareness they may form a coherent focal entity. This is an action, but one that has always an element of *passivity* in it. We can assimilate an object as a tool if we believe it to be actually useful to our purposes and the same holds for the relation of meaning to what is meant and the relation of the parts to a whole. The act of personal knowing can sustain these relations only because the acting person believes that they are apposite: that he has not *made them* but *discovered them*. The effort of knowing is thus guided by a sense of obligation towards the truth: by an effort to submit to reality.

Moreover, since every act of personal knowing appreciates the coherence of certain particulars, it implies also submission to certain standards of coherence. While the athlete or the dancer putting forward their best, act as critics of their own performances, connoisseurs are acknowledged as critics of the goodness of specimens. All personal knowing appraises what it knows by a standard set to itself.

### 10. Summary

Let me sum up my argument so far. I started with the exact sciences, defining them as a mathematical formalism with a bearing on experience. There appeared to be present a personal participation on the part of the scientist in establishing this bearing on experience. This was least noticeable in classical mechanics and I accordingly accepted that chapter of physics as the closest approximation to a completely detached natural science. Its statements could indeed be so formulated as to admit of strict falsification by experience. There followed two sets of examples for a more massive and not conceivably negligible personal participation in the exact sciences. The first of these comprised the knowledge of probabilities in science; and more particularly of the degrees of coincidence involved in assuming that an apparently significant pattern of events had come about as the result of chance. The second set demonstrated the assessment of orderly patterns in the exact sciences and showed that standards of orderliness, though bearing on experience, cannot be conceivably falsified by it. On the contrary, as in the case of statements of probability, they themselves appraise any relevant samples of experience.

Experience can of course offer clues to encourage or disappoint statements of probability or standards of order and this effect is important, but not *much* more important than the factual theme of a novel is for its acceptability. Yet personal knowledge in science is not made but discovered, and as such it claims to establish contact with reality beyond the clues on which it relies. It commits us, passionately and far beyond our comprehension, to a vision of reality. Of this responsibility we cannot divest ourselves by setting up objective criteria of verifiability—or falsifiability, or testability, or what you will. For we live in it as in the garment of our own skin. Like love, to which it is akin, this commitment is a 'shirt of flame', blazing with passion and, also like love, consumed by devotion to a universal demand. Such is the true sense of objectivity in science, which I illustrated in my first chapter. I called it the discovery of rationality in nature, a name which was meant to say that the kind of order which the discoverer claims to see in nature goes far beyond his understanding; so that his triumph lies precisely in his foreknowledge of a host of yet hidden implications which his discovery will reveal in later days to other eyes.

My argument was clearly overflowing already at that stage into domains far beyond the exact sciences. In this chapter I have pursued the roots of personal knowledge towards its most primitive forms which lie behind the operations of a scientific formalism. Tearing away the paper screen of graphs, equations and computations, I have tried to lay bare the inarticulate manifestations of intelligence by which we know things in a purely personal manner. I have entered on an analysis of the arts of skilful doing and skilful knowing, the exercise of which guides and accredits the use of scientific formulae, and which ranges far further afield, unassisted by any formalism, in shaping our fundamental notions of most things which make our world.

Here, in the exercise of skill and the practice of connoisseurship, the art of knowing is seen to involve an intentional change of being: the pouring of ourselves into the subsidiary awareness of particulars, which in the performance of skills are instrumental to a skilful achievement, and which in the exercise of connoisseurship function as the elements of the observed comprehensive whole. The skilful performer is seen to be setting standards to himself and judging himself by them; the connoisseur is seen valuing comprehensive entities in terms of a standard set by him

64

for their excellence. The elements of such a context, the hammer, the probe, the spoken word, all point beyond themselves and are endowed with meaning in this context; and on the other hand a comprehensive context itself, like dance, mathematics, music, possesses intrinsic or existential meaning.

The arts of doing and knowing, the valuation and the understanding of meanings, are thus seen to be only different aspects of the act of extending our person into the subsidiary awareness of particulars which compose a whole. The inherent structure of this fundamental act of personal knowing makes us both necessarily participate in its shaping and acknowledge its results with universal intent. This is the prototype of intellectual commitment.

It is the act of commitment in its full structure that saves personal knowledge from being merely subjective. Intellectual commitment is a responsible decision, in submission to the compelling claims of what in good conscience I conceive to be true. It is an act of hope, striving to fulfil an obligation within a personal situation for which I am not responsible and which therefore determines my calling. This hope and this obligation are expressed in the universal intent of personal knowledge. The sense in which this may be said to be the case will be made more definite as I proceed further and it will be summed up at the close of Part Three.

PART TWO

# THE TACIT COMPONENT

# 5

# ARTICULATION

## 1. INTRODUCTION

GUA the chimpanzee was born in captivity on November 15th, 1930, in Cuba. When she reached the age of seven months and a half she was adopted by Mr. and Mrs. Kellogg of Bloomington, Indiana, to become a companion to their baby Donald, who had just completed the fifth month of his life.[1] During the following nine months the two infants were brought up in exactly the same way and their development was recorded by identical tests. A graph comparing the number of successful intelligence tests passed by them shows a striking parallelism in the development of the two. It is true that the child, though the younger, soon took the lead over the chimpanzee and retained this throughout, but the advantage was slight compared with the child's prospective intellectual superiority which was presently to become apparent. At the age of 15 to 18 months the mental development of the chimpanzee is nearing completion; that of the child is only about to start. By responding to people who talk to it, the child soon begins to understand speech and to speak itself. By this one single trick in which it surpasses the animal, the child acquires the capacity for sustained thought and enters on the whole cultural heritage of its ancestors.

The gap which separates the small feats of animal and infant intelligence from the achievements of scientific thought is enormous. Yet the towering superiority of man over the animals is due, paradoxically, to an almost imperceptible advantage in his original, inarticulate faculties.[2] The situation

---

[1] W. N. and L. A. Kellogg, *The Ape and the Child*, New York, 1933.

[2] The superiority of the child is greater than Kellogg and Kellogg's comparison would suggest in view of the shorter time in which the chimpanzee reaches its maturity. But other observations restrict the range of this advantage. It seems established now, for example, that many animals, and among them notably birds, can be taught to identify numbers. They can recognize the number of objects presented to them and also reproduce a fixed number of consecutive acts. The numbers identified range up to the number eight. Otto Köhler, who most effectively established this fact, found also that human

can be summed up in three points. (1) Man's intellectual superiority is almost entirely due to the use of language. But (2) man's gift of speech cannot itself be due to the use of language and must therefore be due to pre-linguistic advantages. Yet (3) if linguistic clues are excluded, men are found to be only slightly better at solving the kind of problems we set to animals. From which it follows that the inarticulate faculties—the potentialities—by which man surpasses the animals and which, by producing speech, account for the entire intellectual superiority of man, are in themselves almost imperceptible. Accordingly, we shall have to account for man's acquisition of language by acknowledging in him the same kind of inarticulate powers as we observe already in animals.

The enormous increase of mental powers derived from the acquisition of formal instruments of thought stands also in a peculiar contrast with the facts collected in the first part of this book, which demonstrate the pervasive participation of the knowing person in the act of knowing by virtue of an art which is essentially inarticulate. The two conflicting aspects of formalized intelligence may be reconciled by assuming that articulation always remains incomplete; that our articulate utterances can never altogether supersede but must continue to rely on such mute acts of intelligence as we once had in common with chimpanzees of our own age.

Admittedly the scientist's art of knowing, which I have surveyed previously, is on a higher level than the child's or the animal's and can be acquired only in conjunction with a knowledge of science as a formal discipline. Other intellectual skills of a high order are acquired similarly in the course of a continued formal education; and indeed our mute abilities keep growing in the very exercise of our articulate powers. Our formal upbringing evokes in us an elaborate set of emotional responses, operating within an articulate cultural framework. By the strength of these affections we assimilate this framework and uphold it as our culture; yet the comparison of the baby and the chimpanzee will go a long way towards accounting for the vastly superior intelligence of man.

Before turning to our principal task of tracing the relation between articulate and inarticulate intelligence, we may use the present vantage ground to set our course towards an ultimate aim of this enquiry which comes into view at this point.[1] If, as it would seem, the meaning of all our utterances is determined to an important extent by a skilful act of our own—the act of knowing—then the acceptance of any of our own

subjects cannot identify the number of any more numerous group of objects than birds can, provided that the subjects are not allowed time to count them. (Cf. W. H. Thorpe, *Ibis*, **93** (1951), p. 48, who quotes seven papers by O. Köhler published from 1935–50.)

[1] My use of the words 'articulate', 'articulation', etc., in this chapter is wider than the common linguistic usage, in which these terms refer only to the actual enunciation of the sounds of language. The context, however, should make my meaning clear, and it is not without precedent. See for example: A. D. Sheffield, *Grammar and Thinking*, New York and London, 1912, p. 22: 'Psychologically, the simple assertory sentence expresses the articulation of a conceptual whole into such of its elements as are pertinent to the interest guiding the train of thought.'

utterances as true involves our approval of our own skill. To affirm anything implies, then, to this extent an appraisal of our own art of knowing, and the establishment of truth becomes decisively dependent on a set of personal criteria of our own which cannot be formally defined. If everywhere it is the inarticulate which has the last word, unspoken and yet decisive, then a corresponding abridgement of the status of spoken truth itself is inevitable. The ideal of an impersonally detached truth would have to be reinterpreted, to allow for the inherently personal character of the act by which truth is declared. The hope of achieving an acceptable balance of mind in this respect will guide the subsequent enquiry throughout Parts Two and Three of this book.

## 2. INARTICULATE INTELLIGENCE

I shall start on this task systematically by returning to the analysis of the inarticulate manifestations of intelligence in animals and children. I accept for the moment without discussion the usual distinction between the automatic functioning of the organism, including its instinctive performances, and the higher forms of behaviour not specifically included in the animal's native repertoire. Such behaviour will be called learning in which term I shall include also acts of problem solving. Learning will be regarded as a sign of intelligence, in contrast to the functioning of internal organs or to instinctive performances which will be classed as sub-intelligent.[1]

The various modes of learning fall readily into three classes, two of which are more primitive and are rooted respectively in the *motility* and the *sentience* of the animal, while the third handles both these functions of animal life in an *implicit operation of intelligence*. This division follows E. R. Hilgard (*Theories of Learning* (1948; 2nd edition, 1956)), and to some extent also O. H. Mowrer (*Learning Theory and Personality Dynamics* (1950)), who in their turn were guided to a considerable extent by E. C. Tolman (*Purposive Behavior in Animals and Men* (1932)). My presentation differs, however, so much from that of these authors that I can only acknowledge my debt to them here in a summary fashion.

*Type A. Trick Learning.* The best demonstration of motoric learning is B. F. Skinner's.[2] He places a hungry rat in a box equipped with a lever, the depression of which releases a food pellet. The rat will first roam about the box, sniffing and pawing at any prominent object. Having once accidentally depressed the lever, it eats the pellet so released. After a while the rat may happen to depress the lever again and learning will set in, showing

[1] I shall set aside at this stage the question whether learning may be represented within an extended framework of physiology, either as experimental conditioning or as stimulated maturation, for this need not affect the practical distinction between lower and higher performances, of which the former are said to fall below and the latter above the intelligent level.

[2] B. F. Skinner, *The Behavior of Organisms*, New York, 1938.

itself by the fact that the action rapidly becomes more frequent. Finally the rat becomes engaged assiduously in lever-pressing and pellet-eating; and the process of learning is complete.

This amplification of the rat's feeding behaviour is elicited here by providing an object which it can use as a tool, and it consists in discovering and practising the proper use of this tool. We may say that the rat has learned to *contrive* an effect that is useful to it, or else that it has discovered a useful *means-end* relationship. The anthropomorphic imputations implied here—and similarly in the following survey of learning—are deliberate and will be justified later in Part Four against behaviourist objections.

*Type B. Sign-learning.* A dog which is trained to expect an electric shock shortly after a red light is flashed on a screen, has recognized a sign foretelling an event. This type of learning has been sharply illuminated, but also somewhat distorted, by Pavlov's experiments, in which he induced salivation in dogs by giving them definite signs (like ringing a bell) that food was forthcoming. In Pavlov's terms the sound of a bell announcing food, the conditioned stimulus, replaces in its effect the presentation of food, the unconditioned stimulus. Similarly, according to Pavlov, a red light announcing the imminence of an electric shock would be supposed to act like the shock itself on the trained animal. But none of these is quite true: the dog does not jump and snap at the bell as if it were food, nor does a red light cause the kind of muscular contraction which results from an electric shock. In fact, the 'conditioned response' differs quite generally from the original 'unconditioned response', in the same way in which the anticipation of an event differs from the effect of the event itself.[1] This entitles us to say, in contrast to Pavlov's description of the process, that in sign-learning the animal is taught to expect an event by recognizing a sign foretelling the event.

A closer analysis of sign-learning is provided by using a discrimination box, of which there are different types. For example, the animal is faced with two doors leading to two compartments with different markings that can be shifted from one door to the other. The animal, usually a rat, is trained to recognize the marking which signifies the presence of food behind the door as distinct from the other marking, behind which no food will be found. The greater freedom of action given here to the animal allows its behaviour to reveal some of the preliminary stages through which learning is achieved.

The first stage consists in realizing the presence of a problem. To induce this, the animal is presented with a version of the situation so simplified that it grasps it at a glance. Food is first offered openly in one or the other of the two compartments; then the entrances to the compartments are closed and the animal is made to push open the doors, behind which he will find either the food or an empty space. These experiences establish an

---

[1] This criticism of the conditioned reflex theory is well known. See e.g. D. O. Hebb, *The Organization of Behavior*, New York, 1949, p. 175.

awareness that food is hidden in one of the two compartments and that it can be got at by pushing open the right door. The understanding of this problem rouses the animal to search for food by pushing its way into one or the other compartment. It is by such attempts at guessing the right compartment that it will eventually hit on the fact that certain markings on the door signify the presence of food behind the door.

There is evidence that during these attempts the animal's choices are not random, but follow from the start some such system as 'turn always to the right' or 'always to the left' or 'alternately right and left', until it eventually tumbles to the relevance of the markings and then fairly rapidly identifies the true one.[1] The whole process clearly shows the animal's capacity to be intrigued by a situation, to pursue consistently the intimation of a hidden possibility for bringing it under control, and to discover in the pursuit of this aim an orderly context concealed behind its puzzling appearances. The essential features of problem-solving are thus apparent even at this primitive level.

While sign-learning results, like trick learning, in new motoric habits, these are comparatively trivial and only of secondary importance. What the animal will eventually do can be readily varied by slightly modifying the experimental apparatus, so that the learning of the sign event relationship results in quite different motoric actions. *Type B* learning consists therefore not primarily in the contriving of skilful actions, but in the *observing* of a *sign-event* relation on which these actions follow. Such learning is grafted primarily not on motility but on *perception*. Animals like rats and dogs are richly equipped by nature to grasp coherently the things they perceive, and sign learning appears to be an extension of this perceptive faculty by the power of intelligence.[2]

Animals learn only when impelled by desire or fear, and in this sense all learning is purposive. But while in the contriving of a useful trick purpose guides action directly, the observing of a useful sign is guided only by a general alertness of the senses, which is stimulated but not determined by any specific purpose. Thus trick learning, like the performance of human skills, is more completely controlled by purpose than sign-learning, which, like connoisseurship, is primarily the achievement of strained *attention*.

*Type C.* When an animal contrives a new trick it reorganizes its behaviour to serve a purpose by exploiting a particular means-end relationship; and similarly, the animal learning a new sign reorganizes its sensory field by establishing in it a valid and useful coherence between a sign and the event signified by it. Both forms of learning establish a time sequence, whether contrived or observed by the learner. (*Type A* or *B.*) Learning of

[1] See Hilgard, *Theories of Learning* (2nd edition), New York, 1956, pp. 106–7, quoting I. Krechevsky (1932 and 1933) on 'hypotheses' in rats. Lashley had already said that normal animals never behave in random fashion. (*Brain Mechanisms and Intelligence*, Chicago, 1929, p. 138.)

[2] Hilgard, *op. cit.* (1st edn., 1948), p. 333, distinguished between *motor learning* and *perceptual learning* (cf. 2nd edn., p. 466).

Type C occurs when the process of reorganization is achieved not by a particular act of contriving or observing, but by achieving a *true under-standing of a situation which had been open to inspection almost entirely from the start*. Type C has been described as *latent learning*, to suggest that in such cases the animal learns something which it can intelligently manifest in more numerous and less predictable ways than the lessons of trick or sign learning. Thus a rat which has learned to run a maze will show a high degree of ingenuity in choosing the shortest alternative path when one of the paths has been closed to it.[1] This behaviour of the rat is such as would be accounted for by its having acquired a mental map of the maze, which it can use for its guidance when faced with different situations within the maze.[2]

The capacity for deriving from a latent knowledge of a situation a variety of appropriate routes or alternative modes of behaviour amounts to a rudimentary logical operation. It prefigures the use of an articulate inter-pretative framework on which we rely as a representation of a complex situation, drawing from it ever new inferences regarding further aspects of that situation. Latent learning is transformed into pure problem-solving when the situation confronting the subject can be taken in by it from the start, at a glance. This reduces exploration to a minimum and shifts the task altogether to the subsequent process of inference. Learning becomes then an act of 'insight', preceded by a period of quiet deliberation; as we see demonstrated by the performances of Köhler's chimpanzees.

The functioning of a latent understanding as a guide to the act of problem-solving comes out most clearly by contrast, when understanding is only partial. The chimpanzee who piles up packing-cases in a grossly unstable manner (for example by placing them edgewise), shows that he has grasped the principle of gaining height by constructing a tower on which to climb up, without knowing the conditions for making the constructions stable. Its error is a 'good error', as Köhler calls it,[3] for it testifies to an ingenious process of inference which overreaches itself by relying partly on mistaken assumptions. Thus the very rise of inferential power brings with it the conjoint capacity for inferential error. We shall presently see this manifested further in the process of transposing practical problems into verbal terms (p. 93).

The development of inarticulate behaviour up to the point where it approximates and finally achieves an articulate form, can be watched in the maturing child. Observations of this kind, carried out extensively by Piaget, have been analysed by him in terms of such logical operations as

---

[1] This was beautifully demonstrated, for example, by an experiment of Tolman and Honzik described in Hilgard, *op. cit.*, 2nd edn., p. 194 (fig. 26), from E. C. Tolman and C. H. Honzik, *Univ. Calif. Publ. Psychol.*, **4**, (1940), 215–32. Hilgard mentions some recent criticism of this experiment but maintains his account of it.

[2] E. C. Tolman, 'Cognitive maps in rats and men', in *Collected Papers in Psychology*, Berkeley and Los Angeles, 1951, pp. 261–4 (from *Psych. Rev.*, **55** (1948), 189–208).

[3] W. Köhler, *The Mentality of Apes*, 2nd edn., London, 1927, pp. 123, 194.

he finds embodied in the behaviour of the child at consecutive stages of its maturation.[1] At the earliest stage, even more primitive than that usually studied in intelligence tests on animals, the infant can be observed building up a spatial framework. At first he does not recognize objects as permanent, but gives up any attempt to find them as soon as they are covered up. For example, when a watch is hidden by a handkerchief the child, instead of lifting the handkerchief, withdraws his hand. But with growing maturity, he learns that objects continue to exist even while not seen or felt, and learns to see them as having constant sizes and shapes though presented at different distances and from different angles.[2] Further improvements in the capacity for spatial orientation may be tested, for example, by an experiment in which three dolls of different colours strung on a wire are moved behind a screen, and children are asked to predict (1) the direct order of reappearance at the opposite side of the screen, and (2) the reverse order of return. The reverse order is predicted only at about 4–5 years, at the end of what Piaget calls the 'preconceptual period'.[3]

The progress achieved by the child in this manner has been described by Piaget as a development of its intelligence, but it might be more precise to call it an increased mental discipline, achieved by establishing a fixed interpretative framework of growing complexity. An inference guided by a fixed framework can always be traced back to its premises, and such 'reversibility', Piaget points out, may be regarded as a characteristic feature of disciplined thought.[4]

Reversibility can be contrasted with the irreversible processes which predominate throughout an important part of intelligent behaviour. In each case of the three types (A) Trick learning, (B) Sign learning, (C) Latent learning, we may distinguish between the process of learning, which is irreversible, and the performances achieved by learning, which are comparatively reversible. In the first two cases the distinction is clear enough. In case A we have the irreversible act of contriving a trick, as distinct from the subsequent performances, which involve no change in it and in this sense may be said to be reversible. In case B we have the irreversible act of establishing a sign-event relation, as distinct from the subsequent reversible performance of reacting to a sign already recognized as such. In case C the distinction is perhaps not always so clear. The first, irreversible phase may be one of systematic exploration, resulting in the gradual building up of an interpretative framework, but it may also be merely a puzzled contemplation of a situation, leading to a solution in a flash of insight. Again, the amount of ingenuity contributing an irreversible coefficient to the conceptual operations of the second phase, may vary considerably. Yet

[1] J. Piaget, *Psychology of Intelligence*, London, 1950.
[2] Piaget describes the way babies seem to explore the variable appearances of an object at different distances by alternately approaching it to their eyes or moving it away at arm's length. *Ibid.*, pp. 130 ff.
[3] *ibid.*, pp. 161–2.
[4] *ibid.*, p. 62; *Judgement and Reasoning in the Child*, London, 1928, pp. 173, 176.

in spite of this we may distinguish also in case C, clearly enough, between an act of insight, which is irreversible, and the resultant performance, which is comparatively reversible.

In each case the actual process of learning is covered by the first stage, while the second stage consists in displaying the knowledge acquired by learning. We may call the first a *heuristic* act by contrast to the second which is of a more or less *routine* character. For type A the heuristic act is a contriving; for B an observing; for C an understanding. The routine acts are: for A, the repeating of a trick, for B, the continued responding to a sign, and for C, the solving of a routine problem. The capacity for contriving, observing or understanding something for the first time cannot be rated intellectually below that of the performances based on the resultant knowledge. Therefore, we acknowledge already at this primitive level the existence of two kinds of intelligence: one achieving innovations, irreversibly, the other operating a fixed framework of knowledge, reversibly. Although at the inarticulate level of intellectual life this distinction may appear precarious, its more fully established manifestations in the corresponding domains of articulate intelligence are clearly enough prefigured here.

Our three types of animal learning are primordial forms of three faculties more highly developed in man. Trick-learning may be regarded as an act of *invention*; sign-learning as an act of *observation*; latent learning as an act of *interpretation*. The use of language develops each of these faculties into a distinctive science to which the other two contribute subsidiarily.

Thus, invention will include at its highest reaches the whole array of ingenious and useful operations of the kind that are described in patents and form the subjects of engineering and technology. Observation, even when restricted to the kind of things animals are concerned with in learning experiments, may be taken to include at the highest articulate level the whole of natural science. Experimental conditioning amounts from the animal's point of view to a process of inductive inference. An animal recognizing a sign-event relationship is therefore producing a primordial form of observational science.

The transition from inarticulate learning of type C to its articulate counterpart (which I have called Interpretation), has been traced in Piaget's work on the genesis of disciplined thought in children. Eventually, the operational rules implicitly governing the intelligent behaviour of the child growing up to adolescence will comprise a system of logic, together with the elements of mathematics and classical mechanics. The highest articulate forms of this type of intelligence are mathematics, logic and mathematical physics, or more generally, the deductive sciences. While applied mathematics is object-directed, pure mathematics has no outside object; being concerned with objects of its own creation, it may be described as 'object creating'.

At the articulate level of intelligence, heuristic acts fall distinctly apart from mere routine applications of established knowledge. They are the acts of the inventor and discoverer, which require originality and offer scope for genius, differing in this both from the performance of engineers who apply known devices and of teachers demonstrating established results of science. Intellectual acts of a heuristic kind make an *addition* to knowledge and are in this sense irreversible, while the ensuing routine performances operate within an *existing* framework of knowledge and are to this extent reversible. The wider significance of the difference between reversible and irreversible mental processes, and the bearing of this distinction on that between specifiable and unspecifiable forms of knowledge, will become apparent later.

### 3. Operational Principles of Language

I shall now try to define the main principles by which language becomes the instrument for the tremendous feats of articulation.

There are three main kinds of utterances, namely: (1) expressions of feeling, (2) appeals to other persons, (3) statements of fact. To each of these there corresponds a different function of language. The transition from the tacit to the articulate which I am envisaging here is restricted to the indicative forms of speech, as used for statements of fact.[1]

Admittedly, language is primarily and always interpersonal and in some degree impassioned; exclusively so in emotional expression (passionate communication) and imperative speech (action by speech), while even in declaratory statements of fact there is some purpose (to communicate) and passion (to express belief). In fact, it is precisely the ingredient of personal

---

[1] These three forms or functions of language are a matter for general agreement among linguistic theorists. The functions of 'Ausdruck, Appel, Darstellung' distinguished by K. Bühler (*Sprachtheorie*, Jena, 1934), are adopted, for example, by D. V. McGranahan, in 'The Psychology of Language' (*Psychological Bulletin*, 1936, 33, pp. 178–216); or by Bruno Snell in *Der Aufbau der Sprache* (Hamburg, 1952), p. 11. Cf. also George Humphrey, *Thinking*, London, 1951, p. 217. On the other hand, the question whether and how one of the three functions is pre-eminent in the *origin* of language, whether in the individual or the species, is a matter for extensive and sharp disagreement (see e.g. the survey by McGranahan, *loc. cit.*, pp. 179 f., or of types of expressive theory by L. H. Gray, *Foundations of Language*, New York, 1939, p. 40, *cf.* also G. Révész, *Origin and Prehistory of Language*, London, 1955). The present argument lies outside the range of this controversy, and its restriction to the representative function is not meant to endorse, e.g. a 'Representative' as against an 'Expressive' or 'Evocative' theory. I am engaged here not in constructing still another theory of the origin of language, but in an epistemological reflection on the relation of language to its inarticulate roots. Some of the theories of linguists have of course certain affinities with my own: for example, Sapir's analysis of the conceptual role of speech in *Language* (New York, 1921); or A. H. Gardiner's insistence, in the *Theory of Speech and Language* (London, 1932), on the importance of the 'thing-meant' in the speech situation; or W. J. Entwhistle's rebellion against the more extremely behaviouristic linguists: 'The chief error of the mechanical view is to eliminate Man from his own speech, treating the latter as if it were a machine independent of Man (*Aspects of Language*, London, 1953, p. 39). But the linguists are concerned, reasonably enough, with the verbal techniques of speech itself: not primarily, as I am, with the nature of spoken truth in view of its inarticulate and unformalizable grounds.

passion inherent in and necessary to even the least personal forms of speech which my argument seeks to exhibit. But the peculiar intellectual powers conferred by articulation can be recognized more clearly if we disregard this possibility for the moment, and attend principally to the bare indicative solitary use of language.[1] Even so, language should be taken from the start to include writing, mathematics, graphs and maps, diagrams and pictures; in short, all forms of symbolic representation which are used as language in the sense defined by the subsequent description of the linguistic process.[2]

The operational principles of language which account for the entire intellectual superiority of men over animals seem to be twofold. The first controls the process of linguistic *representation*, the second the *operation* of symbols to assist the process of thought. Each principle can be demonstrated by taking its advantages to an extreme and obviously absurd limit of perfection, and thus exhibiting the necessity of a restraint which had thereby been left out of account.

(1) Suppose you wanted to improve a language by increasing its richness indefinitely. We can get an idea of the enormous number of printed or written words that could be formed by different combinations of phonemes or letters, by envisaging the fact that from an alphabet of 23 letters we could construct $23^8$, i.e. about one hundred thousand million eight-letter code words. This should allow us to replace each different sentence ever printed in the English language by a different printed word, so that this code word (which would function as a verb) would cover what that sentence asserts. This millionfold enrichment of the English language would completely destroy it; not only because nobody could remember so many words, but for the more important reason that they would be meaningless. For the meaning of a word is formed and manifested by its repeated usage, and the vast majority of our eight-letter code words would be used only once or at any rate too rarely to acquire and express a definite meaning. It follows that a language must be poor enough to allow the same words to be used a sufficient number of times. We may call this the Law of Poverty.[3]

Of course, if ten thousand words must do duty for making ten thousand

---

[1] Again, this is not to prejudge the case for the importance of *tu* as against *ego* in the beginnings of speech (see Entwistle, *op. cit.*, pp. 15–24), nor to enter into the controversy as to how egocentric or otherwise is the speech of small children (see D. McCarthy, 'Language Development' in Murchison, *Handbook of Child Psychology*, Worcester, Mass., 1933, pp. 278–315). I am merely concerned here to deal with one aspect of language which undoubtedly does exist.

[2] I am here drawing a dividing line at a rather different place from the one usual among psychologists, who have been interested, back to the Würzburg school, in the distinction between verbalized and 'wordless' thought. I should prefer to consider, along with Samuel Butler in the essay already quoted, that Mrs. Bentley's snuff-box was language.

[3] Cf. Locke, *Essay concerning Human Understanding*, Book III, ch. 3, sect. 2–4, where the existence of universal terms is derived from a somewhat similar argument. Cf. also E. Sapir, *op. cit.*, p. 11.

million statements, this can be achieved only if we can form combinations of words so that they jointly express an intended meaning. A fixed vocabulary of sufficient poverty must therefore be used within some fixed modes of combination, which always have the same meaning. Only grammatically ordered word clusters can say with a limited vocabulary the immense variety of things that are apposite to the range of known experience.[1]

The Laws of Poverty and Grammar do not exhaust the first operational principle of language. They refer to words, but words are not words unless they are both identifiably repeated and consistently used. Thus, underlying the Laws of Poverty and Grammar, we have two further requirements: the Laws of Iteration and Consistency.

In order that words may be identifiably repeated in different spoken or written sentences, phonemes and letters must be repeatable. They must be chosen for and defined by some feature possessing the kind of distinctiveness which Gestalt-psychology has described as *prägnanz* and which I have acknowledged before, in Part One, together with other types of order, by contrasting it with random configurations. The process of repeating or identifying words (whether in speech or writing) can of course never be quite free from hazards, whence arise verbal errors which may falsify history[2] or lead to permanent changes in linguistic usage.[3] Mispronunciations and the confusion of similar words are (or at least were until recently) the stock-in-trade of music halls for making fun of less educated people. Phonemes, scripts and words are *good* if they reduce these hazards by their distinctive gestalt.[4]

If their identifiable shape distinguishes words from shapeless utterances, like groans or squeaks, their consistent use distinguishes them from clearly repeatable utterances—like tunes—which have no consistent use for conveying an expression, an appeal or a statement. Only when repeatable utterances are used consistently can they have a definite meaning, and utterances without definite meaning are not language. The poverty of language can fulfil its denotative functions only if utterances are both repeatable and consistent.

'Consistency' is a deliberately imprecise term designating an unspecifiable quality. Since the world, like a kaleidoscope, never exactly repeats any

[1] Cf. E. Sapir, *op. cit.*, p. 39.

[2] As for example when Michael Bruce is reputed to have saved the life of Lafayette, because historians have substituted 'Lafayette' for the obscurer title of the Marquis de Lavalette.

[3] See Snell, *op. cit.*, p. 171, quoting Leumann, *Homerische Wörter*, Basel, 1950, etc. Cf. also S. Ullmann, *The Principles of Semantics*, Glasgow, 1951, pp. 234 ff.

[4] I. A. Richards ('Responsibilities in the teaching of English', Harvard Educational Review, **20** (1950), p. 37), observes that the distinctiveness of a sign consists in its safety from being mistaken for another sign. In the Latin alphabet the three letters o, c, e are least distinctive for they are more or less incomplete forms of each other. Liable to be mistaken for each other are also symmetricals, p b, q b, u n, p q, d b; and among numerals 6 and 9. The article mentions the 'trick of seeing' a letter, which is more difficult to learn for letters that can be more easily mistaken for each other.

previous situation (and indeed, if it did we would not know it, as we would have no means of telling that time had passed in between), we can achieve consistency only by identifying manifestly different situations in respect to some particular feature, and this requires a series of personal judgments. First, we must decide what variations of our experience are irrelevant to the identification of this recurrent feature, as forming no part of it, i.e. we must discriminate against its random background. Secondly, we must decide what variations should be accepted as normal changes in the appearance of this identifiable feature, or should be taken, on the contrary, to discredit this feature altogether as a recurrent element of experience. Thus the Laws of Poverty and Consistency imply that every time we use a word for denoting something, we perform and accredit our performance of an act of generalization and that, correspondingly, the use of such a word is taken to designate a class to which we attribute a substantial character.

Moreover, by being prepared to speak in our language on future occasions, we anticipate its applicability to future experiences, which we expect to be identifiable in terms of the natural classes accredited by our language. These expectations form a theory of the universe, which we keep testing continuously as we go on talking about things. So long as we feel that our language classifies things well, we remain satisfied that it is right and we continue to accept the theory of the universe implied in our language as true.

The nature of this universal theory which we accept by using a language can be more clearly understood as follows. Of the 2000–3000 English words in common usage today, each occurs on the average a hundred million times in the daily intercourse of people throughout Britain and the United States. In a library of a million volumes using a vocabulary of 30,000 words, the same words will recur on the average more than a million times. A particular vocabulary of nouns, adjectives, verbs and adverbs, thus appears to constitute a theory of all subjects that can be talked about, in the sense of postulating that these subjects are all constituted of comparatively few recurrent features, to which the nouns, adjectives, verbs and adverbs refer.[1] Such a theory is somewhat similar to that of chemical compounds. Chemistry alleges that the millions of different compounds are composed of a small number—about a hundred—of persistent and identical chemical elements. Since each element has a name and characteristic symbol attached to it, we can write down the composition of any compound in terms of the elements which it contains. This corresponds to writing down a sentence in the words of a certain language. The parallel can be pushed still further. We may regard the system of brackets used for specifying the internal structure of a compound of given composition, as analogous to grammatical constructions which indicate the internal relations between the things denoted by the words composing a sentence.

[1] The question whether adverbs are 'really' words, or only pseudo-words (see S. Ullmann, *op. cit.*, pp. 58–9) may be left open here.

To talk about things, we have seen, is to apply the theory of the universe implied by our language to the particulars of which we speak. Such talk is therefore continuous with the process described in Part One, by which the theories of the exact sciences are brought to bear on experience. But the connection is still closer with the descriptive sciences to be treated later in Part Four. To classify things in terms of features for which we have names, as we do in talking about things, requires the same kind of connoisseurship as the naturalist must have for identifying specimens of plants or animals. Thus the art of speaking precisely, by applying a rich vocabulary exactly, resembles the delicate discrimination practised by the expert taxonomist.

The lesson derived in Part One from reflecting on the application of the exact sciences to experience can be extended now as follows. We have seen that in all applications of a formalism to experience there is an indeterminacy involved, which must be resolved by the observer on the ground of unspecifiable criteria. Now we may say further that the process of applying language to things is also necessarily unformalized: that it is inarticulate. Denotation, then, is an art, and whatever we say about things assumes our endorsement of our own skill in practising this art. This personal coefficient of all affirmations inherent in the use of language will be presently reconsidered in the wider context of ineffable knowledge and ineffable thought.

(2) The second operational principle of language can be discovered from the absurdity of taking to its limit another manner of perfecting language. I can best exemplify this by the process of mapping. A map is the more accurate the nearer its scale approaches unity, but if it were to reach unity and represent the features of a landscape in their natural size, it would become useless, since it would be about as difficult to find one's way on the map as in the region represented by it. We may conclude that linguistic symbols must be of reasonable size, or more generally that they must consist of easily manageable objects. The manageable size of printed language enables a single shelf holding the *Encyclopaedia Britannica* to contain information ranging over all the largest and the most minute objects in existence. Language can assist thought only to the extent to which its symbols can be reproduced, stored up, transported, re-arranged, and thus more easily pondered, than the things which they denote. Churches and pyramids are symbols but they are not language because they cannot be easily reproduced or handled. We may call this requirement the Law of Manageability.

This requirement has been anticipated already to some extent, by assuming that we can utter the same designation on repeated occasions and that we can compose a great number of different sentences by putting together the same words according to certain rules. But the services of manageability go far beyond this in enlarging the intellectual powers of man.

In the most general terms, the principle of manageability consists in devising a representation of experience which reveals new aspects of it. This principle can be put into operation simply by writing down or otherwise uttering a designation of an experience, from which we can directly read off novel features of it. Alternatively, the manageability of symbols may include their capacity to be manipulated according to rules acknowledged as symbolic operations, or else merely to be handled informally, as when we turn the pages of a book in order to reconsider its subject.

These services of manageability to thought can all be described as taking place in three stages:

1. Primary denotation.
2. Its reorganization.
3. The reading of the result.

Stages 2 and 3 are merged into one when reorganization occurs mentally by a novel reading of the primary denotation.

Each of the three stages may be relatively trivial or else may require various grades of ingenuity up to that of genius (of which I shall treat later). Furthermore, the process of reorganization may be taken to include the transposition of the primary denotation into another set of symbols, as when numerical observations are represented by graphs or verbal statements by equations, a process which may also require considerable ingenuity.

We have seen that in the process of latent learning, described as Type C, animals reorganize their memories of experience mentally. It appears now that the intellectual superiority of man is due predominantly to an extension of this power by the representation of experience in terms of manageable symbols which he can reorganize, either formally or mentally, for the purpose of yielding new information. This enormously increased power of reinterpretation is of course ultimately based on that relatively slight superiority of the tacit powers which constitute our gift of speech. To speak is to *contrive* signs, to *observe* their fitness, and to *interpret* their alternative relations; though the animal possesses each of these three faculties, he cannot combine them.[1]

## 4. The Powers of Articulate Thought

The following examples should illustrate the immense range of mental powers generated by the simple machinery of denoting, reorganizing and reading, and should show at the same time that though our powers of thought be ever so much enhanced by the use of symbols, they still operate ultimately within the same medium of unformalized intelligence which we share with the animals.

[1] It is this intelligence which the infant begins to develop as he begins to speak. See J. Piaget, 'Le Language et la Pensée du point du vue génètique', in G. Révész, *Thinking and Speaking*, Amsterdam, 1954, p. 51; W. F. Leopold, 'Semantic Learning in Infant Language', *Word*, 4 (1948), pp. 173–80.

The use of a geographical map for finding one's way offers, for this exceptionally simple case, a rough numerical estimate of the inferential powers derived from a suitably arranged representation of experience. A rough map of England can be drawn by marking by dots on a sheet of paper the geographical positions of the 200 largest English towns, the Cartesian co-ordinates of each dot being chosen in a constant proportion to the longitude and latitude of one town, and each dot having the name of the corresponding town printed below it. From such a map we can read off at a glance the itineraries by which we can get about from any town to any other, so that our original input of 400 positional data (200 longitudes and 200 latitudes) thus yields $\dfrac{200 \times 200}{2} = 20{,}000$ itineraries. Actually, the information derived from mapping will be much ampler even than this. Each itinerary will comprise on the average some fifty places, amounting to about a million items, that is 2500 times the input.

The original catalogue of 200 towns, listing their longitudes and latitudes, would be comparatively useless, for it does not represent their mutual position in a way which the eye can readily take in. We may regard the transposition of the catalogue into the shape of a map as a formal operation carried out on its data, to be followed by the informal operation of reading off a variety of itineraries from the map. Similarly, the reports coming in during the air battles over England (1940) were continuously pictured at Air Force Command on a large table, offering thereby to the Supreme Commander a representation of the changing situation which he could grasp far better than the reports themselves. We know how the mere plotting of a series of numerical data on paper in the form of a graph, may reveal functional relationships quite unsuspected from our knowledge of the original figures. An example of this is the graphic representation of time-tables used for the direction of railroad traffic, which shows immediately the places and times at which trains overtake or meet each other, a piece of information which is not easily deduced from ordinary time-tables.

In all these instances of the enhancement of our intellectual powers by suitable symbolization, it is clear that the mere manipulation of symbols does not in itself supply any new information, but is effective only because it assists the inarticulate mental powers exercised by reading off their result. This may not be so obvious for the process of deriving new information by means of mathematical computations; but it is true all the same here too. Supposing we know that Paul is one year less than twice the age of Peter, while the difference between their ages is four, and we want to find the ages of each. We have first to set out the situation symbolically; age of Paul $x$, age of Peter $y$, and $x = 2y-1$; $x-y = 4$. Then we operate on the symbols and obtain $x = 9$, $y = 5$; which is finally read out as: Paul is aged 9 and Peter is aged 5. However crudely mechanical this procedure may be, its performance does require a measure of controlling intelligence.

The original situation of Peter and Paul must be understood and the problem involved in it clearly recognized; then its symbolic representation, including the subsequent operations, has to be correctly performed and the result correctly interpreted. All of this requires intelligence, and it is in the course of these tacit feats of intelligence that the formal operations utilized in the process are accredited and their result accepted by the person carrying them out.

The operations of the few simple principles illustrated here can in fact account (in terms of the first approximation defined on page 70) for the expansion of human intelligence from the basic types of inarticulate learning observed in animals to the articulate domains of engineering, of the natural sciences and of pure mathematics.

Take first the natural sciences, both of the exact and the descriptive kind. The numerical denotation of experience, followed by computations yielding new information, can be expanded into the logical machinery of the exact sciences by including in our computation the use of a formula representing a law of nature. I have already dealt at some length with the exact empirical sciences as a system of formalisms in Part One and shall return to the subject in the next chapter.

To the descriptive sciences, like zoology and botany, we can advance, as has been hinted already, from a more primitive level of articulation, relying only on rudimentary or at any rate quite informal logical operations. These sciences are an expansion of ordinary speech by the addition of a scientific nomenclature, while the symbolic operation on which they mainly rely is the systematic accumulation of recorded knowledge, and the rearranging and reconsidering of these records from new points of view.

Yet even here the process of articulation has rendered immensely effective assistance to our native mnemonic powers. Man is not much superior to a rat in finding his way in a maze; and it is not clear that he possesses in other ways either much greater native intelligence than the animal for reorganizing remembered experiences. But the bare unaided memory of animals can only collect scraps of information, unsystematically; nor could man do much better, but for the power of systematization dependent on speech. And even so, not until the invention of printing enormously speeded up the reproduction of records and made them much more compendious, could descriptive zoology and botany expand from the Aristotelian and medieval natural history covering a few hundred types to a systematic science comprising millions of species.

Decisive assistance is rendered to memory by the compilation of manageable records also in the great domains of humanistic scholarship such as history, literature and law, to which I may refer in passing, though my programme has excluded for the time being the class of interpersonal articulation to which these branches of scholarship belong. Their progress depends altogether on the expansion of printed records derived from the renewed exploration of primary sources, which themselves are largely printed records or printed literary works. Books setting out such information concisely and libraries which make the books readily accessible are decisive in enlarging the opportunities of such scholarship.

Allied to the mnemonic services of articulation is their capacity for assisting the speculative imagination of the inventor. The inventor's sketchbook is his laboratory. There is a standard experiment to test inventiveness, in which a person is confronted with two ropes hanging from the ceiling and almost reaching to the floor, the points of suspension being so far apart that while holding the end of either rope in one hand you cannot reach the other rope as it hangs straight down.[1] The task is to tie the two ends of the ropes together. People who failed to discover how to do this, readily found the solution when they drew a picture on paper of the arrangement set up in front of them. Articulation pictures the essentials of a situation on a reduced scale, which lends itself more easily to imaginative manipulation than the ungainly original; it thereby makes possible a science of engineering.

Thus the joint application of the two operational principles of language can be seen to expand speech into the texts of science and technology. But the invention of suitable symbols and their manipulation according to fixed rules can transcend altogether the task of dealing with matters of experience. Processes of inference, conducted by symbolic operations, can be carried out without reference to actually counted or measured entities, and such inferences may be interesting. Hence pure mathematics is possible.

Like chessmen, the symbols of pure mathematics stand not, or not necessarily, for anything denoted by them, but primarily for the use that can be made of them according to known rules. The mathematical symbol embodies the conception of its operability, just as a bishop or a knight in chess embodies the conception of the moves of which it is capable. The invention of new mathematical symbols which can be used in a more interesting or practically more effective manner has been going on through the centuries. The conception of numbers is present already in animals, but, by successive symbolic inventions, man has developed it far beyond its original range of six to eight integers. The advent of positional notation, of arabic numerals, of the zero sign and the decimal point, have facilitated the invention of arithmetical operations which have both greatly enriched our notion of numbers, and made the practical application of numbers for counting and measuring more powerful.

A notation invented by one mathematician may suggest to another some interesting variation of the corresponding conception. Laplace remarks how fortunate was Descartes' notation of the exponent of a power in stimulating speculations about the possibility of other than positive integer powers.[2] Some questions of number theory had long remained unapproachable

---

[1] N. R. F. Maier, 'Reasoning in humans—II', *Journal of Comparative Psychology*, **12** (1931), pp. 181–94.
[2] F. Laplace, *Traité de Probabilité, Ouevres*, Acad. Sc. edn., 1886, **7**, p. 2.

on account of the forbidding labour of the computations required to explore them, until the construction of electronic computers speeded up these manipulations many thousand times. Thus the progress of mathematics depends greatly on the invention of expressive and easily manipulable symbols for the representation of mathematical conceptions.

The rise of formal logic resembles the advances made in pure mathematics by the advent of happy symbolic innovations. Logical symbols allow us to state clearly such complex sentences as would be quite incomprehensible in ordinary language. The range of manageable grammatical structures being thus vastly increased, we can also perform in these terms feats of deductive argument which could not conceivably be attempted otherwise. This has opened up a new domain of inferences of such ingenuity and profundity as to be worth serious cultivation for its own sake.

The surprisingly varied terms in which systems of algebra or geometry can be interpreted, demonstrate the tenuousness of their denotative functions. They do not refer to particular things and may be altogether empty categories, well defined, but applying to nothing. Thus the infinite set $\aleph$ counts all numbers, the next larger infinite sets $\aleph_1$ and $\aleph_2$ count respectively all geometrical points and all conceivable curves, but the sets $\aleph_3$, $\aleph_4$ ... etc. are infinitely larger than any set of objects so far conceived, and so they apply to nothing definite at all—without being disqualified thereby as mathematical entities. These self-contained systems of pure mathematics may tell us something which is important, without primarily referring to anything outside themselves. Thus the second operational principle of articulation predominates here altogether over the first. Indeed, mathematics deploys the highest powers of this principle and testifies to our pleasure in exercising these powers. Of this intellectual passion, which is essential to mathematics, I shall say more in the next chapter.

We have now before us the following sequence of sciences relying decreasingly on the first and increasingly on the second operational principle of language: (1) the descriptive sciences, (2) the exact sciences, (3) the deductive sciences. It is a sequence of increasing formalization and symbolic manipulation, combined with decreasing contact with experience. Higher degrees of formalization make the statements of science more precise, its inferences more impersonal and correspondingly more 're-versible'; but every step towards this ideal is achieved by a progressive sacrifice of content. The immense wealth of living shapes governed by the descriptive sciences is narrowed down to bare pointer-readings for the purpose of the exact sciences, and experience vanishes altogether from our direct sight as we pass on to pure mathematics.

There is a corresponding variation in the tacit coefficient of speech. In order to describe experience more fully language must be less precise. But greater imprecision brings more effectively into play the powers of inarticulate judgment required to resolve the ensuing indeterminacy of speech. So it is our personal participation that governs the richness of concrete experience to which our speech can refer. Only by the aid of this tacit coefficient could we ever say anything at all about experience—a conclusion I have reached already by showing that the process of denotation is itself unformalizable.

## 5. THOUGHT AND SPEECH.  I. TEXT AND MEANING

These recurrent suggestions regarding the participation of the tacit in the process of articulation must remain obscure until we define the process by which the tacit co-operates with the explicit, the personal with the formal. But we are not yet ready for a frontal attack on this question. We must examine first three characteristic areas in which the relation between speech and thought varies from one extreme type to an opposite extreme, through the intermediary of a balanced type, lying midway between them. These three areas are:

(1) The area where the tacit predominates to the extent that articulation is virtually impossible; we may call this the *ineffable domain*.

(2) The area where the tacit component is the information conveyed by easily intelligible speech, so that *the tacit is co-extensive with the text of which it carries the meaning*.

(3) The area in which the tacit and the formal fall apart, since the speaker does not know, or quite know, what he is talking about. There are two extremely different cases of this, namely (*a*) an ineptitude of speech, owing to which articulation encumbers the tacit work of thought; (*b*) symbolic operations that outrun our understanding and thus anticipate novel modes of thought. Both (*a*) and (*b*) may be said to form part of *the domain of sophistication*.

(1) When I speak of ineffable knowledge, this should be taken literally and not as a designation of mystic experience, to which I do not wish to refer at this stage. Even so my attempt to speak of the ineffable may be thought to be logically meaningless,[1] or alternatively, to offend against the Cartesian doctrine of 'clear and distinct ideas' which the early Wittgenstein transposed into terms of semantics in his aphorism: 'Of what cannot be said'—i.e. said exactly, as a sentence in natural science—'thereof one must be silent'.[2] The answer to both objections lies ready in the mass of observations presented in Part One and in the foregoing sections of Part Two, which have demonstrated everywhere the limits of formalization. These observations show that strictly speaking nothing that we know can

[1] Comp. Ernst Topitsch, 'The Sociology of Existentialism', *Partisan Review* (1954), p. 296.
[2] L. Wittgenstein, *Tractatus Logico-Philosophicus*, London, 1922, p. 1889, I shall have something to say later of the attempts to modify the demand for precision by referring to more informal kinds of language. Some of the difficulties encountered in this programme are described by P. L. Heath in 'The Appeal to Ordinary Language', *Philosophical Quarterly*, 2 (1952), pp. 1–12.

be said precisely;[1] and so what I call 'ineffable' may simply mean something that I know and can describe even less precisely than usual, or even only very vaguely. It is not difficult to recall such ineffable experiences, and philosophic objections to doing so invoke quixotic standards of valid meaning which, if rigorously practised, would reduce us all to voluntary imbecility. This will become clearer as we proceed to perform what such objections would condemn as meaningless or impossible.

What I shall say of ineffability will in fact cover largely the same ground which I have previously traversed for the demonstration of the unspeci-fiability of personal knowledge; the difference being that I shall now regard the unspecifiable part of knowledge as the residue left unsaid by a defective articulation. Such defectiveness is common and often glaring. I may ride a bicycle and say nothing, or pick out my macintosh among twenty others and say nothing. Though I cannot say clearly how I ride a bicycle nor how I recognize my macintosh (for I don't know it clearly), yet this will not prevent me from saying that I know how to ride a bicycle and how to recognize my macintosh. For I know that I know perfectly well how to do such things, though I know the particulars of what I know only in an instrumental manner and am focally quite ignorant of them; so that I may say that I know these matters even though I cannot tell clearly, or hardly at all, what it is that I know.

Subsidiary or instrumental knowledge, as I have defined it, is not known in itself but is known in terms of something focally known, to the quality of which it contributes; and to this extent it is unspecifiable. Analysis may bring subsidiary knowledge into focus and formulate it as a maxim or as a feature in a physiognomy, but such specification is in general not exhaust-ive. Although the expert diagnostician, taxonomist and cotton-classer can indicate their clues and formulate their maxims, they know many more things than they can tell, knowing them only in practice, as instrumental particulars, and not explicitly, as objects. The knowledge of such particu-lars is therefore ineffable, and the pondering of a judgment in terms of such particulars is an ineffable process of thought. This applies equally to con-noisseurship as the art of knowing and to skills as the art of doing, where-fore both can be taught only by aid of practical example and never solely by precept.

But the relationship of the particulars jointly forming a whole may be ineffable, even though all the particulars are explicitly specifiable. The subject matter of topographic anatomy is such an ineffable relationship and will serve us as an example to illustrate the principle of this type of ineffability.

The medical student first learns a list of bones, arteries, nerves, and

[1] Cf. A. N. Whitehead, *Essays in Science and Philosophy*, London, 1948, p. 73: 'There is not a sentence which adequately states its own meaning. There is always a background of presupposition which defies analysis by reason of its infinitude.' Whitehead proceeds to illustrate this maxim by the example 'One and one make two'. See below, Part Three, ch. 8.

viscera which constitutes systematic anatomy. This is hard on the memory, but mostly presents no difficulty to the understanding, for the characteristic parts of the body can usually be clearly identified by diagrams. The major difficulty in the understanding, and hence in the teaching of anatomy, arises in respect to the intricate three-dimensional network of organs closely packed inside the body, of which no diagram can give an adequate representation. Even dissection, which lays bare a region and its organs by removing the parts overlaying it, does not demonstrate more than one aspect of that region. It is left to the imagination to reconstruct from such experience the three-dimensional picture of the exposed area as it existed in the unopened body, and to explore mentally its connections with adjoining unexposed areas around it and below it.

The kind of topographic knowledge which an experienced surgeon possesses of the regions on which he operates is therefore ineffable knowledge. In saying this, I disregard altogether the act of personal knowing involved in forming the conceptions of normal anatomy from a great number of actual instances which vary in detail. Let all human bodies be taken as absolutely identical, and let it be assumed that we have unlimited time and patience for mapping out the internal organs of the body. Let the body be cut for this purpose into a thousand thin slices and each cross section be depicted in detail; and let us even grant for full measure that, by a superhuman feat of cramming, a student could memorize precisely the picture of all the thousand cross sections. He would know a set of data which fully determine the spatial arrangement of the organs in the body; yet he would not know that spatial arrangement itself. Indeed, the cross-sections which he knows would be incomprehensible and useless to him, until he could interpret them in the light of this so far unknown arrangement; while on the other hand, had he achieved this topographic understanding, he could derive an indefinite amount of further new and significant information from his understanding, just as one reads off itineraries from a map. Such processes of inference, which may involve sustained efforts of intelligence, are ineffable thoughts.

The shortcomings of the powers of mapping, of which we have here an extreme case, set in already the moment we pass from the mapping of objects lying on a plane to objects on a curved surface. We can map the whole surface of the earth on a flat sheet of paper only in the form of a distorted projection, while its representation by a globe is clumsy and shows only one hemisphere at a time. This inadequacy is increased to the level of an impossibility when we come to an intricate three-dimensional arrangement of closely packed opaque objects. Diagrams or demonstrations of instructive aspects of the aggregate will now merely offer clues to its understanding, while understanding itself must be achieved by a difficult act of personal insight, the result of which must remain inarticulate.[1]

---

[1] The same difficulty of effective representation arises in respect to other aggregates of opaque objects: e.g. the representation of the arrangement of atoms in a crystal

We have now two inadequacies of articulation before us; different and yet closely related. When I am riding a bicycle or picking out my macintosh, I do not know the particulars of my knowledge and therefore cannot tell what they are; when on the other hand I know the topography of a complex three-dimensional aggregate, I know and could describe its particulars, but cannot describe their spatial interrelations. The limitations of articulation are correspondingly different in the two cases. When arts of knowing are explained by maxims, these never disclose fully the subsidiarily known particulars of the art, so that the powers of articulation are already restricted at this stage. No such limitation is imposed on the articulation of a spatial topography, the particulars of which are fully accessible. The difficulty lies here entirely in the subsequent integration of the particulars, and the inadequacy of articulation consists altogether in the fact that the latter process is left without formal guidance. The degree of intelligence required from the student to perform the act of insight which ultimately conveys to him the knowledge of the topography, offers here a measure of the limitations of the articulation representing this topography.

This ineffable domain of skilful knowing is continuous in its inarticulateness with the knowledge possessed by animals and infants, who, as we have seen, also possess the capacity for reorganizing their inarticulate knowledge and using it as an interpretative framework. The anatomist exploring by dissection a complex topography is in fact using his intelligence very much like a rat running a maze; and since he cannot tell any more than the rat what he gets to know in this way, his understanding of topographic anatomy remains similar in this respect as well to that which rats acquire of a maze. We may say in general that by acquiring a skill, whether muscular or intellectual, we achieve an understanding which we cannot put into words and which is continuous with the inarticulate faculties of animals.

What I *understand* in this manner has a meaning for me, and it has this meaning in itself, and not as a sign has a meaning when denoting an object. I have called this earlier on an existential meaning.[1] Since animals have no language which could denote anything, we may describe all meaning of the kind that is understood by animals as existential. The learning of signs, which is the first step towards denotation, would then be only a special case

lattice or of the arrangement of parts in a complex machine. Students of crystallography or of engineering have to think in terms of these elements, the pictorial representation of which must always remain fragmentary. On machines, cf. F. Kainz, 'Vorformen des Denkens', in Révész, *op. cit.*, pp. 61–110: p. 85, 'das mechanische Denken'. The task of mapping geological strata presents similar problems, for which geologists have recently been devising new and imaginative techniques. See L. Dudley Stamp, *The Earth's Crust*, London, 1951; and for the newly developed 'ribbon technique' W. E. Nevill, 'The Millstone Grit and Lower Coal Measures of the Leinster Coalfield', *Proc. Royal Irish Acad.*, **58**, B1 (1956), plates III, IV, and V; or *British Regional Survey*, *Pennines and Adjacent Uplands*, Department of Scientific and Industrial Research: Geological Survey and Museum, 1954.

[1] P. 58.

of existential meaning, but when we come to a deliberately chosen system of signs, constituting a language, we must admit that these have a denotative meaning which is not inherent in a fixed context of things or actions.[1]

Now that I have spoken at some length of the ineffable, it is easier to see why this is neither impossible nor self-contradictory. To assert that I have knowledge which is ineffable is not to deny that I can speak of it, but only that I can speak of it adequately, the assertion itself being an appraisal of this inadequacy. Reflections of the kind that I made a moment ago, when recalling the particular contents of our knowledge which we cannot adequately specify, have served to substantiate the inadequacy of our articulation for the cases in question. Such reflections must of course appeal ultimately to the very sense of inadequacy which they intend to justify. They do not try to eliminate, but only to evoke more vividly our sense of inadequate representation, by persevering in the direction of greater precision and reflecting on the ultimate failure of this attempt.

I believe that we should accredit in ourselves the capacity for appraising our own articulation. Indeed, all our strivings towards precision imply our reliance on such a capacity. To deny or even doubt our possession of it would discredit any effort to express ourselves correctly, and the very conception of words as consistently used utterances would dissolve if we failed to accredit this capacity. This does not imply that this capacity is infallible, but merely that we are competent to exercise it and must ultimately rely on our exercise of it. This we must admit if we are to speak at all, which I believe to be incumbent on us to do.

(2) Having acknowledged our own capacity to distinguish what we know from what we may be saying about it, we are free to distinguish also between hearing a message and knowing what it conveys to us.[2] We may recall in this connection once more how, having just read a letter, I no longer knew in what language it was written, though I knew its content precisely.[3] The knowledge that I had acquired was the meaning of the message. This kind of knowledge, or meaning, resembles in its tacitness the kinds of knowledge that I have described as ineffable, but differs from them profoundly by its verbal origin. While I read the letter, I was consciously aware both of its text and of the meaning of the text, but my awareness of the text was merely instrumental to that of the meaning, so that the text was transparent in respect to its meaning. After putting the letter down, I

---

[1] Our widened use of the word 'understanding' makes it comprise the domain of 'conception' as well as that of 'schema', the term used by Claparède and Piaget for designating a complex motoric faculty. I shall use these words interchangeably, to stand for a kind of latent knowledge, or aspects of such knowledge, as distinct from any overt performances based on this kind of knowledge. Later on 'intuition' or 'insight' will be introduced to describe the act of understanding, particularly in mathematics.

[2] This is reminiscent of the distinction between 'nom' and 'sens' by Saussure (see Ullmann, *op. cit.*, pp. 70-1); but his insistence that he is considering this relation *apart* from the relation to the referent or thing-meant, puts his analysis out of court for the purpose of my argument.

[3] p. 57.

lost my conscious awareness of the text, but remained subsidiarily aware of it in terms of my inarticulate knowledge of its content.[1] Tacit knowledge is manifestly present, therefore, not only when it exceeds the powers of articulation, but even when it exactly coincides with them, as it does when we have acquired it a moment before by listening to or reading a text.[2]

Even *while* listening to speech or reading a text, our focal attention is directed towards the meaning of the words, and not towards the words as sounds or as marks on paper. Indeed, to say that we read or listen to a text, and do not merely see it or hear it, is precisely to imply that we are attending focally to what is indicated by the words seen or heard and not to these words themselves.

But words convey nothing except by a previously acquired meaning, which may be somewhat modified by their present use, but will not as a rule have been first discovered on this occasion. In any case, our knowledge of the things denoted by words will have been largely acquired by experience, in the same way as animals come to know things, while the words will have acquired their meaning by previously designating such experience, either when uttered by others in our presence or when used by ourselves. Therefore, when I receive information by reading a letter and when I ponder the message of the letter, I am subsidiarily aware not only of its text, but also of all the past occasions by which I have come to understand the words of the text, and the whole range of this subsidiary awareness is presented focally in terms of the message. This message or meaning, on which attention is now focussed, is not something tangible: it is the conception evoked by the text. *The conception in question is the focus of our attention, in terms of which we attend subsidiarily both to the text and to the objects indicated by the text.* Thus the meaning of a text resides in a focal comprehension of all the relevant instrumentally known particulars, just as the purpose of an action resides in the co-ordinated innervation of its instrumentally used particulars. This is what we mean by saying that we *read* a text, and why we do not say that we *observe* it.

While focal awareness is necessarily conscious, subsidiary awareness may vary over all degrees of consciousness. When reading a text or listening to speech we have a completely *conscious* subsidiary awareness of it, even while we remain consciously aware of the text also in terms of its message, to which we keep attending focally. The relation between words and thought

---

[1] Experiments have shown the rather obvious fact that the context of a text is learned faster than its words if the text is understood. J. A. McGeoch, *The Psychology of Human Learning*, New York and London, 1942, p. 166. In a more recent experiment at Oxford, when one group of people wrote out a summary of a passage of 300 words from memory immediately after hearing it, and a second group made a précis of the same passage while consulting the text, the remembered versions and the précis were found to be indistinguishable. The experimenter, Dr. Gumulicki, concludes that this indicates 'the operation of an unwittingly abstractive procedure which seems to develop concurrently with the process of understanding the passage as it is being read'. See Harry Kay, in *Experimental Psychology*, Ed. B. A. Farrell, 1955, p. 14.

[2] The classic text for this distinction is the *De Magistro* of St. Augustine.

is the same, therefore, whether we have the words consciously in mind or not. This will allow us to agree with Révész[1] that 'wordless' thought can be, and often is, founded on language, without agreeing at the same time to disqualify all ineffable mental processes as lacking the character of thought. Further comments on this point will follow.

(3) I have shown a domain in which both knowledge and thought are of necessity predominantly tacit, and then a second domain in which the tacit, on which our attention is focussed, is the meaning of speech to which we are listening or have just listened.[2] The domain of sophistication, on which we now enter, is formed by *not fully understood* symbolic operations which can be

(*a*) a fumbling, to be *corrected* later by our tacit understanding

(*b*) a pioneering, to be *followed up* later by our tacit understanding.

More precisely speaking, we should say that we are referring in both these cases to a state of mental uneasiness due to the feeling that our tacit thoughts do not agree with our symbolic operations, so that we have to decide on which of the two we should rely and which we should correct in the light of the other.

The first of these two types of disagreement occurs when children learn to speak. They often show themselves encumbered rather than assisted by their new articulate equipment, the operations of which they have not yet fully mastered. Piaget has observed how often children find verbal problems intractable, though they know, and have known for a long time, how to solve the practical problems corresponding to them. He concludes that all the operations of logic must be learned all over again on the verbal plane of thought.[3]

Although the gains made by casting our thoughts into articulate terms eventually outweigh by far these initial disadvantages, there will always remain certain chances of error—and even of grave error—which arise from our very adoption of an articulate interpretative framework. This risk is therefore inherent in the exercise of all higher forms of human reason. Animals can make mistakes; rabbits fall into traps, fish rise to the angler's fly, and such errors may be fatal. But animals are exempt from the errors due to elaborate systems of false interpretation, which can be established only in verbal terms. Animism, belief in witchcraft, oracles and taboos prevail universally among primitive people, and a tendency

---

[1] *op. cit.*, 'Denken und Sprechen', pp. 3 ff.

[2] There is no third domain where our attention is focussed on the words or other symbols in themselves, so that we utter and operate them altogether without attending to their meaning. Such purely mechanical handling of symbols, guided by no intelligent purpose, would be futile. Even when carrying out a computation by a machine, we turn its handle in confidence in the outcome, and in doing so rely on the operational principles of the machine. No meaningless thing can be acknowledged as a symbol and no meaningless manipulation can be acknowledged as a symbolic operation. In this sense formalization must always remain necessarily incomplete. This has already been repeatedly foreshadowed and will be further elaborated later.

[3] J. Piaget, *Judgment and Reasoning in the Child*, pp. 92, 93, 213, 215.

towards kindred superstitions can be found also in childhood. When superstition is superseded by philosophy and theology, or by mathematics and natural science, we become involved once more in new systems of fallacies from which our practice of mathematics, science, philosophy or theology can never be strictly free. The mind which entrusts itself to the operation of symbols acquires an intellectual tool of boundless power; but its use makes the mind liable to perils the range of which seems also unlimited. The gap between the tacit and the articulate tends to produce everywhere a cleavage between sound common sense and dubious sophistication, from which the animal is quite free.

The linguistic school of philosophy aims at eliminating such uncertainties by bringing the use of words under stricter control. But you cannot benefit from the formalization of thought, unless you allow the formalism which you have adopted to function according to its own operational principles, and to this extent you must abandon yourself to this functioning and risk being led into error. Remember how various new kinds of numbers—irrational, negative, imaginary, transfinite—were produced as a result of extending familiar mathematical operations into unexplored regions, and how these numbers, after having first been repudiated as meaningless, were eventually accepted as denoting important new mathematical conceptions. Such spectacular gains, achieved by the speculative use of mathematical notations for purposes not originally entertained, remind us that the major fruitfulness of a formalism may be revealed in its entirely uncovenanted functions, precisely at points where the peril seems greatest of its drifting into absurdity. Gödel has shown that the scope of mathematical formulae is indeterminate, in the sense that we cannot decide within a deductive system like arithmetic whether any set of axioms comprising the system are consistent or mutually contradictory.[1] We must commit ourselves to the risk of talking complete nonsense, if we are to say anything at all within any such system.

This is true also for ordinary language applying to matters of experience. It contains descriptive terms, each of which implies a generalization affirming the stable or otherwise recurrent nature of some feature to which it refers, and these testimonies to the reality of a set of recurrent features constitute, as we have seen (p. 80), a theory of the universe which is amplified by the grammatical rules according to which the terms can be combined to form meaningful sentences. So far as this universal theory is true, it will be found to anticipate, like other true theories, much more knowledge than was possessed or even surmised by its originators. We may recall as a crude model of this how even a small map multiplies a thousand-fold the original input of information; and add to this that, actually, the number of meaningful and interesting questions one could study by means of such a map is much greater and not wholly foreseeable. Much less can we control in advance the myriads of arrangements in which nouns,

---

[1] K. Gödel, *Monatsh. Math. Phs.*, **38** (1931), 173–98.

adjectives, verbs and adverbs can be meaningfully combined to form new affirmations or questions, thus developing, as we shall see, the meaning of the words themselves ever further in these new contexts. Verbal speculation may therefore reveal an inexhaustible fund of true knowledge and new substantial problems, just as it may also produce pieces of mere sophistry.

How shall we distinguish between the two? The question cannot be fully answered at this stage; but from what has been already said, we can see, at least in outline, by what method the decision will have to be reached. Three things will have to be borne in mind: the *text*, the *conception* suggested by it, and the *experience* on which this may bear. Our judgment operates by trying to adjust these three to each other. The outcome cannot be predicted from the previous use of language, for it may involve a decision to correct, or otherwise to modify, the use of language. On the other hand, we may decide instead to persist in our previous usage and to reinterpret experience in terms of some novel conception suggested by our text, or at least to envisage new problems leading on to a reinterpretation of experience. And in the third place, we may decide to dismiss the text as altogether meaningless.

Thus to speak a language is to commit ourselves to the double indeterminacy due to our reliance both on its formalism and on our own continued reconsideration of this formalism in its bearing on experience. For just as, owing to the ultimately tacit character of all our knowledge, we remain ever unable to say all that we know, so also, in view of the tacit character of meaning, we can never quite know what is implied in what we say.[1]

## 6. FORMS OF TACIT ASSENT

Before proceeding further, I must return for a moment to the point where I set out my programme for Parts Two and Three. I proposed there to bring the conception of truth into accordance with the following three facts which became broadly apparent from the start:

(1) Nearly all knowledge by which man surpasses the animals is acquired by the use of language.

(2) The operations of language rely ultimately on our tacit intellectual powers which are continuous with those of the animals.

(3) These inarticulate acts of intelligence strive to satisfy self-set standards and reach their conclusions by accrediting their own success.

I have already traced back these decisive tacit coefficients of articulation

---

[1] The irreducible indeterminacy inherent in the meaning of all descriptions and the origin and function of this indeterminacy in relating meaning to reality was affirmed and elaborated in my *Science, Faith and Society*, Oxford, 1946, pp. 8–9. Waismann's 'open texture' ('Verifiability', *PAS Suppl.*, **19** (1945), stated part of the same reflections within a context of regulative principles which I find unacceptable (see p. 113 below).

to the three basic types of learning in animals; but this does not account for our intensive personal participation in the search for and conquest of our knowledge. The origin of this intellectual striving which (somewhat paradoxically) both shapes our understanding and assents to its being true, must lie in an *active* principle. It stems in fact from our innate sentience and alertness, as manifested already in the lowest animals in exploratory movements and appetitive drives, and at somewhat higher levels in the powers of perception. Here we find self-moving and self-satisfying impulses of both purpose and attention which antedate learning in animals and themselves actuate learning. These are the primordial prototypes of the higher intellectual cravings which both seek satisfaction in the quest for articulate knowledge and accredit it by their own assent. In reaching out to these prototypes we must proceed from the higher to the lower forms of intellectual strivings and, accordingly, shall come to perception first and deal with drives afterwards.

Perception is manifestly an activity which seeks to satisfy standards which it sets to itself. The muscles of the eye adjust the thickness of its lens, so as to produce the sharpest possible retinal image of the object on which the viewer's attention is directed, and the eye presents to him as correct the picture of the object seen in this way. This effort anticipates the manner in which we strive for understanding and satisfy our desire for it, by seeking to frame conceptions of the greatest possible clarity.

But sharpness of contour does not always predominate in the shaping of what we see. Ames and his school have shown that when a ball set against a featureless background is inflated, it is seen as if it retained its size and was coming nearer.[1] This illusion seems to be due to the fact that in this case we accommodate our eyes to a closer range, even though in consequence the object gets out of focus. Worse still, we simultaneously increase the convergence of our eyes so that the two retinal images are displaced from corresponding positions, which would normally make us see the object double. These defects of the quality and position of our retinal images are accepted here by the eye, in the urge to satisfy the more pressing requirement of seeing the object behave in a reasonable way. Since tennis balls are not known to blow themselves up to the size of foot-balls, a ball which does so must be seen as approaching us, even though in shaping this perception the eye must override standards of correctness which it would otherwise accept as binding.

The rule that we follow in shaping the sight of the inflated ball is one that we taught ourselves as babies, when we first experimented with approaching a rattle to our eyes and moving it away again. We had to choose then between seeing the rattle swelling up and shrinking alter-

[1] A. H. Hastorf, 'The Influence of Suggestion on the Relationship between Stimulus Size and Perceived Distance', *J. Psychol.*, **29** (1950), pp. 195–217. Cf. W. H. Ittelson and A. Ames, 'Accommodation, Convergence and their Relation to Apparent Distance', *J. Psychol.*, **30** (1950), pp. 43–62; W. H. Ittelson, *The Ames Demonstration in Perception*, Princeton, 1952.

nately, or seeing it change its distance while retaining its size, and we adopted the latter assumption. By this way of seeing things we eventually constructed a universal interpretative framework that assumes the ubiquitous existence of objects, retaining their sizes and shapes when seen at different distances and from different angles, and their colour and brightness when seen under varying illuminations. (See p. 80 *ante*.)

This tremendous generalization, on which we base our understanding of the universe, we have in common with the higher animals. Their native sensory equipment and ours both set us similar standards of correct seeing, and it is this primordial standard which induces us to flout, in the case of the inflated ball, the contrary evidence of our retinal images. Indeed, it induces us to intervene actively in producing false evidence of the ball's approach to our eyes, by a misplaced effort of visual accommodation, undeterred by the fact that this destroys the sharpness and the binocular correspondence of our retinal images. The process illustrates clearly the active principle which seeks to establish a coherence between all the clues of visual perception, so that our subsidiary awareness of them in terms of what we see shall satisfy us of having truly comprehended the things seen.[1]

In a larger perspective, the present experience of seeing the inflated ball come nearer to our eyes appears merely as the last of a life-long chain of experiences encountered and shaped by us, to each of which we reacted to make sense of it as best we could, and which are now all subsidiarily effective in the shaping and comprehension of our present experience. The sensory clues offered by the inflated ball thus appear to be evaluated together with an immense array of past clues, gone beyond recall—but *not* without effective trace.

This process, by which the meaning of sensory clues is established in terms of our perceptions, is closely analogous to that by which we shape the meaning of denotative words in the lifelong course of applying them to a long series of identifiable instances. These linguistic identifications are in fact based primarily on the sensory identification of objects at varying distances, under varying angles and varying illumination, and merely extend the theory of the universe implied in our sensory interpretations to the wider theory, implied in the vocabulary by which we talk about things.

We owe to Gestalt psychology much of the available evidence showing that perception is a comprehension of clues in terms of a whole. But perception usually operates automatically, and gestalt psychologists have tended to collect preferentially examples of the type in which perception

---

[1] The normal way of seeing objects 'right way up' satisfies our self-set standards of coherence between visual and tactile, as well as proprioceptive, clues. Spectacles which invert our retinal pictures make us see objects 'upside down'. But after a few days' habituation to the spectacles the eye restores coherence once more by seeing things now *right way up through the spectacles*. On removing the spectacles, objects now appear *upside down, without the spectacles*, but eventually coherence is restored again by the re-establishment of *normal vision* (I. Köhler, *Die Pyramide*, 5 (1953), 92–5, 6 (1953), 109–13).

goes on without any deliberate effort on the part of the perceiver and is not even corrigible by his subsequent reconsideration of the result. Optical illusions are then classed with true perceptions, both being described as the equilibration of simultaneous stimuli to a comprehensive whole. Such an interpretation leaves no place for any intentional effort which prompts our perception to explore and assess in the quest of knowledge the clues offered to our senses. I believe this is a mistake, and shall say more in Part Four of the reasons for recognizing persons who use their senses as centres of intelligent judgment. At this stage it is enough to recall some features of this active personal participation.[1] We recognize it in the signs of watchfulness which distinguish an alert animal from one rendered listless by exhaustion or neurotic disturbance. A sign-learning experiment can succeed only if we can arouse the animal's interest in his situation and make him aware of a problem which can be solved by straining his powers of observation. This may of course be done by offering a reward. But once he has learned a trick, the animal's inclination to repeat it without reward, for the mere fun of it, shows that his pleasure in solving a problem has a purely intellectual component. It has been proved also that the learning of a maze goes on even when no reward is offered. The animal's intelligence is spontaneously alive to the problem of making sense of its surroundings.[2]

I shall presently return to these primordial signs of intellectual passions in the animal. As to ourselves, we should know well the joy of seeing things; the curiosity aroused by novel objects; the straining of our senses to make out what it is that we see and the vast superiority of some people in quickness of eye and penetrating powers of observation. I believe that we should acknowledge these sensory actions as proper strivings which we both share and rely on. This endorsement of our native powers of making sense of our experience according to our own standards of rationality should also make it possible for us to acknowledge the ubiquitous contributions made by sense perception to the tacit components of articulate knowledge. And, eventually, it should duly condition our manner of acknowledging truth in its articulate forms.

This analysis of perception bears on the traditional question whether an object is to be equated to the aggregate of the impressions made by it on our senses. Ryle's linguistic analysis dismisses this question as nonsensical on the ground that sense impressions cannot be observed and that everything we observe is an object.[3] That is true, but the problem remains. For we can 'see' objects without observing them as such. Babies probably always see them like that. The newborn child experiences the world without controlling it intellectually, for he lacks the integrative control of the organs

[1] For a similar critique of Gestalt theory see D. Katz, *Gestaltpsychologie.*, Basel, 1944; and M. Scheerer, *Die Lehre von der Gestalt*, Berlin & Leipzig, 1931, p. 142.
[2] The vivid alertness of perception could be exhibited also by reference to the pre-linguistic development of the child's intelligence. See pp. 74–5 above.
[3] G. Ryle, *The Concept of Mind*, London, 1949, pp. 234–40.

which direct the examination and identification of external objects. His ocular fixation being deficient, his eyes stare uncomprehendingly at the things around him. In this way he can see only coloured patches of no definite shape or size, appearing at no particular distance and undergoing perpetual changes of shade and colour. If faced with objects that are effectively camouflaged or altogether novel to them, adults too see only patches of colour. Patients born blind and gaining sight after an operation must laboriously learn to recognize objects; and similarly, chimpanzees reared in the dark need several weeks of practice to see even so interesting an object as their nursing bottle.[1] Besides, a deliberate act of contemplation can dissolve objects into patches of colour.[2] In passing from the visionary contemplation of an object to its observation, we do make an affirmation, therefore, of something which lies beyond what we had seen before. This is an act involving a commitment which can prove misguided. It establishes a conception of reality experienced in terms of a subsidiary awareness of the coloured patches which had previously been experienced as such in an act of contemplation.

If perception prefigures all our knowing of things, drive satisfaction prefigures all practical skills, and the two are always interwoven. The efforts made to satisfy our cravings and to avoid pain are guided by perception; and insofar as they lead to the satisfaction of our appetites this is in its turn a manner of ascertaining a fact, namely, that certain things are satisfying to our appetites. The pursuit of drives is a mute exploration, which in the event of success leads to a mute affirmation; and—as in the case of sensory perception—the process by which information is gained itself selects and correlates from its own point of view the things to which it refers, and judges them in their relation to its own motivation. Though the information that we thus acquire—for example by eating, smoking, or making love—is necessarily centred on ourselves as agents, such information does in fact enter emphatically into our articulate picture of the world. To a disembodied intellect, entirely incapable of lust, pain or comfort, most of our vocabulary would be incomprehensible. For the majority of nouns and verbs refer either to living beings, whose behaviour can be appreciated only from an experience of the drives which actuate them, or to things made by man for his own use, which again can be appreciated only through an understanding of the human needs to which they minister.

Drive satisfaction and perception are the primordial rudiments of two classes of intelligent behaviour which manifest themselves—at a higher, though still inarticulate level—in learning of two types, the practical and the cognitive. The first (*type A*) enlarges on innate sensory-motor faculties by the grasping of new means-ends relationships, while the second (*type B*)

---

[1] A. H. Riesen, 'The Development of visual perception in man and chimpanzee', *Science*, **106** (1947), pp. 107-8; M.v. ⁀enden, *Raum und Gestaltauffasung bei operierten Blindgeborenen vor und nach der Operation*, Leipzig, 1932.

[2] See Part Two. ch. 6, pp. 197-200 below.

deploys the animal's innate sensory powers in the learning of new sign-event relations.

Learning of *type C*, by which an animal comes to understand and control a complex situation, uses both its motor and sensory faculties as part of a primitive conceptual operation. We may recognize the earliest rudiments of such combined operations in the animal's exploratory behaviour and in the incessant adjustment of its balance, culminating, e.g., in its stratagems for restoring its normal position when turned upside-down. These drives, which preserve the animal's rational coherence both within itself and in relation to its environment, foreshadow the learning of alternative part-whole relations at a more highly developed level of intelligence.

All these inarticulate achievements are guided by self-satisfaction. The adaptation of our sense organs, the urge of our appetites and fears, our capacity for locomotion, balancing and righting, as well as the processes of learning which an inarticulate intelligence develops from these strivings, can be said to be what they are and to achieve what they are said to achieve, only to the extent to which we accredit their implied assent to their own performance, shaped by them in accordance with standards set by themselves to themselves. Therefore, at each of the innumerable points at which our articulation is rooted in our sub-intellectual strivings, or in any inarticulate feats of our intelligence, we rely on tacit performances of our own, the rightness of which we implicitly confirm.

## 7. Thought and Speech. II. Conceptual Decisions

We may now begin to recognize the nature of the tacit faculty which accounts in the last resort for all the increase in knowledge achieved by articulation, and the nature of the urge to exercise it. We have seen this faculty revealed in somewhat different ways in all three characteristic relations between thought and speech. In the ineffable domain it made sense of the scanty clues conveyed by speech; in listening to a readily intelligible text and remembering its message, the conception grasped by it formed the focus of our attention; and lastly, it was seen to be the centre of operations for readjusting the tacit and the formal components of thought, which had fallen apart by a process of sophistication. The faculty on which we relied in all these situations was our power for comprehending a text and the things to which the text refers, within a conception which is the meaning of the text.

We have seen how the urge to look out for clues and to make sense of them is ever alert in our eyes and ears, and in our fears and desires. The urge to understand experience, together with the language referring to experience, is clearly an extension of this primordial striving for intellectual control. The shaping of our conceptions is impelled to move from obscurity to clarity and from incoherence to comprehension, by an

100

intellectual discomfort similar to that by which our eyes are impelled to make clear and coherent the things we see. In both cases we pick out clues which seem to suggest a context in which they make sense as its subsidiary particulars.

This may resolve the paradox that we intellectually owe so much to articulation, even though the focus of all articulation is conceptual, with language playing only a subsidiary part in this focus. For since the conceptions conveyed by speech, when speech is properly understood, make us aware both of the way our speech refers to certain things and of the way these things are constituted in themselves, we can never learn to speak except by learning to know what is meant by speech. So that even while our thoughts are of things and not of language, we are aware of language in all thinking (so far as our thinking surpasses that of the animals) and can neither have these thoughts without language, nor understand language without understanding the things to which we attend in such thoughts.

An illustration—akin to that of topographic anatomy by which we exemplified the ineffable—may exhibit this dual movement of comprehension in learning a language. Think of a medical student attending a course in the X-ray diagnosis of pulmonary diseases. He watches in a darkened room shadowy traces on a fluorescent screen placed against a patient's chest, and hears the radiologist commenting to his assistants, in technical language, on the significant features of these shadows. At first the student is completely puzzled. For he can see in the X-ray picture of a chest only the shadows of the heart and the ribs, with a few spidery blotches between them. The experts seem to be romancing about figments of their imagination; he can see nothing that they are talking about. Then as he goes on listening for a few weeks, looking carefully at ever new pictures of different cases, a tentative understanding will dawn on him; he will gradually forget about the ribs and begin to see the lungs. And eventually, if he perseveres intelligently, a rich panorama of significant details will be revealed to him: of physiological variations and pathological changes, of scars, of chronic infections and signs of acute disease. He has entered a new world. He still sees only a fraction of what the experts can see, but the pictures are definitely making sense now and so do most of the comments made on them. He is about to grasp what he is being taught; it has clicked. Thus, at the very moment when he has learned the language of pulmonary radiology, the student will also have learned to understand pulmonary radiograms. The two can only happen together. Both halves of the problem set to us by an unintelligible text, referring to an unintelligible subject, jointly guide our efforts to solve them, and they are solved eventually together by discovering a conception which comprises a joint understanding of both the words and the things.

But this duality of speech and knowledge is *asymmetrical*, in a sense anticipated on the inarticulate level, in the distinction (apparent already in the learning of animals) between knowledge and the performances based on knowledge. We have seen there that the acquisition of the type of knowledge called latent learning can be manifested by an indefinite range of performances, depending on the situation with which the animal is confronted after his training. Indeed, once it has learned something new, the animal's every subsequent reaction may be affected to some extent by its previously acquired knowledge: a fact which is known as the transfer of learning. And it is easy to see that knowledge, even when acquired verbally, has a 'latent' character; to express it in words is a performance based on our possession of such latent knowledge.

Take the knowledge of medicine. While the correct use of medical terms cannot be achieved in itself, without the knowledge of medicine, a great deal of medicine can be remembered even after one has forgotten the use of medical terms. Having changed my profession and moved from Hungary to England, I have forgotten most of the medical terms I learned in Hungary and have acquired no others in place of them; yet I shall never again view—for example—a pulmonary radiogram in such a totally uncomprehending manner as I did before I was trained in radiology. The knowledge of medicine is retained, just as the message of a letter is remembered, even after the text which had conveyed either kind of knowledge has passed beyond recall. To speak of this message, or of medical matters, is therefore a performance based on knowledge, and it is indeed only one of an indefinite range of conceivable performances by which such knowledge can be manifested. We grope for words to tell what we know and our words hang together by these roots. 'The true artists of speech', writes Vossler,[1] 'remain always conscious of the metaphorical character of language. They go on correcting and supplementing one metaphor by another, allowing their words to contradict each other and attending only to the unity and certainty of their thought.' Humphrey[2] rightly aligns the capacity to express knowledge in an unlimited variety of spoken terms, with the capacity of a rat to manifest its knowledge of a maze in an unlimited number of different actions.

## 8. THE EDUCATED MIND

By the time these pages appear in print, Donald Kellogg may have completed his studies at a university. He may be on his way to becoming a capable doctor, lawyer or clergyman, destined perhaps to become an authority on medicine, law or theology, or even a pioneer whose greatness will only dawn on generations yet to come—while Gua the chimpanzee, his playmate and intellectual rival until the age of a year and a half, will never have got beyond the stage of intelligence which they both reached

[1] K. Vossler, *Positivismus und Idealismus in der Sprachwissenschaft*, Heidelberg (1904), pp. 25–6. Cf. also I. Murdoch, 'On Thinking and Language', *PAS*. Suppl. vol. 25 (1951), p. 25. [2] G. Humphrey, *Thinking*, London, 1951, p. 262.

as infants. Donald has acquired all his superior knowledge by the exercise of inarticulate powers—exceeding those of Gua mainly in the capacity for combining the practical, observational and interpretative endowments which they both shared—for setting in motion the operational principles of speech, of print and of other linguistic symbols, and perhaps even enlarging this heritage of knowledge by discoveries of his own.

Knowledge acquired by education may be of various kinds. It may be medical knowledge, legal knowledge, etc., or simply the general knowledge of an educated person. We are clearly aware of the extent and special character of our knowledge, even though focally aware of hardly any of its innumerable items. Of these particulars we are aware only in terms of our mastery of the subject of which they form part. This sense of mastery is similar in kind to the inarticulate knowledge of knowing one's way about a complex topography, but its range is enhanced by the aid of verbal and other linguistic pointers, the peculiar manageability of which enables us to keep track of an immense amount of experience and to rest assured of having access, when required, to many of its countless particulars. Consciousness of our education resides ultimately, therefore, in our conceptual powers, whether applied directly to experience or mediated by a system of linguistic references. Education is *latent* knowledge, of which we are aware subsidiarily in our sense of intellectual power based on this knowledge.

The power of our conceptions lies in identifying new instances of certain things that we know. This function of our conceptual framework is akin to that of our perceptive framework, which enables us to see ever new objects as such, and to that of our appetites, which enables us to recognize ever new things as satisfying to them. It appears likewise akin to the power of practical skills, ever keyed up to meet new situations. We may comprise this whole set of faculties—our conceptions and skills, our perceptual framework and our drives—in one comprehensive power of anticipation.

Owing to the unceasing changes which at every moment manifestly renew the state of things throughout the world, our anticipations must always meet things that are to some extent novel and unprecedented. Thus we find ourselves relying jointly on our anticipations and on our capacity ever to re-adapt these to novel and unprecedented situations. This is true in the exercise of skills, in the shaping of our perception and even in the satisfaction of appetites; every time our existing framework deals with an event anticipated by it, it has to modify itself to some extent accordingly. And this is even more true of the educated mind; the capacity continually to enrich and enliven its own conceptual framework by assimilating new experience is the mark of an intelligent personality. Thus our sense of possessing intellectual control over a range of things, always combines an anticipation of meeting certain things of this kind which will be novel in some unspecifiable respects, with a reliance on ourselves to interpret them successfully by appropriately modifying our framework of anticipations.

This is no truism, but the very heart of our subject. The oddity of our thoughts in being much deeper than we know and in disclosing their major import unexpectedly to later minds, has been acknowledged in my first chapter as a token of objectivity. Copernicus anticipated in part the discoveries of Kepler and Newton, because the rationality of his system was an intimation of a reality incompletely revealed to his eyes. Similarly, John Dalton (and long before him the numerous precursors of his atomic theory) beheld and described the dim outline of a reality which modern atomic physics has since disclosed in precisely discernible particulars. We know also that mathematical conceptions often disclose their deeper significance only to later generations, by revealing yet unsuspected implications or undergoing a surprising generalization. Moreover, a mathematical formalism may be operated in ever new, uncovenanted ways, and force on our hesitant minds the expression of a novel conception. These major intellectual feats demonstrate on a large scale the powers which I have claimed for all our conceptions, namely of making sense beyond any specifiable expectations in respect to unprecedented situations.

Why do we entrust the life and guidance of our thoughts to our conceptions? Because we believe that their manifest rationality is due to their being in contact with domains of reality, of which they have grasped one aspect. This is why the Pygmalion at work in us when we shape a conception is ever prepared to seek guidance from his own creation; and yet, in reliance on his contact with reality, is ready to re-shape his creation, even while he accepts its guidance. We grant authority over ourselves to the conceptions which we have accepted, because we acknowledge them as intimations—derived from the contact we make through them with reality —of an indefinite sequence of novel future occasions, which we may hope to master by developing these conceptions further, relying on our own judgment in its continued contact with reality. The paradox of self-set standards is re-cast here into that of our subjective self-confidence in claiming to recognize an objective reality. This brings nearer by a great step the final conception of truth within which I shall seek to establish my balance of mind. But let me proceed with the subject on hand a little further for the moment.

## 9. The Re-interpretation of Language

I have shown that the educated mind relies for most of its knowledge on verbal clues. It follows, then, that its conceptual framework will be developed mostly by listening or speaking, and that its conceptual decisions will usually entail also a decision to understand or use words in a novel fashion. In any case, every use of language to describe experience in a changing world applies language to a somewhat unprecedented instance of its subject matter, and thus somewhat modifies both the meaning of language and the structure of our conceptual framework.[1] I implied this already, when I spoke of denotation as an art (p. 81), and likened the life-long process by which the meaning of words is established to that by which we interpret and re-interpret our sensory clues (pp. 98–100). These hints can now be consolidated and developed, within a fuller analysis of the way in which the reiteration of linguistic utterances with reference to identifiable occasions carries with it a change of their meaning, each time we either listen to speech or utter it ourselves.

The re-interpretation of language can take place at a number of different levels. (1) The child learning to speak practises it *receptively*. (2) Poets, scientists or scholars can propose linguistic *innovations, and teach others* to use them. (3) Re-interpretation takes place also at an *intermediate* level in the everyday use of language which modifies it imperceptibly, without any conscious effort at innovation.

I shall deal with all three of these cases in turn; but I must first pick up yet another thread which may serve us as a guide here. Piaget has des-cribed the subsumption of a new instance under a previously accepted con-ception as a process of *assimilation*; while he describes as *adaptation* the formation of new or modified conceptions for the purpose of dealing with novel experience.[2] I shall use these terms to describe the two associated movements by which we both apply and re-shape our conceptions at the same time; for I regard this combination as essential to all conceptual decisions, even though one of the two characters may predominate in any particular case.

The distinction between assimilation of experience by a fixed inter-pretative framework and the adaptation of such a framework to comprise the lessons of a new experience, gains a new and more precise meaning when the framework in question is articulate. The first represents the ideal of using language impersonally, according to strict rules; the second relies on a personal intervention of the speaker, for changing the rules of language to fit new occasions. The first is a routine performance, the second a heuristic act. A paradigm of the first is counting, which leaves its inter-pretative framework—the numbers used in counting—quite unchanged; the ideal of the second is found in the originality of poetic phrasing or of new mathematical notations covering new conceptions. Ideally, the first is strictly reversible, while the second is essentially irreversible. For to modify our idiom is to modify the frame of reference within which we shall hence-forth interpret our experience; it is to modify ourselves. In contrast to a formal procedure which we can recapitulate at will and trace back to its premises, it entails a conversion to new premises not accessible by any

---

[1] See W. Haas ('On Speaking a Language,' *PAS*, **51** (1951), pp. 129-66) on living language.

[2] J. Piaget, *Psychology of Intelligence*; cf. also *Plays, Dreams and Imitation in Child-hood*, London, 1951, p. 273. Piaget's terms are 'assimilation' and 'accommodation'. I am using 'adaptation' as a more acceptable English synonym of the latter. Piaget himself uses 'adaptation' in a more general sense, covering both types of process.

strict argument from those previously held. It is a decision, originating in our own personal judgment, to modify the premises of our judgment, and thus to modify our intellectual existence, so as to become more satisfying to ourselves.

But again, this urge to satisfy ourselves is not purely egocentric. Our craving for greater clarity and coherence, both in our speech and in the experience of which we speak, seeks a solution to a problem, a solution on which we may henceforth rely. It tries to discover something and to establish it firmly. We seek self-satisfaction here only as a token of what should be universally satisfying. The modification of our intellectual identity is entered upon in the hope of achieving thereby closer contact with reality. We take a plunge only in order to gain a firmer foothold. The intimations of this prospective contact are conjectural and may prove false, but they are not therefore mere guesses like betting on a throw of dice. For the capacity for making discoveries is not a kind of gambler's luck. It depends on natural ability, fostered by training and guided by intellectual effort. It is akin to artistic achievement and like it is unspecifiable, but far from accidental or arbitrary.

This is the sense in which I called denotation an art. To learn a language or to modify its meaning is a tacit, irreversible, heuristic feat; it is a transformation of our intellectual life, originating in our own desire for greater clarity and coherence, and yet sustained by the hope of coming by it into closer touch with reality. Indeed, any modification of an anticipatory framework, whether conceptual, perceptual or appetitive, is an irreversible heuristic act, which transforms our ways of thinking, seeing and appreciating in the hope of attuning our understanding, perception or sensuality more closely to what is true and right. Though each of these non-linguistic adaptations will affect our language, I shall continue to deal here only with the interaction between modifications of the conceptual and linguistic frameworks, to which I referred at the opening of this section.

(1) The first of the three levels on which I have proposed to illustrate the re-interpretation of language is that of the child learning to speak. Its early guesswork may appear floundering and foolish to adults, but the conjectural character of linguistic usage which it reveals is necessarily inherent in all speech and remains inherent in ours to the end. A child will point at the washing fluttering in the wind and call it 'weather', and call the pegs fastening the washing 'small weather', and the windmill 'big weather'. Such infantile false generalizations in guessing the meaning of words are known as 'childish verbalism',[1] but the mistakes persisting in adult life are quite similar. Few people seem to know, for example, that the common adjective 'arch' means 'cunning' or 'playfully roguish'. Even exceptionally well educated persons may tell you that it means 'oily', 'ingratiating', 'ironical', or 'pretending to be aristocratic'. For some years past the *Readers Digest* has published every week a different list of ten words known

[1] J. Piaget, *Judgment and Reasoning*, p. 115.

to most people, asking the reader to identify, out of three classes mentioned, that to which the notion designated by each word belongs; and rarely does anyone get the whole list right. We have a comparatively safe knowledge of the most frequently used words, but this assured vocabulary is surrounded by a swarm of half-understood expressions which we hardly ever venture to use at all. This hesitation reflects a sense of intellectual uneasiness, which induces us to grope for greater clarity and coherence.

I have already expressed my belief that we must credit ourselves with the capacity for appraising the inadequacy of our own articulation (p. 91). I shall now claim this capacity myself for saying that verbal mistakes go hand in hand with misunderstandings of the subject matter by which we feel puzzled. A child who uses the same word 'weather' for rain, clothespegs and windmills, has an unsatisfying and hence unstable conception of weather in which all these disparate things are amalgamated. I can still remember a puzzling conception I had as a child, in which buns and luggage were fused together, in view of the fact that I could not distinguish between the German words 'Gebäck' and 'Gepäck', which applied to the two respectively. Dylan Thomas tells how he fused in childhood the two meanings of 'front', designating the entrance to the house and the battlefields in France,[1] and wondered at the curious consequences flowing from this hybridization. Rarer words like 'epicene' or 'cynosure' evoke in most of us confused and uncertain conceptions combining disjointed clues, borrowed mostly from the meaning of similarly sounding words. Scholars continue to conjecture on precisely what conceptions are covered by such Greek terms as 'arete' and 'sophrosyne'; their guesses are guided by criteria of fitness similar to those on which the child relies in its fumblings to understand speech.

(2) Confusion may prevail for a long time also in some branch of the natural sciences, and be finally resolved only in conjunction with a clarification of terms. The atomic theory of chemistry was established by John Dalton in 1808 and generally accepted almost at once. Yet for about fifty years, in which the theory was universally applied, its meaning remained obscure. It came as a revelation to scientists when in 1858 Cannizaro distinguished precisely the three closely related conceptions of atomic weight, molecular weight and equivalent weight (weight per valence), which had been used until then in an indeterminately interchangeable manner. The appositeness of Cannizaro's interpretative framework brought new clarity and coherence into our understanding of chemistry. Such clarification is irreversible; it is as difficult to reconstruct today the confused conceptions which chemists used during the previous half century (and which for example induced Dalton to reject Avogadro's Law as contrary to the atomic theory of chemistry), as it is to be baffled once more by a puzzle after having discovered its solution. Remember also how, during almost a century after the first appearance of Mesmer, men of

[1] D. Thomas, 'Reminiscences of Childhood', *Encounter*, 3 (1954), p. 3.

science felt that they had either to accept the false claims of 'animal magnetism', or reject all the evidence in its favour as illusory or fraudulent, until at last Braid resolved the false dilemma by suggesting the new conception of 'hypnotism'.[1] Great pioneers of hypnotism, like Elliotson, had fallen tragic victims to the confusion previously prevailing, for lack of a conceptual framework in which their discoveries could be separated from specious and untenable admixtures.

Cannizaro and Braid made conceptual discoveries, which they consolidated by an improvement of language; their better understanding of their subject matter enabled them to speak about it more appositely. Such linguistic innovation is linked to the shaping of new conceptions in the same way as the learning of an established language is linked to the acquiring of current conceptions of its subject matter. As in the case of childish verbalism, the confusions of which we have seen examples in the natural sciences consist in a deficiency of intellectual control, which causes uneasiness, and is remediable by conceptual and linguistic reform.

I must digress here briefly to consider more closely the process by which confusion is eliminated in these different cases and others related to them. The falling apart of text and meaning, whether in the child or in the scientist, is the sign of a problematic state of mind. The seat of such confusion is always conceptual. There is independent evidence from the study of animals that confusion can arise on the purely inarticulate level.[2] Human confusion may be verbal, in the sense that it could not be produced without the use of language: a man speculating on the possibility of predicting his own actions is puzzled in a way no chimpanzee can be. Yet his perplexity is similar to that in which he would be involved by speculating on the possibility of lifting himself by his own shoe-laces; though this perplexity *could* be experienced inarticulately by a child or a chimpanzee trying to lift itself up in this manner.

[1] Part One, ch. 4, p. 51.
[2] The following is an example, observed by Köhler in a chimpanzee. The chimpanzee is at liberty and armed with a stick. A banana lies on the floor inside a cage; the cage is boarded on three sides; there is a gap in the horizontal boards on the side to which the objective is nearest, and there are vertical bars on the opposite side. Thus the arrangement is such that the animal can get at the banana only by pushing it away from the boarded-up side by means of its stick towards the bars on the opposite side, where she can reach it by walking round the cage. The animal has discovered this solution and has practised it before. She is now about to repeat it, and has started pushing the banana away from herself across the floor of the cage; but suddenly, interrupted by a noise, she apparently forgets her intention and yields to the more primitive impulse of pulling the banana towards herself (which is useless since the boards prevent her from reaching the prize)—then, having finished this useless pulling, *she walks round to the other side of the cage*, intending apparently to reach for the banana as usual, though it is of course quite inaccessible from here. 'Nobody', writes Köhler, 'could look more nonplussed than Chica, when she peered into the cage and saw the objective as far as it could be from the bars' (*op. cit.*, p. 267). The motoric schema 'pushing banana with stick away, followed by walking round to get it through the bars' gets mixed up here with the schema 'hauling banana towards yourself'. The animal goes on operating the first, though its lapse into the second had cancelled its premises. The uneasiness shown by animals puzzled by a problem will be described later more fully.

108

When a child confuses homonyms or fuses the meanings of similar-sounding words, or when it is perplexed by verbally formulated problems the answer to which it has long known how to find in practice, its use of language is obscuring what had previously been clear to its tacit understanding. Such childish sophistication can be cured by teaching children to understand and use speech in accordance with their anterior inarticulate understanding of the subject-matter. Modern analytic philosophy has demonstrated that this may hold also in philosophy. Philosophic problems may sometimes be dissolved by defining the meaning of their terms in accordance with our unsophisticated understanding of their subject matter.

But purely speculative problems are not always so fruitless. Speculations about lifting oneself by one's own shoelaces, for example, coincide essentially with speculations on mechanical devices of perpetual motion, which were resolved only by the discovery of mechanics, to which they effectively contributed. The paradox raised by Einstein as a schoolboy about the behaviour of light in a laboratory moving with the speed of light, was resolved only by Einstein's reform of the concept of simultaneity, and his conjoint establishment of special relativity. The fundamental part played by various logical and semantic paradoxes in stimulating the recent conceptual development in logic is equally notorious. I believe that the solution of philosophic puzzles, like that raised by the question whether we can predict our own actions, may also lead to important conceptual discoveries.[1] In fact, my present book rests on precisely these grounds; I am attempting to resolve by conceptual reform the apparent self-contradiction entailed in believing what I might conceivably doubt.

I have suggested before (p. 95) that when text and meaning fall apart we must choose whether to

(1) (*a*) Correct the meaning of the text.
    (*b*) Re-interpret the text.
(2) Re-interpret experience.
(3) Dismiss the text as meaningless.

Case (1*a*) is now seen to cover both the receptive process by which we improve our knowledge of a language, and the elimination of verbal puzzles by a stricter control of language, as practised by modern philosophy. Combinations of (1*b*) and (2) are exemplified by conceptual discoveries in science; analogous discoveries, not referring to experience, are possible in mathematics, to which I shall yet return. The dismissal of a text as meaningless, and of the problem raised by it as a pseudo-problem (Case (3)), may result from the philosophic clarification of its terms (Case (1*a*)).

Every one of these choices involves the shaping of meaning in the

[1] See M. Cranston, *Freedom: A new Analysis*, London, 1953, p. 163.

light of our standards of clarity and reason. Such a choice constitutes a heuristic act which may display the highest degree of originality. This I have illustrated just now by recalling the examples of Cannizaro and Braid. But I want also to mention again the case of Ernst Mach, who denounced as meaningless Newton's 'absolute space': a conception which the discovery of relativity later proved to be not meaningless but false.[1] For this error recalls others of its kind. When Poincaré said that a proportionate change in the linear dimensions of *all* solid bodies would be unobservable, and therefore empty of meaning,[2] he overlooked a host of consequences arising from the corresponding change in the relation of volumes and sizes. For a time it was thought that the Lorentz-Fitzgerald contraction was essentially unobservable,[3] which is false. The paradox of the Liar was long regarded as a mere sophism, without importance in logic,[4] in which it was later recognized as a fundamental problem. The interpretative act by which a question is dismissed as a pseudo-problem is inevitably fraught with all the risks of a heuristic decision.

(3) Language is continuously re-interpreted in its everyday use without the sharp spur of any acute problem, and some kindred questions of nomenclature are usually settled in a similarly smooth fashion in science. The general principle which governs these occasions has already been stated; I shall re-assert it now as follows. In this changing world, our anticipatory powers have always to deal with a somewhat unprecedented situation, and they can do so in general only by undergoing some measure of adaptation. More particularly: since every occasion on which a word is used is in some degree different from every previous occasion, we should expect that the meaning of a word will be modified in some degree on every such occasion. For example, since no owl is exactly like any other, to say 'This is an owl', a statement which ostensibly says something about the bird in front of us, also says something new about the term 'owl', that is, about owls in general.

This raises an awkward question. Can we safely sanction the practice of adapting the meaning of words so that what we say shall be true? If we can say of an unprecedented owl, belonging perhaps to a new species: 'This is an owl', using this designation in an appropriately modified sense, why should we not equally well say of an owl: 'This is a sparrow', meaning a new kind of sparrow, not known so far by that name? Indeed, why should we ever say one thing rather than another, and not pick our descriptive terms at random? Or alternatively, if our terms are to be

---

[1] See Part One, ch. 1 above.

[2] H. Poincaré, *Science and Method*, London, 1914, pp. 94 f.

[3] Cf. e.g. Physics Staff of the University of Pittsburgh, *Outline of Atomic Physics*, 2nd edn. London, 1937, p. 313. This error too was given currency by Poincaré in *Science and Method*. Its fallacy was revealed by observations of the bi-refringence of crystals and of the capacity of condensers, both of which should be, but proved not, measurably affected by the Lorentz-Fitzgerald contraction.

[4] H. Weyl, *Philosophy of Mathematics and Natural Science*, Princeton, 1949, p. 220.

defined by conformity to their present applications, would any statement say more than 'this is this', which is clearly useless?

I shall try to answer this by an illustration from the exact sciences. When heavy hydrogen (deuterium) was discovered by Urey in 1932, it was described by him as a new isotope of hydrogen. At a discussion held by the Royal Society in 1934 the discoverer of isotopy, Frederic Soddy, objected to this on the grounds that he had originally defined the isotopes of an element as chemically inseparable from each other, and heavy hydrogen was chemically separable from light hydrogen.[1] No attention was paid to this protest and a new meaning of the term 'isotope' was tacitly accepted instead. The new meaning allowed heavy hydrogen to be included among the isotopes of hydrogen, in spite of its unprecedented property of being chemically separable from its fellow isotopes. Thus the statement 'There exists an element deuterium which is an isotope of hydrogen' was accepted in a sense which re-defined the term isotope, so that this statement, which otherwise would be false, became true. The new conception abandoned a previously accepted criterion of isotopy as superficial, and relied instead only on the identity of nuclear charges in isotopes.

Our identification of deuterium as an isotope of hydrogen thus affirms two things: (1) that there exists in the case of hydrogen and deuterium an instance of a new kind of chemical separability, pertaining to two elements of equal nuclear charge, (2) that these elements are to be regarded as isotopes in spite of their separability, merely on the grounds of their equal nuclear change. The new observations referred to in (1) necessitated the conceptual and linguistic reforms decreed in (2). They rendered the linguistic rule 'all "isotopes" are chemically inseparable' untenable, and compelled its replacement by a new usage reflecting the truer conception of isotopy derived from these observations. For to retain the original conception of isotopy, by which the chemical differences between light and heavy hydrogen would be classed with the chemical differences between two elements filling different places of the periodic system, would have been misleading to the point of absurdity. This demonstrates the principle which must guide us when adapting the meaning of words, so that what we say shall be true: the corresponding conceptual decisions must be right—their implied allegations true.

Thus we call a new kind of owl an owl, rather than a sparrow, because the modification of the conception of owls by which we include the bird in question as an instance of 'owls' makes sense; while a modification of our conception of sparrows, by which we would include this bird as an instance of 'sparrows', makes nonsense. The former conceptual decision is right and its implications true, in the same sense in which the decision to accept deuterium and hydrogen as isotopes in a modified sense of this term, is right and its implications true. Similarly, in both cases—of owls

[1] *Proc. Roy. Soc.* (A), **144** (1934), pp. 11–14.

and of isotopes—the alternative decisions are wrong and their implications untrue. There is only this difference between the two cases: that the adaptation of the isotope concept to accommodate the observations on deuterium and hydrogen can be specified in terms of an amendment to the definition of isotopy, while the adaptations of a morphological concept like that of 'owl', by which it is made to include novel specimens, can usually not be so specified. I shall expand these observations in Part Four within a wider context.

The adaptation of our conceptions and of the corresponding use of language to new things that we identify as new variants of known kinds of things is achieved subsidiarily, while our attention is focussed on making sense of a situation in front of us. Thus we do this in the same way in which we keep modifying, subsidiarily, our interpretation of sensory clues by striving for clear and coherent perceptions, or enlarging our skill without focally knowing how by practising them in ever new situations. The meaning of speech thus keeps changing in the act of groping for words without our being focally aware of the change, and our gropings invest words in this manner with a fund of unspecifiable connotations. Languages are the product of man's groping for words in the process of making new conceptual decisions, to be conveyed by words.[1]

Different languages are alternative conclusions, arrived at by the secular gropings of different groups of people at different periods of history. They sustain alternative conceptual frameworks, interpreting all things that can be talked about in terms of somewhat different allegedly recurrent features. The confident use of the nouns, verbs, adjectives and adverbs, invented and endowed with meaning by a particular sequence of groping generations, expresses their particular theory of the nature of things.[2] In learning to speak, every child accepts a culture constructed on the premises of the traditional interpretation of the universe, rooted in the idiom of the group to which it was born, and every intellectual effort of the educated mind will be made within this frame of reference. Man's whole intellectual life would be thrown away should this interpretative framework be wholly false; he is rational only to the extent to which the conceptions to which he is committed are true. The use of the word 'true' in the preceding sentence is part of a process of re-defining the meaning of truth, so as to make it truer in its own modified sense.

Different vocabularies for the interpretation of things divide men into groups which cannot understand each other's way of seeing things and of acting upon them. For different idioms determine different patterns of possible emotions and actions. If, and only if, we believe in witches may

[1] Changes in the meaning of words have of course been extensively studied by linguists. My examples taken from science should demonstrate that the accompanying conceptual decisions rank with other discoveries of science. This suggests that changes of meaning may quite generally have implications which are true or false.
[2] This is the aspect of meaning stressed by the 'context' school, notably Weisgerber and Trier. See summary by S. Ullmann, *op. cit.*, pp. 75 and 155 ff.

we burn people as witches; if, and only if, we believe in God will we build churches; if we believe in master races we may exterminate Jews and Poles; if in class war, we may join the Communist Party; if in guilt, we may feel remorse and punish offenders; if in guilt-complexes, we may apply psychoanalysis instead; and so on.

Modern writers have rebelled against the power exercised by words over our thoughts, and have expressed this by deprecating words as mere conventions, established for the sake of convenient communication. This is just as misleading as to say that the theory of relativity is chosen for convenience. We may properly ascribe convenience only to a minor advantage in the pursuit of a major purpose. It is nonsense, for example, to compare the convenience of interpreting sudden death in the idiom of witchcraft with the convenience of using instead a medical terminology, or to compare the convenience of describing political opponents as such, with that of calling them spies, monsters, enemies of the people, etc. Our choice of language is a matter of truth or error, of right or wrong—of life or death.

The understatement that language is a set of convenient symbols used according to the conventional rules of a 'language game' originates in the tradition of nominalism, which teaches that general terms are merely names designating certain collections of objects—a doctrine which, in spite of the difficulties admittedly attached to it, is accepted today by most writers in England and America, in abhorrence of its metaphysical alternatives. The question how the same term can apply to a series of indeterminately variable particulars is avoided by admitting that terms have an 'open texture'.[1] 'Open' terms, however, lack any definite meaning; they may mean anything, unless some intervention is admitted which is competent to control the range of their meaning. My own view admits this controlling principle by accrediting the speaker's sense of fitness for judging that his words express the reality he seeks to express. Without this, words having an open texture are totally meaningless, and any text written in such words is meaningless. Refusing to make this admission, the nominalist has either to refrain from enquiring how such words can be applied, except arbitrarily, to experience; or else to invoke a set of vague regulative principles—without asking on what authority these rules are to be accepted and how they can be applied, unless arbitrarily, in view of their own vagueness.[2] All these deficiencies are overlooked in an overriding desire to avoid reference to metaphysical notions or at least to cover these up under a cloak of nominalist respectability.

Alternatively, the study of linguistic rules is used as a pseudo-substitute for the study of the things referred to in its terms. For example, Wittgenstein says, ' "I don't know whether I am in pain or not" is not a

---

[1] See F. Waismann, 'Verifiability', in *Logic and Language I*, Oxford, 1952, pp. 117 ff.
[2] *ibid.*, cf. I. Murdoch, *loc. cit.* See Part Three, ch. 10, p. 307 for the further analysis of 'regulative principles' and their use.

significant proposition.[1] But the experience of pediatricians shows that children are often in doubt whether they are in pain or are uncomfortable for other reasons. Thus the pseudo-character of the substitution becomes apparent here because the implied statement is mistaken. Had Wittgenstein said. 'It is in the nature of pain that I can always tell whether I feel it or not' he would have been mistaken in fact. By using the pseudo-substitute: 'It is contrary to accepted usage to speak of pain whether I feel it or not', he said something true which was irrelevant to the nature of pain, about which he was actually mistaken.

Correspondingly, disagreements on the nature of things cannot be expressed as disagreements about the existing use of words. Whether an alleged machine of perpetual motion is such a machine or not cannot be decided by studying the use of the terms in question. Whether the law is but 'the will of the stronger' or 'the command of the sovereign' or . . . etc., cannot be decided by linguistic investigations, which are irrelevant to the issue. These controversial questions can be attended to only if we use language as it exists to direct our attention to its subject matter and not the other way around, selecting instances of relevant cases to direct our attention to our use of language. 'Grammar' is precisely the total of linguistic rules which can be observed by using a language *without* attending to the things referred to. The purpose of the philosophic pretence of being merely concerned with grammar is to contemplate and analyse reality, while denying the act of doing so.[2]

There are of course 'Scheinprobleme', and none of these could have arisen without the use of language. Newton could not have formulated his axioms on time and space without talking about absolute rest. But the conception of absolute rest was not suggested by any abuse of language, nor could it be eliminated by reference to ordinary experience and everyday usage, for it was actually rooted in these. Nor were Mach's speculations unavailing for being misguided by Newton's conceptual error; for they raised a problem which led to a great discovery.

I suggest that we should be more frank in facing our situation and acknowledge our own faculties for recognizing real entities, the designations of which form a rational vocabulary. I believe that a classification made according to rational criteria should form groups of things which we may expect to have an indefinite number of properties in common, and that accordingly the terms designating such classes will have an intension referring to an indefinite range of uncovenanted common properties shared by the members of a class. The ampler the intensions of a key feature, the more rational should be as a rule the identification of things in its terms and the more truly should such classification reveal the nature of the classified objects; while classifications made according to terms having no intension should be rejected as purely artificial, unreal, nonsensical;

---

[1] L. Wittgenstein, *Philosophical Investigations*, Oxford, 1953, p. 408.
[2] The same criticism applies to Wittgenstein's use of the term 'language-game'.

unless indeed they *are* designed purely for convenience as, e.g., an alphabetic register of words.

This belief in our capacity for conceiving objective classifications may be acknowledged here on the ground of its continuity with my previous endorsement of personal knowledge and of the personal coefficient of articulation in a great variety of aspects—even though the kind of personally grounded objectivity which it entails must still remain unelucidated at this stage. I shall go on therefore to elaborate this belief further.

There are three successively deeper strata of intensions, of which the *first* comprises the readily specifiable properties which a class of things are known to share apart from their common key-feature; such manifest intension forms the patent evidence for the reality of the classification. The *second* stratum comprises the known but not readily specifiable properties which these things share. The range of such properties subsumed under a term is the measure to which its analysis may lead to a deeper understanding of the things it designates. Words of great human significance accumulate through the centuries an unfathomable fund of subsidiarily known connotations, which we can bring partly into focus by reflecting on the use of such words—in the same way as we may recognize the characteristic elements of a physiognomy, or the tricks conducive to a skill—by a process of analytic reflection. Hence the fruitfulness of a Socratic enquiry into the meaning of words like 'justice' or 'truth' or 'courage', etc.

Understood in these terms, definition is a formalization of meaning which reduces its informal elements and partly replaces them by a formal operation (the reference to the definiens). This formalization will be incomplete also in the sense that the definiens can be understood only by those conversant with the definiendum. Even so, the definition may still throw new light on the definiendum, in the way a guiding maxim illuminates the practice of an art, though its application must rely on the practical knowledge of the art. Such definitions (like 'causation is necessary succession', 'life is continuous adaptation') are, if true and new, analytic discoveries. Such discoveries are among the most important tasks of philosophy.

To take cognizance focally of a subsidiary element of a comprehension is a new experience, and an act which is usually hazardous. The conclusion thus reached is in the nature of an explanation. We see combined here the characteristics of an empirical observation with those of an analytic proposition. This is due ultimately to the fact that the dichotomy between analytic propositions that are necessary, and synthetic statements that are contingent, no longer holds when we can know the same thing in two different ways which cannot be transposed into each other by logical operations, but can be identified only by an enquiry of the Socratic type.

Such an enquiry must be guided by the fact that to speak of 'justice', 'truth', 'courage', etc., is but a performance based on our understanding of the subject matter of these terms. Only if we are confident that we can identify what is just, true or courageous, can we reasonably undertake to analyse our own practice of applying the terms 'justice', 'truth' or 'courage', and hope that such an analysis will reveal to us more clearly what is just, true or courageous.

It is the same as if we studied the motions involved in using a hammer effectively with a view to improving our hammering. For this we must wield a hammer as efficiently as we can, even while watching our motions to discover the best way of hammering. Similarly, we must *use* the word 'justice', and use it as correctly and thoughtfully as we can, while watching ourselves doing it, if we want to analyse the conditions under which the word properly applies. We must look, intently and discriminatingly, *through* the term 'justice' at justice itself, this being the proper use of the term 'justice', the use which we want to define. To look instead *at* the word 'justice' would only destroy its meaning. Besides, to study the recurrence of the word 'justice' as a mere noise in its repeated occurrence in appropriate situations is *impossible*, for only the meaningful use of the term can indicate to us what situations we are to look at.

Speaking more generally: in order to analyse the use of a descriptive term we must use it for the purpose of contemplating its subject matter, and an analysis of this contemplation will inevitably extend to the contemplated object. It will thus amount to an analysis of the conception by which we are jointly aware both of the term and the subject matter, or more precisely, to an analysis of the particulars covered by this conception: from which we may derive both a more rational use of the term and a better understanding of the things which it designates.

The third and deepest level of intensions is formed by the indeterminate range of anticipations expressed by designating something. When we believe that we have truly designated something real, we expect that it may yet manifest its effectiveness in an indefinite and perhaps wholly unexpected manner. This intension comprises a range of properties which only future discoveries may reveal—confirming thereby the rightness of the conception conveyed by our term.[1]

I have already affirmed that these indeterminate anticipatory powers of an apposite vocabulary are due to its contacts with reality. We may extend the conception of reality implied here to account also for the capacity of formal speculations to raise new problems and lead on to new discoveries. A new mathematical conception may be said to have reality if its assumption leads to a wide range of new interesting ideas. Geometries

---

[1] Such a process of classification implies an empirical generalization of a kind usually disregarded in the attempted formalizations of induction. This is pointed out by H. Jeffreys ('The Present Position in Probability Theory', *Brit. J. Phil. Sc.*, **5** (1954–5), pp. 275 ff., 282): 'It seems to me that the epistemology of this process has been unduly neglected. It is of far wider application than Laplacian induction, and the principles involved in framing and arranging the questions seem to me the sort of thing that philosophers might have something to say about.'

based on alternatives to Euclid's postulate of parallels were explored by Saccheri a century before Lobatschevski, but he failed to realize that they could be true. It was only the range of interesting ideas developed by Lobatschevski and Bolyai from non-Euclidean assumptions that eventually convinced a reluctant public. They had then to admit that such conceptions had the same degree of reality as had hitherto been ascribed to Euclid's system. We may extend this conception of reality to the arts, by recalling for example E. M. Forster's distinction between 'flat' and 'round' personages in a novel. A character is called flat if its actions are almost wholly predictable, while we say that a character is round if it can 'convincingly surprise' the reader. The fruitfulness of a new mathematical conception betokens its superior reality; and so, in a novel, does the internal spontaneity by which a 'round' personage may unexpectedly reveal new features which nevertheless flow from its original character, and are therefore convincing.

We come up here once more against the paradox of our self-reliance in seeking contact with a reality of which we believe that it will yet manifest itself in unexpected ways. This paradox we must carry forward unrelieved until we find a balance to it within the framework of commitment.

## 10. UNDERSTANDING LOGICAL OPERATIONS

When we find our bearings by aid of a map we gain an understanding of the region represented by the map, and from this conception we can derive an indefinite number of itineraries. We are conscious of our mastery of the region without attending focally either to the map or to the landmarks around us; for our knowledge of these particulars enters subsidiarily into a conception comprising jointly both the map and the area that it represents. We find our way about the country by reorganizing this conception so as to reveal the particular itineraries in which we are interested. Such a conceptual decision, not induced by new experience, but by a new kind of interest in what we know already, is a speculative act of a kind prefigured primordially in rats running mazes. It is a recognition of alternative part-whole relations of the kind achieved by learning of type C.

Though the conceptual reorganization in question is based on articulation, it is itself informal. Yet it may require a mental effort and might be said to solve a problem. If so, it is a process of deductive inference, as it leads to a new conception wholly implied by our original conception, yet different from it. Such deduction, being informal, is predominantly irreversible; but it can be acknowledged as reversible to the extent to which it follows fixed rules of procedure, be they focally known or not.

The process of reorganizing a conception for drawing new inferences from it can be formalized, by accepting as inferential operations certain rules for manipulating the symbols representing a state of affairs. Although such manipulations are symbolic, they denote not a state of affairs, but the transformation of one conception of a state of affairs into another conception of it that is implied in the first. They evoke the conceptual transformation which they symbolize in the same way as a descriptive term like 'cat' evokes the conception for which it stands. The tacit component of a formalized process of reasoning is broadly analogous to that of a denotation. It conveys both our understanding of the formal manipulations, and our acceptance of them as right.

It may be thought that the difficulty of understanding a formalized chain of reasoning, as for example a mathematical proof, lies in its unfamiliar symbolism. But a verbal sentence may be as difficult to understand as any mathematical formula. Take the sentence constructed by Professor Findlay[1] to paraphrase verbally the result of Gödel's (first) theorem. It runs:

> We cannot prove the statement which is arrived at by substituting for the variable in the statement form 'We cannot prove the statement which is arrived at by substituting for the variable in the statement form $Y$ the name of the statement form in question' the name of the statement form in question.

When you substitute for the variable $Y$ the name of the statement form in question which is the text in quotes, you see that Findlay's sentence says of itself that it cannot be demonstrated, and that hence the sentence is true, just as a Gödelian sentence is true if it cannot be demonstrated.

Even with this explanation to aid them, most people may read Findlay's sentence twenty times over without making head or tail of it; indeed, it may never convey any meaning to them, for they keep losing track of the process of comprehension by which it would make sense. Natural aptitude and training make all the difference in this matter. Lord Russell, to whom I showed Findlay's text in the summer of 1949, took in its meaning at a glance.

No one can be convinced by a proof which he does not understand, and to learn up a mathematical proof which has not convinced us adds nothing to our knowledge of mathematics. Indeed, no teacher will be satisfied with imparting a chain of formulae connected by formal operations as constituting a mathematical proof, and no student of mathematics should be satisfied with memorizing such sequences. To look at a mathematical proof by merely verifying each consecutive step—says Poincaré—is like watching a game of chess, noting only that each step obeys the rules of chess. The least that is required is a grasp of the logical sequence as a purposeful procedure: what Poincaré describes as 'the something which constitutes the unity of the demonstration'.[2] It is this 'something'—per-

---

[1] J. Findlay, 'Gödelian Sentences, a non-numerical approach', *Mind*, **51** (1942), pp. 259-65, 262.

[2] H. Poincaré, 'L'intuition et la logique en mathematique' (1900). Reproduced in *La Valeur de la Science*, Ch. 1, quoted by Daval et Guilbaud, *Le Raisonnement Mathematique*, Paris, 1945, p. 43. See also K. Koffka, *Principles of Gestalt Psychology*, London,

haps in the form of an outline embodying the main steps in the proof— for which the student will grope, if baffled by a sequence of operations which convey no sense to him, and it is again this outline, embodying the general principle or general structure of the mathematical proof, which will be remembered when the details of the proof are forgotten. I can still recall the general procedure followed in evaluating the wave equation of a hydrogen atom on which I lectured about ten years ago, though I could no longer write down any part of the actual demonstration; and this comprehensive recollection satisfies me that I still understand wave mechanics, and sustains in me the conviction of its cogency. On the other hand, though I have repeatedly gone over all the consecutive steps in the formal proof of Gödel's aforementioned theorem, they have conveyed nothing to me, for I have not been able to grasp their sequence as a whole.

Even among mathematicians an argument which seems entirely convincing to one person may not even be comprehensible to another.[1] Hence the striving to remove any occasion for the exercise of personal judgment by a strict formalization of the deductive sciences; a striving which can now be seen to aim at defeating itself. For the meaning of a formalism lies in our subsidiary awareness of it within a conceptual focus sustained in terms of this formalism, and is necessarily absent therefore in operations carried out on symbols seen quite impersonally, as objects. This limiting case is adumbrated when the complete formalization of a proof is attempted in mathematics: the gain in exactitude, resulting from a stricter elimination of ambiguities, is accompanied by a loss in clarity and intelligibility.[2]

I have suggested that the formal operation of symbols conveys the conception of a logical entailment, even as the word 'cat' conveys the conception of a cat. But while the subject-matter denoted by a mathematical proof is less tangible than a cat, a proof does more than denote its subject matter: it brings it about. As the second operational principle of language is set in motion, we pass on from the making of signs to the devising of a formal process, which first contrives what it subsequently conveys. This contriving is guided by the specific purpose of establishing a particular implication and compelling its acceptance. It endorses this purpose as worthy of a great effort, and sets up standards of economy and beauty for the manner of its achievement. We have here a far more elaborate sequence of purposive actions than that involved in the naming of recurrent features of experience; it is a linguistic operation in the stricter sense of an ingenious contrivance.

1935, pp. 555–6, for the transformation caused by understanding a mathematical proof. For a supporting statement by a mathematician see Van der Waerden, 'Denken ohne Sprache', in Révész, *op. cit.*, p. 165.

   [1] A. Tarski, *Introduction to Logic*, New York, 1946, p. 132.
   [2] A. Tarski, *op. cit.*, p. 134. We may recall here also that legal documents and government regulations, which are carefully worded to achieve the greatest precision, are notoriously unintelligible.

## 11. Introduction to Problem-Solving

There is a purposive tension from which no fully awake animal is free. It consists in a readiness to perceive and to act, or more generally speaking, to make sense of its own situation, both intellectually and practically. From these routine efforts to retain control of itself and of its surroundings, we can see emerging a process of problem solving, when the effort tends to fall into two stages, a first stage of perplexity, followed by a second stage of doing and perceiving which dispels this perplexity. We may say that the animal has seen a problem, if its perplexity lasts for some time and it is clearly trying to find a solution to the situation which puzzles it. In doing so, the animal is searching for a hidden aspect of the situation, the existence of which it surmises, and for the finding or achieving of which the manifest features of the situation serve it as tentative clues or instruments.

To see a problem is a definite addition to knowledge, as much as it is to see a tree, or to see a mathematical proof—or a joke. It is a surmise which can be true or false, depending on whether the hidden possibilities of which it assumes the existence do actually exist or not. To recognize a problem which can be solved and is worth solving is in fact a discovery in its own right. Famous mathematical problems have descended from generation to generation, leaving in their wake a long trail of achievements stimulated by the attempt at solving them. And even at the level of animal experiments, we can see the psychologist demonstrating to the animal the presence of a problem in order to start it off in search of a solution. A rat in a discrimination box is made to realize that there is food hidden in one of two compartments, both of which are accessible, and only if it has grasped this will it start searching for something which discriminates the door or screen with food behind it from that of the empty compartment. Similarly, animals will not start solving a maze unless they are made aware of the fact that there exists a path through it, with some reward at its outlet. In Köhler's insight experiments his chimpanzees grasped their problem from the start, and marked their appreciation of the task by composing themselves quietly to concentrate on it.

Accident usually plays some part in discovery and its part may be predominant. Learning experiments can be so arranged that, in the absence of any clearly understood problem, discovery can only be accidental.[1] Mechanistically minded psychologists who devise such experiments would

[1] Guthrie and Horton placed a cat in a cage in which a small pole, placed in the middle of the floor, acted as release mechanism. Cats who had touched the pole by accident and found themselves freed in consequence, quickly realized the connection and proceeded to repeat their releasing action in an exactly stereotyped manner. The situation in which the cat was placed offered no intelligible problem to the cat and the solution, found accidentally, showed no clear understanding of the release mechanism; the role played by intelligence in the whole process was negligible (comp. Hilgard, *op. cit.*, pp. 65–8).

explain all learning as the lucky outcome of random behaviour. This conception of learning underlies also the cyberneticist model of a machine which 'learns' by selecting the 'habit' which had proved successful in a series of random trials. I shall disregard now this model of heuristics, and continue to explore the process of discovery resulting from intelligent effort, irrespective of the neural model that may be proposed for it.

Intelligent problem-solving is manifested among animals most effectively in Köhler's experiments on chimpanzees, whose behaviour already presents the characteristic stages through which, according to Poincaré, discovery is achieved in mathematics. I have already mentioned the first: the appreciation of a problem. A chimpanzee in a cage within sight of a bunch of bananas out of its reach neither makes any futile effort to get hold of it by sheer force, nor abandons its desire of acquiring the prize. It settles down instead to an unusual calm, while its eyes survey the situation all round the target; it has recognized the situation as problematical and is searching for a solution.[1] We may acknowledge this (using the terminology of Wallas based on Poincaré) as the stage of Preparation.[2]

In the most striking cases of insight observed by Köhler, this preparatory stage is suddenly followed by intelligent action. Sharply breaking its calm, the animal proceeds to carry out a stratagem by which it secures its aim, or at least shows that it has grasped a principle by which this can be done. Its unhesitating manner suggests that it is guided by a clear conception of the whole proposed operation. This conception is its discovery, or at least—since it may not always prove practicable—its tentative discovery. We may recognize in its coming the stage of Illumination. Since the practical realization of the principle discovered by insight often presents difficulties which may even prove insurmountable, the manipulations by which the animal puts his insight to the test of practical realization may be regarded as the stage of Verification.

Actually, Poincaré observed four stages of discovery: Preparation, Incubation, Illumination, Verification.[3] But the second of these, Incubation, can be observed only in a rudimentary form in chimpanzees. Yet the observation described in some detail by Köhler, in which one of his animals sustained its effort of solving a problem even while otherwise occupied for a time,[4] anticipates to a remarkable extent the process of

---

[1] 'The greatest impression on the visitor (writes Köhler) was made when Sultan made a pause, scratching his head leisurely and not moving anything but his eyes and very slightly his head, scrutinizing the situation around him in the minutest detail', *The Mentality of Apes*, London, 1927, p. 200.

[2] G. Wallas, *The Art of Thought*, London, 1946, pp. 40 ff.

[3] 'Verification' means, of course, in mathematics, 'demonstration', and is therefore closer to what I describe later (Part Two, ch. 6, p. 202) as 'validation' rather than to the 'verification' of experimental science.

[4] An ape which for a while had been searching for a tool to rake in a bunch of bananas lying outside its cage. and had made various fruitless attempts in this direction—such as trying to break off a board from the lid of a wooden case or hitting out with a stalk of straw in the direction of the prize—had apparently abandoned the task altogether. It went on playing with one of its fellows for about 10 minutes without turning again to

incubation: that curious persistence of heuristic tension through long periods of time, during which the problem is not consciously entertained.

An extensive preoccupation with a problem imposes an emotional strain, and a discovery which releases from it is a great joy. The story of Archimedes rushing out from his bath into the streets of Syracuse shouting 'Eureka' is a witness to this; and the account I have quoted from Köhler, of the way his chimpanzees behaved before and after solving a problem, suggests that they also experience such emotions. I shall show this more definitely later. I mention it now only to make it clear that nothing is a problem or discovery in itself; it can be a problem only if it puzzles and worries somebody, and a discovery only if it relieves somebody from the burden of a problem. A chess problem means nothing to a chimpanzee or to an imbecile, and hence does not puzzle them; a great chess master on the other hand may fail to be puzzled by it because he finds its solution without effort; only a player whose ability is about equal to the problem will find intense preoccupation in it. Only such a player will appreciate its solution as a discovery.[1]

It appears possible to appraise the comparative hardness of a problem, and to test the intelligence of subjects by their capacity for solving problems of a certain degree of hardness. The intelligence of chimpanzees and the hardness of certain problems were both successfully assessed by Köhler when he devised a series of problems, which some of his apes could solve with some effort while others among them usually failed altogether to do so. The success of Yerkes in setting problems to earthworms (which these could solve after about a hundred trials), shows that he could assess even such extremely low powers of intelligence as were required here from the earthworm.[2] Editors of a crossword column undertake a similar feat in supplying their readers with a steady stream of always equally difficult problems. We may conclude that while a problem must always be regarded as being a problem to some kind of person, it is possible for an observer reliably to recognize it as such in respect to identifiable persons.

If an animal who has solved a problem is placed once more in the original situation, it proceeds unhesitatingly to apply the solution which it

the bananas outside the cage. Then suddenly, its attention having been diverted from its game by a shout nearby, its eyes happened to fall on a stick attached to the roof of the cage, and at once it went for the stick and by jumping up a number of times finally secured it and hauled in the bananas by its aid. We may take this to show that even while otherwise occupied the animal kept its problem alive 'at the back of its mind', keeping it ready to pounce on the instruments of a solution when they happened to meet its eye. Köhler, *op. cit.*, p. 184.

[1] Kurt Lewin has observed that we do not become emotionally involved either in a task that is too easy or in one that is too difficult, but only in tasks that we can master at our best. Hoppe, following Lewin, calls this the measure of ego-involvement. See Hilgard, *op. cit.*, p. 277.

[2] R. M. Yerkes, 'The Intelligence of Earthworms', *Journ. Anim. Behav.*, 2 (1912), pp. 332–52. Cf. N. R. F. Maier and T. Schneirla, *Principles of Animal Psychology*, New York and London, 1935, pp. 98–101.

had originally discovered at the cost of much effort and perhaps after many unsuccessful trials. This shows that by solving the problem the animal has acquired a new intellectual power, which prevents it from being ever again puzzled by the problem. Instead, it can now deal with the situation in a routine manner involving no heuristic tension and achieving no discovery. The problem has ceased to exist for it. Heuristic progress is irreversible.

The irreversible character of discovery suggests that no solution of a problem can be accredited as a discovery if it is achieved by a procedure following definite rules. For such a procedure would be reversible in the sense that it could be traced back stepwise to its beginning and repeated at will any number of times, like any arithmetical computation. Accordingly, any strictly formalized procedure would also be excluded as a means of achieving discovery.

It follows that true discovery is not a strictly logical performance, and accordingly, we may describe the obstacle to be overcome in solving a problem as a 'logical gap', and speak of the width of the logical gap as the measure of the ingenuity required for solving the problem. 'Illumination' is then the leap by which the logical gap is crossed. It is the plunge by which we gain a foothold at another shore of reality. On such plunges the scientist has to stake bit by bit his entire professional life.

The width of the logical gap crossed by an inventor is subject to legal assessment. Courts of law are called upon to decide whether the ingenuity displayed in a suggested technical improvement is high enough to warrant its legal recognition as an invention, or is merely a routine improvement, achieved by the application of known rules of the art. The invention must be acknowledged to be unpredictable, a quality which is assessed by the intensity of the surprise it might reasonably have aroused. This unexpectedness corresponds precisely to the presence of a logical gap between the antecedent knowledge from which the inventor started and the consequent discovery at which he arrived.

Established rules of inference offer public paths for drawing intelligent conclusions from existing knowledge. The pioneer mind which reaches its own distinctive conclusions by crossing a logical gap deviates from the commonly accepted process of reasoning, to achieve surprising results. Such an act is original in the sense of making a new start, and the capacity for initiating it is the gift of originality, a gift possessed by a small minority.

Since the Romantic movement originality has become increasingly recognized as a native endowment which alone enables a person to initiate an essential innovation. Universities and industrial research laboratories are founded today on the employment of persons with original minds. Permanent appointments are given to young scientists who are credited with signs of originality, in the expectation that they will continue to produce surprising ideas for the rest of their lives.

Admittedly, there are minor heuristic acts within the power of ordinary intelligence which are indeed continuous with the adaptive capacities of life down to its lowest levels. We have seen already that whenever we make (or believe we have made) contact with reality, we anticipate an indeterminate range of unexpected future confirmations of our knowledge derived from this contact. The interpretative framework of the educated mind is ever ready to meet somewhat novel experiences, and to deal with them in a somewhat novel manner. In this sense all life is endowed with originality and originality of a higher order is but a magnified form of a universal biological adaptivity. But genius makes contact with reality on an exceptionally wide range: seeing problems and reaching out to hidden possibilities for solving them, far beyond the anticipatory powers of current conceptions. Moreover, by deploying such powers in an exceptional measure—far surpassing ours who are looking on—the work of a genius offers us a massive demonstration of a creativity which can neither be explained in other terms, nor taken unquestioningly for granted. By paying respect to another person's judgment as superior to our own, we emphatically acknowledge originality in the sense of a performance the procedure of which we cannot specify. Confrontation with genius thus forces us to acknowledge the originative power of life, which we may and commonly do neglect in its ubiquitous lesser manifestations.

In choosing a problem the investigator takes a decision fraught with risks. The task may be insoluble or just too difficult. In that case his effort will be wasted and with it the effort of his collaborators, as well as the money spent on the whole project. But to play safe may be equally wasteful. Mediocre results are no adequate return for the employment of high gifts, and may not even repay the money spent on achieving them. So the choice of a problem must not only anticipate something that is hidden and yet not inaccessible, but also assess the investigator's own ability (and those of his collaborators) against the anticipated hardness of the task, and make a reasonable guess as to whether the hoped for solution will be worth its price in terms of talent, labour and money. To form such estimates of the approximate feasibility of yet unknown prospective procedures, leading to unknown prospective results, is the day-to-day responsibility of anyone undertaking independent scientific or technical research. On such grounds as these he must even compare a number of different possible suggestions and select from them for attack the most promising problem. Yet I believe that experience shows such a performance to be possible and that it can be relied upon to function with a considerable degree of reliability.

## 12. Mathematical Heuristics

There are three major fields of knowledge in which discoveries are possible: natural science, technology and mathematics. I have referred to examples from each of these fields to illustrate the anticipatory powers

124

which guide discovery. These are clearly quite similar in all three cases. Yet the efforts of philosophers have been almost wholly concentrated on the process of empirical discovery which underlies the natural sciences—i.e. on an attempt to define and justify the process of induction, while by contrast, nobody seems to have tried to define and justify the process by which technical innovations are made, as for example when a new machine is invented. The process of discovery in mathematics had received some attention, and has recently been attacked both from the logical and psychological point of view, but neither approach has raised any epistemological questions parallel to those so sedulously pursued for centuries in connection with empirical induction. It seems to me that any serious attempt to analyse the process of discovery should be sufficiently general to apply to all three fields of systematic knowledge, and I should like to contribute to this programme here by identifying and acknowledging the powers on which we rely in solving mathematical problems. I shall exclude for the moment the history of major discoveries which involve modifications in the foundations of mathematics and shall attend only to the type of problems that are set to students in teaching them mathematics. Since the solution of these problems is not known to the student, the process of finding it bears the marks of a discovery, even though it involves no fundamental change of outlook.

The fact that the teaching of mathematics relies to a large extent on practice shows that even this most highly formalized branch of knowledge can be acquired only by developing an art. This is true not only of mathematics and formal logic, but equally also of all mathematical sciences, like mechanics, electrodynamics, thermodynamics and the mathematical branches of engineering; you cannot master any of these subjects without working out concrete problems in them. The skill you strive for in all these cases is that of converting a language, which you so far had assimilated only receptively, into an effective tool for handling new subjects— and more particularly in mathematics, for solving problems.

Since the solving of mathematical problems is a heuristic act which leaps across a logical gap, any rules that can be laid down for its guidance can be but vague maxims, the interpretation of which must rely on the very art to which they apply. We shall see that such is indeed the case.[1]

The simplest heuristic effort is to search for an object you have mislaid. When I am looking for my fountain pen I know what I expect to find; I can name it and describe it. Though I know much more about my fountain pen than I can ever recall, I do not know exactly where I left it; but the pen is clearly known to me and I know also that it is somewhere within a certain region though I do not know where. I have much less knowledge of the

---

[1] A distinguished effort to lay down heuristic maxims has been made in our time by the mathematician G. Polya (*How to Solve It*, Princeton, 1945, and *Mathematics and Plausible Reasoning*, 2 vols., London, 1954), mainly with a view to giving guidance for the teaching of mathematics. Penetrating observations on problem solving have also been contributed by the psychologists, mainly by Duncker and by Wertheimer.

thing I am looking for when I am searching for a word to fit into a cross-word puzzle. This time I know only that the missing word has a certain number of letters and designates, for example, something that is badly needed in the Sahara or flows out of a central chimney. These characteristics are merely clues to a word that I definitely do *not* know: clues from which I must try to gain an intimation of what the unknown word may be. Again, a name which I know well but cannot recall at the moment, lies somewhere halfway between these two cases. It is more closely present to my mind than the unknown solution of a crossword puzzle, but less closely perhaps than the mislaid fountain pen and its unknown location. Mathematical problems are in the class of crossword puzzles, for to solve such a problem we must find (or construct) something that we have never seen before, with the given data serving us as clues to it.

A problem may admit of a systematic solution. By ransacking my flat inch by inch, I may make sure of eventually finding my fountain pen which I know to be somewhere in it. I might solve a chess problem by trying out mechanically all combinations of possible moves and countermoves. Systematic methods apply also to many mathematical problems, though usually they are far too laborious to be carried out in practice.[1] It is clear that any such systematic operation would reach a solution without crossing a logical gap and would not constitute a heuristic act.[2]

The difference between the two kinds of problem solving, the systematic and the heuristic, reappears in the fact that while a systematic operation is a wholly deliberate act, a heuristic process is a combination of active and passive stages. A deliberate heuristic activity is performed during the stage of Preparation. If this is followed by a period of Incubation, nothing is done and nothing happens on the level of consciousness during this time. The advent of a happy thought (whether following immediately from Preparation or only after an interval of Incubation) is the fruit of the investigator's earlier efforts, but not in itself an action on his part; it just happens to him. And again, the testing of the 'happy thought' by a former process of Verification is another deliberate action of the investigator. Even so, the decisive act of discovery must have occurred before this, at the moment when the happy thought emerged.

Though the solution of a problem is something we have never met before, yet in the heuristic process it plays a part similar to the mislaid fountain pen or the forgotten name which we know quite well. We are looking for it as if it were there, pre-existent. Problems set to students are of course known to have a solution; but the belief that there exists a hidden solution

---

[1] A. M. Turing (*Science News*, **31** (1954)) has computed the number of arrangements that would have to be surveyed in the process of solving systematically a very common form of puzzle, consisting of sliding squares to be rearranged in a particular way. The number is 20,922,789,888,000. Working continuously day and night and inspecting one position per minute the process would take 4 million years.

[2] This is to disregard the minimum logical gap involved in a formal process of inference. See Part Three, ch. 8, p. 260.

which we may be able to find is essential also in envisaging and working at a never yet solved problem. It determines also the manner in which the 'happy thought' eventually presents itself as something inherently satisfying. It is not one among a great many ideas to be pondered upon at leisure, but one which carries conviction from the start. We shall see in a moment, from a closer analysis of this process, that this is a necessary consequence of the way a heuristic striving evokes its own consummation.

A problem is an intellectual desire (a 'quasi-need' in K. Lewin's terminology) and like every desire it postulates the existence of something that can satisfy it; in the case of a problem its satisfier is its solution. As all desire stimulates the imagination to dwell on the means of satisfying it, and is stirred up in its turn by the play of the imagination it has fostered, so also by taking interest in a problem we start speculating about its possible solution and in doing so become further engrossed in the problem.

Obsession with one's problem is in fact the mainspring of all inventive power. Asked by his pupils in jest what they should do to become 'a Pavlov', the master answered in all seriousness: 'Get up in the morning with your problem before you. Breakfast with it. Go to the laboratory with it. Eat your lunch with it. Keep it before you after dinner. Go to bed with it in your mind. Dream about it.'[1] It is this unremitting preoccupation with his problem that lends to genius its proverbial capacity for taking infinite pains. And the intensity of our preoccupation with a problem generates also our power for reorganizing our thoughts successfully, both during the hours of search and afterwards, during a period of rest.[2]

But what is the object of this intensive preoccupation? How can we concentrate our attention on something we don't know? Yet this is precisely what we are told to do: 'Look at the unknown!'—says Polya—'Look at the end. Remember your aim. Do not lose sight of what is required. Keep in mind what you are working for. *Look at the unknown. Look at the conclusion.*'[3] No advice could be more emphatic.

The seeming paradox is resolved by the fact that even though we have never met the solution, we have a conception of it in the same sense as we have a conception of a forgotten name. By directing our attention on a focus in which we are subsidiarily aware of all the particulars that remind us of the forgotten name, we form a conception of it; and likewise, by fixing our attention on a focus in which we are subsidiarily aware of the data by which the solution of a problem is determined, we form a conception of this solution. The admonition to look at the unknown really means that we should *look at the known data, but not in themselves, rather as*

---

[1] J. R. Baker, *Science and the Planned State*, London, 1945, p. 55.

[2] 'Only such problems come back improved after a rest whose solution we passionately desire and for which we have worked with great tension' writes Polya (*How to Solve it*, Princeton, 1945, p. 172).

[3] *ibid.*, p. 112, italics in the original. K. Duncker, *Zur Psychologie des produktiven Denkens*, Berlin, 1935, p. 13: '. . . eine Lösung entsteht aus der Beanspruchung des Gegebenen durch das jeweils Geforderte.'

*clues to the unknown; as pointers to it and parts of it.* We should strive persistently to feel our way towards an understanding of the manner in which these known particulars hang together, both mutually and with the unknown. By such intimations do we make sure that the unknown is really there, essentially determined by what is known about it, and able to satisfy all the demands made on it by the problem.

All our conceptions have heuristic powers; they are ever ready to identify novel instances of experience by modifying themselves so as to comprise them. The practice of skills is inventive; by concentrating our purpose on the achievement of success we evoke ever new capacities in ourselves. A problem partakes of both these types of endeavour. It is a conception of something we are striving for. It is an intellectual desire for crossing a logical gap on the other side of which lies the unknown, fully marked out by our conception of it, though as yet never seen in itself. The search for a solution consists in casting about with this purpose in mind. This we do by performing two operations which must always be tried jointly. We must (1) set out the problem in suitable symbols and continuously reorganize its representation with a view to eliciting some new suggestive aspects of it, and concurrently (2) ransack our memory for any similar problem of which the solution is known.[1] The scope of these two operations will usually be limited by the student's technical facility for transforming the given data in different ways, and by the range of germane theorems with which he is acquainted. But his success will depend ultimately on his capacity for sensing the presence of yet unrevealed logical relations between the conditions of the problem, the theorems known to him, and the unknown solution he is looking for. Unless his casting about is guided by a reliable sense of growing proximity to the solution, he will make no progress towards it. Conjectures made at random, even though following the best rules of heuristics, would be hopelessly inept and totally fruitless.

The process of solving a mathematical problem continues to depend, therefore, at every stage on the same ability to anticipate a hidden potentiality which will enable the student first to see a problem and then to set out to solve it. Polya has compared a mathematical discovery consisting of a whole chain of consecutive steps with an arch, where every stone depends for its stability on the presence of others, and he pointed out the paradox that the stones are in fact put in one at a time. Again, the paradox is resolved by the fact that each successive step of the incomplete solution is upheld by the heuristic anticipation which originally evoked its invention: by the feeling that its emergence has narrowed further the logical gap of the problem.

The growing sense of approaching to the solution of a problem can be commonly experienced, as I said above, when we grope for a forgotten name. We all know the exciting sense of increasing proximity to the missing word, which we may confidently express by saying: 'I shall remember it

[1] Polya, *op. cit., passim.*

128

in a moment' and perhaps later 'It is on the tip of my tongue'. The expectation expressed by such words is often confirmed by the event. I believe that we should likewise acknowledge our capacity both to sense the accessibility of a hidden inference from given premises, and to invent transformations of the premises which increase the accessibility of the hidden inference. We should recognize that this foreknowledge biasses our guesses in the right direction, so that their probability of hitting the mark, which would otherwise be zero, becomes so high that we can definitely rely on it simply on the ground of a student's intelligence; or for higher performances, on the ground of the special gifts possessed by the professional mathematician.

The feeling that the logical gap separating us from the solution of a problem has been reduced, means that less work should remain to be done for solving it. It may also mean that the rest of the solution will be comparatively easy, or that it may present itself without further effort on our part, after a period of repose. The fact that our intellectual strivings make effective progress during a period of Incubation without any effort on our part is in line with the latent character of all knowledge. As we continuously know a great many things without always thinking of them, so we naturally also keep on desiring or fearing all manner of things without always thinking of them. We know how a set purpose may automatically result in action later on, as when we go to bed resolved to wake up at a certain hour. Post-hypnotic suggestions can set going latent processes which compulsively result after a number of hours in the performance requested of the subject.[1] Mrs. Zeigarnik has shown that unfinished tasks continue likewise to preoccupy us unconsciously; their memory persists after finished tasks are forgotten.[2] The fact that the tension set up by the unfinished task continues to make progress towards its fulfilment, is shown by the well known experience of athletes that a period of rest following on a spell of intensive training produces an improvement of skill. The spontaneous success of the search for a forgotten name or for the solution of a problem, after a period of quiescence, falls in line with this experience.

These antecedents explain also the manner in which the final success of problem solving will suddenly set in. For each step—whether spontaneous or contrived—that brings us nearer to the solution, increases our premonition of the solution's proximity and brings a more concentrated effort to bear on a reduced logical gap. The last stage of the solution may therefore be frequently achieved in a self-accelerating manner and the final discovery may be upon us in a flash.

I have said that our heuristic cravings imply, like our bodily appetites, the existence of something which has the properties required to satisfy

[1] N. Ach, 'Determining Tendencies; Awareness', in D. Rapaport, *Organization and Pathology of Thought*, New York, 1951, pp. 17 ff.

[2] W. D. Ellis, *A Source Book of Gestalt Psychology*, London and New York, 1938, pp. 300–14.

us, and that the intimations which guide our striving express this belief. But the satisfier of our craving has in this case no bodily existence; it is not a hidden object, but an idea never yet conceived. We hope that as we work at the problem this idea will come to us, whether all at once or bit by bit; and only if we believe that this solution exists can we passionately search for it and evoke from ourselves heuristic steps towards its discovery. Therefore, as it emerges in response to our search for something we believe to be there, discovery, or supposed discovery, will always come to us with the conviction of its being true. It arrives accredited in advance by the heuristic craving which evoked it.

The most daring feats of originality are still subject to this law: they must be performed on the assumption that they origina ᐧ nothing, but merely reveal what is there. And their triumph confirms this assumption, for what has been found bears the mark of reality in being pregnant with yet unforeseeable implications. Mathematical heuristics, though aiming at conceptual reorganization without reference to new experience, once more exemplifies in its own terms that an intellectual striving entails its conviction of anticipating reality. It illustrates also how this conviction finds itself confirmed by the eventual solution, which 'solves' precisely because it successfully claims to reveal an aspect of reality. And we can see once more also, how the whole process of discovery and confirmation ultimately relies on our own accrediting of our own vision of reality.

To start working on a mathematical problem, we reach for pencil and paper, and throughout the stage of Preparation we keep trying out ideas on paper in terms of symbolic operations. If this does not lead straight to success, we may have to think the whole matter over again, and may perhaps see the solution revealed unexpectedly much later in a moment of Illumination. Actually, however, such a flash of triumph usually offers no final solution, but only the envisagement of a solution which has yet to be tested. In the verification or working out of the solution we must again rely therefore on explicit symbolic operations. Thus both the first active steps undertaken to solve a problem and the final garnering of the solution rely effectively on computations and other symbolic operations, while the more informal act by which the logical gap is crossed lies between these two formal procedures. However, the intuitive powers of the investigator are always dominant and decisive. Good mathematicians are usually found capable of carrying out computations quickly and reliably, for unless they command this technique they may fail to make their ingenuity effective—but their ingenuity itself lies in producing ideas. Hadamard says that he used to make more mistakes in calculation than his own pupils, but that he more quickly discovered them because the result did not *look* right; it is almost as if by his computations he had been merely drawing a portrait of his conceptually prefigured conclusions.[1] Gauss is widely quoted

[1] J. Hadamard, *An Essay on the Psychology of Invention in the Mathematical Field*, Princeton, 1945, p. 49.

as having said: 'I have had my solutions for a long time but I do not yet know how I am to arrive at them.' Though the quotation may be doubtful it remains well said.[1] A situation of this kind certainly prevails every time we discover what we believe to be the solution to a problem. At that moment we have the vision of a solution which *looks* right and which we are therefore confident to *prove* right.[2]

The manner in which the mathematician works his way towards discovery, by shifting his confidence from intuition to computation and back again from computation to intuition, while never releasing his hold on either of the two, represents in miniature the whole range of operations by which articulation disciplines and expands the reasoning powers of man. This alternation is asymmetrical, for a formal step can be valid only by virtue of our tacit confirmation of it. Moreover, a symbolic formalism is itself but an embodiment of our antecedent unformalized powers—an instrument skilfully contrived by our inarticulate selves for the purpose of relying on it as our external guide. The interpretation of primitive terms and axioms is therefore predominantly inarticulate, and so is the process of their expansion and re-interpretation which underlies the progress of mathematics. The alternation between the intuitive and the formal depends on tacit affirmations, both at the beginning and at the end of each chain of formal reasoning.

---

[1] G. Polya writes: 'When you have satisfied yourself that the theorem is true, you start proving it.' (*Mathematics and Plausible Reasoning*, Vol. 2, p. 76.)
[2] Archimedes describes in his 'Method' a mechanical process of geometrical demonstration which carried conviction for him, though he regards its results as still requiring proof, which he then proceeded to supply. B. L. Van der Waerden, *Science Awakening*, Groningen, 1954, p. 215.

# 6

## INTELLECTUAL PASSIONS

### 1. SIGN-POSTING

THE previous chapter was a digression. Having acknowledged in Part One the ubiquitous participation of the scientist in upholding the affirmations of science, I wanted to investigate the origins of this personal coefficient by tracing it back to the very act of uttering speech. In order to find this juncture the enquiry had to penetrate beyond it, to the inarticulate levels of intelligence of the animal and the infant, in which the personal coefficient of spoken knowledge is primordially preformed. Pursuing the roots of this tacit intelligence even further, we recognized an active principle which controls and sustains it. As far down the scale of life as the worms and even perhaps the amoeba, we meet a general alertness of animals, not directed towards any specific satisfaction, but merely exploring what is there; an urge to achieve intellectual control over the situations confronting it. Here at last, in the logical structure of such exploring—and of visual perception—we found prefigured that combination of the active shaping of knowledge with its acceptance as a token of reality, which we recognize as a distinctive feature of all personal knowing. This is the principle which guides all skills and connoisseurship, and informs all articulate knowing by way of the ubiquitous tacit coefficient on which spoken utterances must rely for their guidance and confirmation.

The tracing of personal knowledge on these lines, through all spoken utterances and further back to the active principles of animal life, has shown that the tacit intellectual powers which we share with animals and infants suffice to account in a first approximation for the immense expansion in the scope of human knowledge opened up by the acquisition of speech. This approximation has, at any rate, the advantage of representing separately those aspects of articulate thought which require no striking expansion of tacit powers beyond those common to animals. But there are other constituents of thought, and of science itself, which are guided by

132

tacit powers far surpassing the range of animal intelligence, and to these I must now turn.

The powers in question would have inevitably forced themselves upon our attention before this, had I not limited my study so far to the affirmative use of language, in which such powers are least prominent. The expressive and interactive uses of language obviously present us with tacit powers beyond those of animals or infants. Works of art or social imperatives have a manifest emotional force, evoked within an articulate culture to which no speechless being has access. But we have met with such powers even in the affirmative use of language itself. The affirmation of a great scientific theory is in part an expression of delight. The theory has an inarticulate component acclaiming its beauty, and this is essential to the belief that the theory is true. No animal can appreciate the intellectual beauties of science.

It is true that the active principle of animal life, in which I have found prefigured all intellectual strivings of man, already sounds a passionate note. Köhler clearly demonstrated that chimpanzees derive pleasure from the discovery of a new ingenious manipulation, quite apart from the practical benefit they derive from it; he described how they will repeat the performance for its own sake, as a kind of play. W. N. and L. A. Kellogg found that a young chimpanzee is just as much inclined as a child of the same age to repeat in play a manipulation involving the use of a tool which it had first invented for some practical purpose. The animal was also as keen as the child to climb into the place where it was usually confronted with the task of solving problems. These intellectual tastes of the animal prefigure, no doubt, the joys of discovery which our articulate powers can attain for man, but in the animal they do not remotely approach these joys in scope and elevation. As language enlarges the range of our thought, the ape's pleasure in playing with a stick is expanded to a complex system of emotional responses by which scientific value and ingenuity of many kinds are appreciated throughout natural science, technology and mathematics. This is the kind of feeling described in the title of this chapter as 'Intellectual Passions'. Before going into it further, let me notice the new context into which science is shifted by attending to this aspect of it. A scientific theory which calls attention to its own beauty, and partly relies on it for claiming to represent empirical reality, is akin to a work of art which calls attention to its own beauty as a token of artistic reality. It is akin also to the mystical contemplation of nature: a kinship shown historically in the Pythagorean origins of theoretical science. More generally, science, by virtue of its passionate note, finds its place among the great systems of utterances which try to evoke and impose correct modes of feeling. In teaching its own kinds of formal excellence science functions like art, religion, morality, law and other constituents of culture.

This alignment greatly amplifies the perspective of our enquiry. Though we had noticed before that science claims to appraise order and probability and accredits scientific skill and connoisseurship, these evaluative components of science were emotionally colourless compared with the intellectual passions by which science appreciates its own beauty. If the upholding of scientific truth requires that we justify such passionate valuations, our task expands inevitably also to the justification of those equally passionate valuations on which the affirmation of the several domains of culture is predicated. Science can then no longer hope to survive on an island of positive facts, around which the rest of man's intellectual heritage sinks to the status of subjective emotionalism. It must claim that certain emotions are right; and if it can make good such a claim, it will not only save itself but sustain by its example the whole system of cultural life of which it forms part.

Yet while accepting the inescapable solidarity of science with other cultural provinces, I shall have to strike a compromise in this book between the claims of this connection and the limitations of my space. Though it may eventually prove easier to uphold a fuller truth on broader grounds, I cannot attempt this task here in its entirety. I propose, therefore, to continue my enquiry into the conditions for upholding factual truth, while digressing only from time to time to indicate the wider implications of this project.

## 2. SCIENTIFIC VALUE

From the start of this book I have had occasion, in various contexts, to refer to the overwhelming elation felt by scientists at the moment of discovery, an elation of a kind which only a scientist can feel and which science alone can evoke in him. In the very first chapter I quoted the famous passage in which Kepler announced the discovery of his Third Law: '. . . nothing holds me; I will indulge my sacred fury . . .'[1] The outbreak of such emotions in the course of discovery is well known, but they are not thought to affect the outcome of discovery. Science is regarded as objectively established in spite of its passionate origins. It should be clear by this time that I dissent from that belief; and I have now come to the point at which I want to deal explicitly with passions *in* science. I want to show that scientific passions are no mere psychological by-product, but have a logical function which contributes an indispensable element to science. They respond to an essential quality in a scientific statement and may accordingly be said to be right or wrong, depending on whether we acknowledge or deny the presence of that quality in it.

What is this quality? Passions charge objects with emotions, making them repulsive or attractive; positive passions affirm that something is precious. The excitement of the scientist making a discovery is an *intellectual* passion, telling that something is *intellectually* precious and, more particularly, that it is *precious to science*. And this affirmation forms part

---

[1] See Part One, ch. 1, p. 7.

134

of science. The words of Kepler which I quoted were not a statement of fact, but neither were they merely a report of Kepler's personal feelings. They asserted as a valid affirmation of science something else than a fact: namely the scientific interest of certain facts, the facts just discovered by Kepler. They affirmed, indeed, that these facts are of immense scientific interest and will be so regarded as long as knowledge lasts. Nor was Kepler deceived in this majestic sentiment. The passing centuries have paid their cumulative tribute to his vision, and so, I believe, will the centuries yet to come.

The function which I attribute here to scientific passion is that of distinguishing between demonstrable facts which are of scientific interest, and those which are not. Only a tiny fraction of all knowable facts are of interest to scientists, and scientific passion serves also as a guide in the assessment of what is of higher and what of lesser interest; what is great in science, and what relatively slight. I want to show that this appreciation depends ultimately on a sense of intellectual beauty; that it is an emotional response which can never be dispassionately defined, any more than we can dispassionately define the beauty of a work of art or the excellence of a noble action.

Scientific discovery reveals new knowledge, but the new vision which accompanies it is not knowledge. It is *less* than knowledge, for it is a guess; but it is *more* than knowledge, for it is a foreknowledge of things yet unknown and at present perhaps inconceivable. Our vision of the general nature of things is our guide for the interpretation of all future experience. Such guidance is indispensable. Theories of the scientific method which try to explain the establishment of scientific truth by any purely objective formal procedure are doomed to failure. Any process of enquiry unguided by intellectual passions would inevitably spread out into a desert of trivialities. Our vision of reality, to which our sense of scientific beauty responds, must suggest to us the kind of questions that it should be reasonable and interesting to explore. It should recommend the kind of conceptions and empirical relations that are intrinsically plausible and which should therefore be upheld, even when some evidence seems to contradict them, and tell us also, on the other hand, what empirical connections to reject as specious, even though there is evidence for them— evidence that we may as yet be unable to account for on any other assumptions. In fact, without a scale of interest and plausibility based on a vision of reality, nothing can be discovered that is of value to science; and only our grasp of scientific beauty, responding to the evidence of our senses, can evoke this vision.

We shall get a firmer hold on this concept of scientific value by representing it as the joint outcome of three contributing factors. An affirmation will be acceptable as part of science, and will be the more valuable to science, the more it possesses:

(1) certainty (accuracy)

(2) systematic relevance (profundity)

(3) intrinsic interest.

The first two of these criteria are inherently scientific, the third is extra-scientific.

The three criteria apply jointly, so that deficiency in one is largely compensated for by excellence in the others. Take for example the evolution of species. Neo-Darwinism is firmly accredited and highly regarded by science, though there is little direct evidence for it, because it beautifully fits into a mechanistic system of the universe and bears on a subject—the origin of man—which is of the utmost intrinsic interest. In other cases we see great accuracy of facts compensating for comparative lack of systematic relevance or intrinsic interest. Manne Siegbahn was awarded the Nobel prize in physics for a greatly increased accuracy in measuring the wave length of certain X-ray spectra, though his results revealed little that is otherwise interesting. Yet there is a limit to the appreciation of accurate facts. Professor T. W. Richards was awarded the Nobel prize in 1914 for a very accurate determination of atomic weights, and his results have never been contested. Yet in 1932 Frederick Soddy could write of this kind of measurement that it appears now 'of as little interest and significance as the determination of the average weight of a collection of bottles, some of them full and some of them more or less empty'.[1] It had been realized meanwhile that the value of atomic weights results from the accidental proportion in which the constituent isotopes happen to be present in the elements as found in nature. A magnitude that had seemed to characterize a deep seated feature of the universe, had turned out to have no such bearing. Though factually correct it had proved deceptive, because—contrary to expectation—it did not correspond to anything substantial in nature. When the exact atomic weight of an element ceased to be of interest to science, what had seemed important turned out to be trivial.

Though not definable in precise terms, scientific value can as a rule be reliably assessed. Its appraisal is required and depended upon every day in the process of advancing and disseminating science. Referees consulted by journals have to judge whether the scientific interest of a contribution would justify the expense of its publication. Others have to decide whether the award of a research grant is worth while. Scientists must be able to recognize what is manifestly trivial, just as what is manifestly false. When the distinguished German physicist Friedrich Kohlrausch (1840–1910) declared, in a discussion about the aims of natural science, that he would be pleased to determine accurately the speed of water running through the gutter,[2] he was talking nonsense. He completely misjudged the nature of scientific value; for the accuracy of an observation does not in itself make it valuable to science.

The foolish promise made by Kohlrausch was of course not his true

[1] F. Soddy, *The Interpretation of the Atom*, London, 1932, p. 50.
[2] E. Warburg, *Verhandlungen der Deutschen Phys. Gesellschaft* **12** (1910), p. 920.

intention. He was merely expounding a false theory of science more consistently than is usual; relying on it, no doubt, that—as Hume said—the errors of philosophy are only ridiculous and its extravagances do not influence our lives.[1] But in doing so, he demonstrated involuntarily that such absurd conclusions can be avoided—without inconsistency—only by abandoning altogether the ideal of a strictly objective science.

It is often said that science (unlike history) is concerned only with regularities and not with unique events. This is true only up to the point covered already by my two first criteria. An event is regular if it is either reproducible or predictably recurrent. The reproducibility of a fact makes its observation exceptionally reliable, while its recurrence reveals that it forms part of a natural system. Indeed, systematic interest can outweigh even complete absence of regularity. Tycho Brahe's observation in 1572 of a new fixed star of exceptional brightness was of the greatest interest to science, for it tended to invalidate the Aristotelian system of an unchangeable empyrean.[2] Similarly, Wöhler's synthesis of urea in 1828 weakened the traditional belief in the uniqueness of living matter. Nor did the scientific interest in the discovery of a living coelacanth depend on the prospect of finding a recurrent supply of such animals, but lay in the great systematic interest of that species as a common ancestor of all the land vertebrates. Discoveries like these are valued for the breadth of their implications, even though they establish no new general laws. They offer something more vague and also more profound; namely, a truer understanding of a large domain of experience. Generality is indeed but an aspect of profundity in science, and profundity itself, as we shall see, but an intimation that we are making a new, more extensive contact with reality.

The difference between scientific and historical interest, moreover, arises, not from the uniqueness of historical events but from their interpersonal appeal, of which I shall speak later. The historical interest of past events depends, like the scientific interest of facts, on their bearing on a scholarly context: in this case the context of history. Admittedly, the appeal which this context makes to the historian is again interpersonal and thus different from the claim made by a mathematical theory on the attention of the scientist.[3]

A great many facts are unanimously regarded by scientists as irrelevant to science, while in respect to some they differ. Pasteur, in his memoir on spontaneous generation, presented to the French Academy in 1860, relates how Biot and Dumas discouraged him from undertaking work on

[1] D. Hume, *A Treatise on Human Nature*, Part IV, Section VII.

[2] The immobility of the new star had already been observed, but Tycho established the absence of a parallax. (Information due to Prof. Z. Kopal, Manchester Univ.)

[3] Isolated facts are of course devoid of scientific significance, just as an isolated event of the past is devoid of historical significance; but this is a tautology, since the isolated character of a fact is logically incompatible with its having an important bearing on the outlook of the scientist or the historian. My point is that uniqueness does not entail isolation.

such a subject.[1] Today, only very few scientists consider it worth while to test the facts of extra-sensory perception or of psychokinesis, since most of them would regard this as a waste of time and an improper use of their professional facilities. It is the normal practice of scientists to ignore evidence which appears incompatible with the accepted system of scientific knowledge, in the hope that it will eventually prove false or irrelevant. The wise neglect of such evidence prevents scientific laboratories from being plunged forever into a turmoil of incoherent and futile efforts to verify false allegations. But there is, unfortunately, no rule by which to avoid the risk of occasionally disregarding thereby true evidence which conflicts (or seems to conflict) with the current teachings of science. During the eighteenth century the French Academy of Science stubbornly denied the evidence for the fall of meteorites, which seemed massively obvious to everybody else. Their opposition to the superstitious beliefs which a popular tradition attached to such heavenly intervention blinded them to the facts in question.[2]

As the two criteria of right perception—namely sharpness of contour and reasonableness of the image—combine in determining what the eye will see, so the claims of the two first criteria of scientific value, which I have called 'certainty' and 'systematic relevance', combine in determining the scientific value of a fact. Just as the eye sees details that are not there if they fit in with the sense of the picture, or overlooks them if they make no sense, so also very little inherent certainty will suffice to secure the highest scientific value to an alleged fact, if only it fits in with a great scientific generalization, while the most stubborn facts will be set aside if there is no place for them in the established framework of science.

There is rivalry and compensation also between the intrinsic interest of a subject matter, which I have listed as the third variable in determining scientific value, and the first two, i.e. accuracy and systematic relevance. In science, as in ordinary perception, our attention is attracted by things that are useful or dangerous to us, even though they present themselves less distinctly and coherently. This sets up a competition between practical and theoretical interests, with which I shall deal more fully when defining the relation of science to technology. But things are also interesting in themselves, and their intrinsic interest varies greatly. Living animals are more interesting than their dead bodies; a dog more interesting than a fly; a man more interesting than a dog. In man himself his moral life is more interesting than his digestion; and, again, in human society the most interesting subjects are politics and history, which are the theatres of great

[1] J. B. Conant, *Pasteur's and Tyndall's Study of Spontaneous Generation*, Harvard *Case Histories in Experimental Science*, No. 7, Cambridge, 1953, p. 25.

[2] 'Scientists in other countries were anxious not to be considered as backward compared with their famous colleagues in Paris', writes F. Paneth ('Science and Miracles', *Durham University Journal*, **10** (1948–9), p. 49). '. . . many public museums threw away whatever they possessed of these precious meteorites; it happened in Germany, Denmark, Switzerland, Italy and Austria.'

moral decisions—while, at the same time, closely interwoven with these human concerns, there is great intrinsic interest also in the subjects which affect man's contemplation of the universe and his conception of himself, his origin and destiny.

The subjects which are most interesting in themselves do not lend themselves best to accurate observation and systematic study. But the two kinds of gradings can compensate for each other over a wide range of disciplines, in which they combine in variable proportions, and thus uphold throughout a steady level of scientific value. The supreme exactitude and scientific coherence of physics compensate for the comparative dullness of its inanimate subject matter, while the scientific value of biology is maintained at the same level as that of physics by the greater intrinsic interest of the living things studied, though the treatment is much less exact and coherent. The Freudian system is perhaps not wholly accepted by science, yet its enormous influence, based on its scientific claims, shows clearly that even a largely conjectural and rather vague doctrine may gain great scientific interest if it deals with man's morality and happiness. Marxism, equipped with only the flimsiest scientific characteristics, has become a force of destiny, by dealing with politics in a manner claiming to be scientific.

In relying for its own interest on the antecedent interest of its subject matter, science must accept to an important extent the pre-scientific conception of these subject matters. The existence of animals was not discovered by zoologists, nor that of plants by botanists, and the scientific value of zoology and botany is but an extension of man's pre-scientific interest in animals and plants. Psychologists must know from ordinary experience what human intelligence is, before they can devise tests for measuring it scientifically, and should they measure instead something that ordinary experience does not recognize as intelligence, they would be constructing a new subject matter which could no longer claim the intrinsic interest attached to that which they originally chose to study. Admittedly, the pursuits of biology, medicine, psychology and the social sciences, may rectify our everyday conceptions of plants and animals, and even of man and society; but we must set against any such modification its effect on the interest by which the study of the original subject matter had been prompted and justified. If the scientific virtues of exact observation and strict correlation of data are given absolute preference for the treatment of a subject matter which disintegrates when represented in such terms, the result will be irrelevant to the subject matter and probably of no interest at all.[1]

The paradigm of a conception of science pursuing the ideal of absolute detachment by representing the world in terms of its exactly determined particulars was formulated by Laplace. An intelligence which knew at

---

[1] See R. B. Perry, *Realms of Value*, Cambridge, Mass., 1954, p. 357: 'If as in the case of sociology the subject matter does not permit of exactness and conclusiveness then it does not suffice to be exact and conclusive about some other subject matter.'

one moment of time—wrote Laplace—'all the forces by which nature is animated and the respective positions of the entities which compose it, . . . would embrace in the same formula the movements of the largest bodies in the universe and those of the lightest atom: nothing would be uncertain for it, and the future, like the past, would be present to its eyes.'[1] Such a mind would possess a complete scientific knowledge of the universe.[2]

This ideal of universal knowledge is mistaken, since it substitutes for the subjects in which we are interested a set of data which tell us nothing that we want to know. Written down mathematically, Laplacean universal knowledge consists in the prediction of the co-ordinates $p$ and the impulses $q$ of all the $n$ atoms of the world, $p^{(1)} \ldots p^{(n)}$; $q^{(1)} \ldots q^{(n)}$, at a time $t$, from the co-ordinates $p_0$ and $q_0$ ($p_0^{(1)} \ldots p_0^{(n)}$, $q_0^{(1)} \ldots q_0^{(n)}$) given at the starting time $t = 0$. The prediction is made by the aid of a series of functions

$$f\,(p_0^{(1)},\, q_0^{(1)} \ldots p_0^{(n)},\, q_0^{(n)}) = p^{(1)}$$
$$f\,(p_0^{(1)},\, q_0^{(1)} \ldots p_0^{(n)},\, q_0^{(n)}) = q^{(1)}$$
$$\vdots$$

ctc.

which determines the whole set of $2n$ values of the $p$'s and the $q$'s at the time $t$. Suppose now that we could actually observe these atomic magnitudes at the time $t$. While it might be interesting to check up on this prediction, this would merely answer a question raised by the theory itself. It would be therefore without any interest except to the hypothetical scientists who had computed and would subsequently observe the $p$'s and $q$'s at the time $t$.

That such virtually meaningless information was identified by Laplace with a knowledge of all things past and all things to come, and that the stark absurdity of this claim has not been obvious to succeeding generations since his day, can be accounted for only by a hidden assumption by which this information was tacitly supplemented. It was taken for granted that the Laplacean mind would not stop short at the list of $p$'s and $q$'s at the time $t$, but proceed by virtue of its unlimited powers of computation to evaluate from this list the events, and indeed all the events, that we might be interested to know.

But this assumption is actually much larger and quite different in character from that explicitly made by Laplace. It neither demands, nor is it satisfied by our having an unlimited capacity for carrying out complex

[1] Laplace, *Traité de Probabilité*, *Œuvres* (Acad. Sc.), Paris, 1886, **7**, pp. vi–vii.

[2] Though quantum mechanics modifies the terms in which the Laplacean mind operates it does not effectively reduce its scope. The time-dependent wave-equation of the world determines for all times the wave-equation of the world, which, in quantum mechanics, represents our ultimate knowledge of all the particles in the world. It fixes the statistical distribution of all possible observables throughout the world, leaving open within this framework only variations which are strictly random.

computations concerning a mechanical system, but requires instead that we should explain *all* kinds of experience *in terms of atomic data*. This is of course the programme of a mechanistic world view, which in modern times was first speculatively mooted by Galileo; but this programme has never been carried out even in principle and we shall see in Part Four that it cannot be carried out at all. The tremendous intellectual feat conjured up by Laplace's imagination has diverted attention (in a manner commonly practised by conjurers) from the decisive sleight of hand by which he substitutes a knowledge of all experience for a knowledge of all atomic data. Once you refuse this deceptive substitution, you immediately see that the Laplacean mind understands precisely nothing and that whatever it knows means precisely nothing.

Yet the spell of the Laplacean delusion remains unbroken to this day. The ideal of strictly objective knowledge, paradigmatically formulated by Laplace, continues to sustain a universal tendency to enhance the observational accuracy and systematic precision of science, at the expense of its bearing on its subject matter. This issue will be dealt with systematically in Part Four by reflecting on our knowledge of living beings. I mention it here only as an intermediate stage in a wider intellectual disorder: namely the menace to all cultural values, including those of science, by an acceptance of a conception of man derived from a Laplacean ideal of knowledge and by the conduct of human affairs in the light of such a conception.

Whitehead wrote that the conflicts of thought in the eighteenth and nineteenth centuries were governed by the fact 'that the world had got hold of a general idea which the world could neither live with nor live without.'[1] Scientific stringency, inflexibly resolved to denature the vital facts of our existence, continues to sustain this conflict, which may yet issue in a sweeping reaction against science as a perversion of truth. This happened before, with much less justification, in the fourth century, when St. Augustine denied the value of a natural science which contributed nothing to the pursuit of salvation. His ban destroyed interest in science all over Europe for a thousand years.

For the time being, however, the peril to the true values of science does not lie in any overt reaction against science. It lies in the very acceptance of a scientific outlook based on the Laplacean fallacy as a guide to human affairs. Its reductive programme, applied to politics, entails the idea that political action is necessarily shaped by force, motivated by greed and fear, with morality used as a screen to delude the victims. This materialistic view of politics can be traced—like the mechanistic conception of man to which it is allied—far back beyond Laplace and all the way to antiquity. But a complex historical movement has since then led, along a number of mutually related lines, to the establishment in our time of the scientific method as the supreme interpreter of human affairs. This

[1] A. N. Whitehead, *Science and the Modern World*, Cambridge, 1926, p. 63.

movement has created a pervasive tension throughout our culture, similar to that generated at an earlier time by the rebellion of reason against religion, but even more comprehensive in its scope. Here I shall deal only with one form of this movement and only with its effects on the appreciation of scientific value.

Applied to human affairs, the Laplacean universal mechanics induces the teaching that material welfare and the establishment of an unlimited power for imposing the conditions of material welfare are the supreme good. But our age overflows with inordinate moral aspirations. By absorbing this zeal the objectives of power and wealth acquire a moral sanctity which, added to their supposed scientific necessity, enforces their acceptance as man's supreme and total destiny. The comprehensive claims of this movement leave no justification to public liberties, and demand that all cultural activities should subserve the power of the State in transforming society for the achievement of welfare. A discovery will then no longer be valued by the satisfaction which it gives to the intellectual passions of scientists, but will be assessed according to its probable utility for strengthening public power and improving the standard of living. Scientific value will be discredited and its appreciation suppressed.

This is how a philosophic movement guided by aspirations of scientific severity has come to threaten the position of science itself. This self-contradiction stems from a misguided intellectual passion—a passion for achieving absolutely impersonal knowledge which, being unable to recognize any persons, presents us with a picture of the universe in which we ourselves are absent. In such a universe there is no one capable of creating and upholding scientific values; hence there is no science.

The story of the Laplacean fallacy suggests a criterion of consistency. It shows that our conceptions of man and human society must be such as to account for man's faculty in forming these conceptions and to authorize the cultivation of this faculty within society. Only by accrediting the exercise of our intellectual passions in the act of observing man, can we form conceptions of man and society which both endorse this accrediting and uphold the freedom of culture in society. Such self-accrediting, or self-confirmatory, progression will prove an effective guide to all knowledge of living beings.

### 3. Heuristic Passion

So far I have been describing only the *selective* function of intellectual passions. This function is continuous with another which also comes out clearly in the same text of Kepler. We may recall here once more his words:

> What I prophesied two-and-twenty years ago, as soon as I discovered the five solids among the heavenly orbits—what I firmly believed long before I had seen Ptolemy's Harmonics—what I had promised my friends in the title

of this fifth book, which I named before I was sure of my discovery—what sixteen years ago I urged to be sought—that for which I have devoted the best part of my life to astronomical contemplation, for which I joined Tycho Brahe . . . at last I have brought it to light, and recognized its truth beyond all my hopes . . . So now since eighteen months ago the dawn, three months ago the proper light of day, and indeed a very few days ago the pure Sun itself of the most marvellous contemplation has shone forth . . . nothing holds me. . . .[1]

Intellectual passions do not merely affirm the existence of harmonies which foreshadow an indeterminate range of future discoveries, but can also evoke intimations of specific discoveries and sustain their persistent pursuit through years of labour. The appreciation of scientific value merges here into the capacity for discovering it; even as the artist's sensibility merges into his creative powers. Such is the *heuristic* function of scientific passion.

Scientists—that is, creative scientists—spend their lives in trying to guess right. They are sustained and guided therein by their heuristic passion. We call their work creative because it changes the world as we see it, by deepening our understanding of it. The change is irrevocable. A problem that I have once solved can no longer puzzle me; I cannot guess what I already know. Having made a discovery, I shall never see the world again as before. My eyes have become different; I have made myself into a person seeing and thinking differently. I have crossed a gap, the heuristic gap which lies between problem and discovery.

Major discoveries change our interpretative framework. Hence it is logically impossible to arrive at these by the continued application of our previous interpretative framework. So we see once more that discovery is creative, in the sense that it is not to be achieved by the diligent performance of any previously known and specifiable procedure. This strengthens our conception of originality. The application of existing rules can produce valuable surveys, but does not advance the principles of science. We have to cross the logical gap between a problem and its solution by relying on the unspecifiable impulse of our heuristic passion, and must undergo as we do so a change of our intellectual personality. Like all ventures in which we comprehensively dispose of ourselves, such an intentional change of our personality requires a passionate motive to accomplish it. Originality must be passionate.

But the words of Kepler show us also that this truth-bearing passion is far from infallible. Kepler rejoiced in his discovery of the five solids among the heavenly orbits; he thought that the solar distances of the six planets known to him corresponded to the sizes of the successive Platonic bodies, as measured by the radii of inscribed and circumscribed spheres. This is nonsense, and we would regard it as nonsense today, however accurately it corresponded to the facts; merely and simply because we no longer believe that the fundamental harmonies of the universe are disclosed in

[1] See p. 7 above.

143

such simple geometrical relations.[1] But though his view of reality led Kepler astray in this case, it was close enough to the truth to guide him aright to the discovery of his three laws of planetary motion. Therefore Kepler remains a great scientist to us, in spite of his erroneous reference to the Platonic bodies. It is only when he talks of such things as the mind residing in the sun which listens to the planets, and puts down in musical notation the several tunes of the planets, that we no longer regard him as a scientist, but as a mystic. We draw here a distinction between two kinds of error, namely, *scientific guesses* which have turned out to be *mistaken*, and *unscientific guesses* which are not only false, but *incompetent*.

Intellectual passions, then, may be altogether misdirected, as were those of Laplace in formulating his objectivist ideal; and even those which lead aright, as in the case of Kepler, may be interwoven with others that are inherently erroneous. A further example will confirm this conclusion, by showing us once more how closely mingled are truth-bearing and fallacious components in the intellectual passions of even the greatest scientists, and in what sense we can yet distinguish between the two.

In his speculations which led to the discovery of relativity, Einstein was guided by his aspiration—stimulated by Mach—to liberate himself from the misleading assumptions inherent in the hitherto current traditional conceptions of space and time, and to replace these by a frankly artificial framework in which the assumption of absolute rest was replaced by that of an absolutely constant velocity of light. Brushing aside the protest of common sense as the complaint of mere habit, he adopted a vision in which the electro-dynamics of moving bodies were set beautifully free from all the anomalies imposed on them by the traditional framework of absolute space and time. Accepting this intellectual beauty as a token of reality, Einstein went on to generalize his vision further and to derive from it a series of new and surprising consequences. This was an unfamiliar beauty in science, for it accepted a new conception of reality. Electro-magnetic vibrations without any vibrating medium were an affront to the mechanical conception of things which had prevailed in physics since Galileo and Newton. The new beauty inaugurated the modern view of a mathematically defined reality.

Yet Einstein's Argus-eyed sensitivity was open to other clues which did not allow him to accept this vision as his only guide. In the same year (1905) in which his work on relativity was published, he solved the riddle of Brownian motion by adopting a concrete mechanical conception of molecular agitation. This theory, soon to be confirmed experimentally by Perrin, re-established the reality of atoms as material particles, which

[1] I have quoted previously (in *The Logic of Liberty*, Chicago and London, 1951, p. 17) a table of figures published in *Nature* (**146** (1940), p. 620) purporting to show that the days of gestation of different rodents is an integer multiple of the number $\pi$, and have said that no amount of such evidence would convince us today that this relationship is real. Number rules to which no scientist will pay attention are frequently put forward today by the adherents of occult sciences.

had been peremptorily denied by Mach's operationalism. But the mechanical character of molecular interactions, so triumphantly upheld in this case, became a stumbling block when Einstein refused on these grounds to accept the ultimate reality of quantum-mechanical probabilities. His insistence that individual molecular events must be determined by specific causes, seems to have been mistaken.[1]

So we see that both Kepler and Einstein approached nature with intellectual passions and with beliefs inherent in these passions, which led them to their triumphs and misguided them to their errors. These passions and beliefs were theirs, personally, even though they held them in the conviction that they were valid, universally. I believe that they were competent to follow these impulses, even though they risked being misled by them.[2] And again, what I accept of their work as true today, I accept personally, guided by passions and beliefs similar to theirs, holding in my turn that my impulses are valid, universally, even though I must admit the possibility that they may be mistaken.

## 4. ELEGANCE AND BEAUTY

But I have yet to deal with a serious objection which will prove a Demosthenic pebble, as its conquest will finally release the major force of my argument. The mathematical theories of physics are formal systems which are applied to experience by symbolic operations. Great discoveries can be made in this manner, as when Adams and Leverrier computed the position of Neptune from Newtonian mechanics, or van't Hoff derived the laws of chemical equilibrium from the Second Law of Thermodynamics. Such operations, however, can be greatly facilitated by re-casting a formal system into more manageable terms. This adds to the beauty and power of the system without enlarging its theoretical scope; it can tell more fluently what it says about nature, but cannot say more than it could say before. So we can achieve greater economy and simplicity in our interpretative framework, and keenly enjoy this as the display of intellectual elegance, without saying anything substantially new.

This admission seems to imperil our claim that the intellectual beauty of a theory is a token of its contact with reality. Could it not be true then after all—as Mach taught—that the advantage of any theory is merely to give an economical account of observed facts?

This question was actually raised and fiercely discussed already by the contemporaries of Copernicus, in respect to the Copernican system, which

---

[1] This is the predominant view today; it is contradicted by D. Bohm, *Quantum Theory*, New York, 1951.

[2] The distinction between incompetence and error is of great importance in the administration of scientific research. The following directions to referees of the Royal Society make this plain: 'A paper should not be recommended for rejection' (it says) 'merely because the referee disagrees with the opinions or conclusions it contains, unless fallacious reasoning or experimental error is unmistakably evident.'

I have used at the opening of this book to exemplify intellectual beauty as a token of reality. Some years before the publication of his theory, a Lutheran Minister in Nüremberg, called Andreas Osiander, had urged on Copernicus that his system was merely a formal advance on Ptolemy, and Osiander actually succeeded—we do not know quite by what means—in getting this view stated in an Address to the Reader at the opening of Copernicus's book. Nor, since Copernicus was dying when his work appeared in print (1543), do we know what he thought of this passage. Osiander was prompted to his intervention by the view, frequently defended in the later Middle Ages and now familiar to us as the teaching of positivism, that 'hypotheses are not articles of faith' but merely 'foundations for calculation', so that it does not matter whether they are true or false, provided that they reproduce exactly the phenomena of motions. These quotations are taken from a letter written by Osiander in 1541, entreating Copernicus to avoid a conflict with the current Aristotelian and theological orthodoxy, by accepting a conventionalist interpretation of his own theory.[1]

This interpretation was violently rejected by the Copernicans of the next generation. Giordano Bruno called it the work of an ignorant and presumptuous ass. Galileo agreed. Kepler declared: 'It is indeed a most absurd fiction to explain natural phenomena by false causes.'[2] The issue engaged the contestants passionately to the death. Believing Copernicanism to be true, Giordano Bruno enlarged it to a vision of an indefinite multiplicity of solar systems, which anticipated the modern conception of stellar spaces. Fifty-seven years after the death of Copernicus, Bruno was burnt alive for his convictions. Galileo suffered lesser persecution over a number of years for adhering to the view (which he never sincerely abandoned) that the Copernican system was true and not merely an economical hypothesis.

[1] See G. Abetti, *History of Astronomy*, London, 1954, p. 73. A. C. Crombie, *Augustine to Galileo*, London, 1952, p. 60-1, describes the use of similar distinctions in the controversies about Aristotelean (or physical) and Ptolemaic (or mathematical) astronomy in the Middle Ages. Thus the statement of St. Thomas, *Summa Theologica*, Part 1, ques. 32, provides the background for Osiander's type of argument: 'For anything a system may be induced in a double fashion. One way is for proving some principle as in natural science where sufficient reason can be brought to show that the motions of the heavens are always of uniform velocity. In the other way, reasons may be adduced which do not sufficiently prove the principle, but which may show that the effects which follow agree with that principle, as in astronomy a system of eccentrics and epicycles is posited because this assumption enables the sensible phenomena of the celestial motions to be accounted for. But this is not a sufficient proof; because possibly another hypothesis might also be able to account for them.' This conventionalist theory of science, current in the late Middle Ages, was meant to deny science access to reality; consequently its wording coincides with the positivist analysis of science which seeks to purify it from metaphysics by avoiding any reference to reality. So, as Crombie points out ('Galilée devant les Critiques de la Posterité', University of Paris, 1956, p. 10), Duhem could say 'that it was Bellarmin and Osiander, and not Galileo and Kepler, who had grasped the precise significance of the experimental method'.

[2] G. Abetti, *op. cit.*, p. 74.

Was the issue which engaged such minds and roused such fateful passions merely verbal: a question of the proper use of the word 'true'? Actually, both sides agreed on what they meant by 'true'; namely, that truth lies in the achievement of a contact with reality—a contact destined to reveal itself further by an indefinite range of yet unforeseen consequences. I believe accordingly—in view of the subsequent history of astronomy—that the Copernicans were right in affirming the truth of the new system, and the Aristotelians and theologians wrong in conceding to it merely a formal advantage.[1]

But the long controversy between the two views shows also that this distinction is as difficult as it is vital.[2] The difficulty is merely covered up by suggesting that a true discovery is characterized by its fruitfulness, which a purely formal advance lacks. You cannot define the indeterminate veridical powers of truth in terms of fruitfulness, unless 'fruitful' is itself qualified in terms of the definiendum. The Ptolemaic system was a fruitful source of *error* for one thousand years; astrology has been a fruitful source of *income* to astrologers for two thousand five hundred years; Marxism is today a fruitful source of *power* for the rulers of one third of mankind. When we say that Copernicanism was fruitful, we mean that it was a fruitful source of *truth*, and we cannot distinguish its kind of fruitfulness from that of the Ptolemaic system, or of astrology, or Marxism, except by such a qualification. To use the word 'fruitful' in this sense, without acknowledging it, is a deceptive substitution, a pseudo-substitution, a Laplacean sleight of hand.

But even when 'fruitfulness' is taken to mean the capacity for leading to new truths, it is an insufficient characterization of truth. Copernicanism could well have been a source of truth—as the apocryphal text of Esdras was for Columbus—even if it had been false. But the Copernican system did not anticipate the discoveries of Kepler and Newton accidentally: it led to them *because* it was true. In saying this we are using the term 'true' to acknowledge the indeterminate veridical quality of Copernicanism: the quality which the Copernicans affirmed against Osiander's interpretation of the Copernican system. They believed that this system was fruitful in this and in no other sense.

Thus the substitution of 'fruitful' for 'true' is specious. But it is also nonsensical, for it implies the manifestly absurd suggestion that fruitfulness is a more concrete and limited quality which could be ascertained without going all the way to establish truth. But at the stage when we have

---

[1] Sir Edmund Whittaker ('Obituary Notice on Einstein', *Biogr. Mem. Roy. Soc.*, 1955, p. 48) points out that, contrary to widespread opinion, the physical significance of Copernicanism is not impaired by relativity. For the Copernican axes are inertial, while the Ptolemaic are not, and the earth rotates with respect to the local inertial axes.

[2] G. de Santillana, *The Crime of Galileo*, Chicago, 1955, p. 164 n. 'On Wolynski's count, there were 2,330 works published on astronomy between 1543 and 1887 . . ., of those, only 180 were Copernican (see Archivo Storico Italiano, 1873, p. 12).'

to make up our minds about the merits of a discovery its future repercussions are still unknown. By the time Newton published his *Principia* anyone could see that Copernicus had been right; but Copernicus and the Copernicans, and among these Newton himself, were convinced of it long before this. The veridical quality of the system, of which they were then convinced, could not be its *subsequently* observed fruitfulness. The attempt to replace the quality of truth, in which they believed, by the observation of the fruitfulness which this belief anticipated, is like the Bellman's advice for spotting a Snark by its habit of dining the following day. The mark of true discovery is not its fruitfulness but the *intimation* of its fruitfulness.

Admittedly—since the convenient reformulation of a theory may considerably enhance the rate at which new consequences are derived from it—a formal advantage also yields fruits, though not in the same sense in which a new discovery is expected to prove fruitful by those who accept it as true. But this objection to the anticipation of fruitfulness as a criterion of true discovery dissolves altogether when the formal advantage is of such magnitude that its contrivance does amount to a discovery. The restatements of Newtonian mechanics by d'Alembert, Maupertuis, Lagrange, and Hamilton embodied great discoveries of this kind. Such discoveries are usually accompanied or prepared by advances in mathematics, and their appreciation rests on qualities which they share with mathematical discoveries, rather than with discoveries in the natural sciences. So the recognition of their presence in some theories of physics does not obliterate, though it somewhat complicates, the distinction between a merely formal elegance and the intellectual beauty of a theory, a beauty which establishes a new contact with external reality.

However, my analysis has by now proliferated too widely from its roots in the history of Copernicanism and Relativity. I must go no further without first supplementing these paradigmatic cases by other instances substantiating the nature of theoretical discovery in physics, and the difference between such discovery and an advance of a purely formal character.

The suggestion made by Louis de Broglie (1923), purely on grounds of intellectual beauty, to ascribe wave nature to ponderable particles, is a case in point. The professors (who included Paul Langevin) to whom he presented the work for the Doctorate were doubtful whether to accept it and wrote to Einstein for advice. The latter recognized its scientific merit and the degree was duly awarded to its author.[1] But no one realised

[1] This account, based on my own recollection of the events in question, has been substantially confirmed by Louis de Broglie, *Le Dualisme des Ondes et des Corpuscules dans L'Œuvre de Albert Einstein*, Institut de France, Academie des Sciences, 1955, p. 16–17. De Broglie adds that his work would presumably have remained unknown for a long time without Einstein's subsequent reference to it in a paper published in 1925 (*ibid.*, p. 18). The only surviving examiner, Charles Mauguin, reports that while he recognized the originality and depth of thought of the candidate, 'when the thesis was

at the time that de Broglie's formulae implied that electronic beams would give diffraction patterns similar to those of X-rays, a consequence which was first envisaged by W. Elsasser in 1925.[1]

Again, the mathematical framework by which Dirac succeeded (1928) in reconciling quantum mechanics with relativity, showed some incomprehensible features which were to turn out eventually to be a description of the positive electron, when this particle was discovered, independently, by Anderson in 1932. Among earlier examples of this kind is the work of Willard Gibbs which was regarded as purely formal, until Bakhuis Roozeboom discovered the wide and illuminating applicability of the Phase Rule. More recently, the voluminous thermodynamic speculations of de Donder, published without gaining any response in the 1920's, came into their own within the new thermodynamics of irreversible systems which they were found to have partly anticipated. But the history of science records only happy endings; far more frequent are formal speculations which lead nowhere. The innumerable papers of van Laar on the thermodynamic potential, published about the same time as de Donder's papers, may be remembered among a vast multitude of such unfortunate cases. This dilution of the meritorious by floods of triviality makes the recognition of true scientific value particularly difficult.

Let me stop here for a moment. I believe that by now three things have been established beyond reasonable doubt: the power of intellectual beauty to reveal truth about nature; the vital importance of distinguishing this beauty from merely formal attractiveness; and the delicacy of the test between them, so difficult that it may baffle the most penetrating scientific minds.

I might have closed this section with these conclusions and carried them forward for later reflection; but I fear that in many readers' minds the convincing power of my evidence may have evoked only a growing feeling of uneasiness. I have watched many a university audience listening to my account of intuitive discoveries silently, with sullen distaste. Then an ironical voice would ask whether the speaker thought that there was any use at all in making experiments; and yet another, whether on such grounds as these, explanations in terms of astrology would not be equally justified. These are important questions.

presented I did not believe in the physical reality of the waves associated with the particles of matter. I saw in them rather pure creations of the mind. . . . Only after the experiments of Davisson and Germer (1927), of G. P. Thomson (1928) and only when I held in my hand the beautiful photographs (Electron diffraction patterns from thin layers of ZnO), which Ponte had succeeded in making in the École Normale, did I understand how inconsistent, ridiculous and nonsensical my attitude was' (*Louis de Broglie und die Physiker*, German edition of *Louis de Broglie: Physicien et Penseur*, Hamburg, 1955, p. 192).

[1] W. Elsasser, *Die Naturwissenschaften* **13** (1925), p. 711. Davisson and Germer's first observations of electron diffraction go back to 1925 but their interpretation of their observations as electron diffraction and the publication of this result followed only two years later, in 1927.

The answer to the first is that experience is an indispensable clue to the understanding of nature, even though it does not determine its understanding. Einstein speaks of 'ein intuitives Heranfühlen an die Tatsachen', which I should call a groping for the meaning of the facts. In this empirical guidance of our groping lies all the difference—elusive and yet utterly decisive—between a merely formal advance and a new insight into the nature of things. Whence this elusiveness? It is a reflection on the canvas of the highest scientific achievement of the fact that we can never tell exactly what we mean, or even whether we mean anything at all. Indeterminacy of meaning is not eliminated, but only restricted, when we eventually decide to accept a theory as a true statement of something new about nature. For, while we heavily commit ourselves thereby to a belief concerning certain things, such a belief can have no bearing on reality unless its scope is still left indeterminate.

The answer to the second question, why we should prefer science to astrology, cannot be given briefly. In the next section I shall approach it by one step and a fairly conclusive reply will be reached at the end of Part Three; but the whole of this book is but a quest for a substantial reply to a question of this kind. In the end I should be able to say as a statement that will appear neither dogmatic nor trivial: 'I do not entertain explanations in terms of astrology, for I do not believe them to be true.'

## 5. SCIENTIFIC CONTROVERSY

Heuristic passion seeks no personal possession. It sets out not to conquer, but to enrich the world. Yet such a move is also an attack. It raises a claim and makes a tremendous demand on other men; for it asks that its gift to humanity be accepted by all. In order to be satisfied, our intellectual passions must find response. This universal intent creates a tension: we suffer when a vision of reality to which we have committed ourselves is contemptuously ignored by others. For a general unbelief imperils our own convictions by evoking an echo in us. Our vision must conquer or die.

Like the heuristic passion from which it flows, the *persuasive passion* too finds itself facing a logical gap. To the extent to which a discoverer has committed himself to a new vision of reality, he has separated himself from others who still think on the old lines. His persuasive passion spurs him now to cross this gap by converting everybody to his way of seeing things, even as his heuristic passion has spurred him to cross the heuristic gap which separated him from discovery.

We can see, therefore, why scientific controversies never lie altogether within science. For when a new system of thought concerning a whole class of alleged facts is at issue, the question will be whether it should be accepted or rejected in principle, and those who reject it on such comprehensive grounds will inevitably regard it as altogether incompetent and

150

unsound. Take, for example, four contemporary issues: Freud's psycho-analysis, Eddington's *a priori* system of physics, Rhine's 'Reach of the Mind', or Lysenko's environmental genetics. Each of the four authors mentioned here has his own conceptual framework, by which he identifies his facts and within which he conducts his arguments, and each expresses his conceptions in his own distinctive terminology. Any such framework is relatively stable, for it can account for most of the evidence which it accepts as well established, and it is sufficiently coherent in itself to justify to the satisfaction of its followers the neglect for the time being of facts, or alleged facts, which it cannot interpret. It is correspondingly segregated from any knowledge or alleged knowledge rooted in different conceptions of experience. The two conflicting systems of thought are separated by a logical gap, in the same sense as a problem is separated from the discovery which solves the problem. Formal operations relying on *one* framework of interpretation cannot demonstrate a proposition to persons who rely on *another* framework. Its advocates may not even succeed in getting a hearing from these, since they must first teach them a new language, and no one can learn a new language unless he first trusts that it means something. A hostile audience may in fact deliberately refuse to entertain novel conceptions such as those of Freud, Eddington, Rhine or Lysenko, precisely because its members fear that once they have accepted this framework they will be led to conclusions which they—rightly or wrongly —abhor. Proponents of a new system can convince their audience only by first winning their intellectual sympathy for a doctrine they have not yet grasped. Those who listen sympathetically will discover for themselves what they would otherwise never have understood. Such an acceptance is a heuristic process, a self-modifying act, and to this extent a conversion. It produces disciples forming a school, the members of which are separated for the time being by a logical gap from those outside it. They think differ-ently, speak a different language, live in a different world, and at least one of the two schools is excluded to this extent for the time being (whether rightly or wrongly) from the community of science.

We can now see, also, the great difficulty that may arise in the attempt to persuade others to accept a new idea in science. We have seen that to the extent to which it represents a new way of reasoning, we cannot con-vince others of it by formal argument, for so long as we argue within their framework, we can never induce them to abandon it. Demonstration must be supplemented, therefore, by forms of persuasion which can induce a conversion. The refusal to enter on the opponent's way of arguing must be justified by making it appear altogether unreasonable.

Such comprehensive rejection cannot fail to discredit the opponent. He will be made to appear as thoroughly deluded, which in the heat of the battle will easily come to imply that he was a fool, a crank or a fraud. And once we are out to establish such charges we shall readily go on to expose our opponent as a 'metaphysician', a 'Jesuit', a 'Jew', or a 'Bolshevik', as the case may be—or, speaking from the other side of the Iron Curtain—as an 'objectivist', an 'idealist' and a 'cosmopolitan'. In a clash of intellectual passions each side must inevitably attack the opponent's person.

Even in retrospect such conflicts can often be appreciated only in these terms. They do not appear as scientific arguments, but as conflicts between rival scientific visions, or else between scientific values and extraneous interests interfering illegitimately with the due process of scientific enquiry. I shall recall here four controversies to illustrate this. The first is the Copernican, to which I have already had occasion to refer. The other three occurred in the nineteenth century and their outcome, like that of the earlier one, had an effective part in developing our present sense of scientific value.

The Ptolemaic and Copernican theories opposed each other for a long time as two virtually complete systems separated by a logical gap. The facts known at any time during the 148 years from the publication of Copernicus' *De Revolutionibus* to the appearance of Newton's *Principia* could be accounted for by either theory. By 1619 the discovery of Kepler's third law may have tipped the balance in favour of Copernicanism,[1] but the non-appearance of any seasonal variation in the angle at which the fixed stars are seen continued to present a serious difficulty to this system. The mistaken argument that falling bodies would not descend vertically to earth if it were in motion was disproved by Galileo; but his explanation of the tides, which he regarded as a crucial proof of terrestrial rotation, fell into a similar error. His discovery of Jupiter's moons was perhaps suggestive, but its significance hardly justified his scornful invective against those who refused to look at these moons through his telescope.[2] The real ground of Galileo's conviction lay in his passionate appreciation of the greater scientific value of the heliocentric view: a feeling which was accentuated by his angry rebellion against Aristotle's authority over science. His opponents had on their side the common-sense view which sees the earth at rest, and, above all, a vivid consciousness of man's uniqueness as the only particle of the universe that feels responsible to God. Their craving to retain for man a location which corresponds to his importance in the universe was the emotional force opposed to the intellectual appeal of Copernicanism.[3]

---

[1] Galileo never made use of this argument, which was the strongest available to him. He seems never to have accepted Kepler's elliptical planetary paths, presumably because his Pythagoreanism was even more rigid than Kepler's. (See G. de Santillana, *op. cit.*, pp. 106 (note 29) and 168–70).

[2] Admittedly, the phases of Venus discovered by Galileo could not be accounted for by the Ptolemaic system, but they were compatible with Tycho Brahe's assumption that the planets circled round the Sun which itself circled round the Earth. Fortunately, no experiment of the Michelson Morley type was carried out at the time, for its negative result would have served as decisive proof that the Earth was at rest.

[3] See Goethe, *Geschichte der Farbenlehre*, Vierte Abteilung, 2te Zwischenbemerkung: 'Vielleicht ist nie eine grössere Forderung an die Menschheit geschehen: denn was ging nicht alles durch diese Anerkennung in Dunst und Rauch auf: ein zweites Paradies, eine Welt der Unschuld, Dichtkunst und Frömmigkeit, das Zeugnis der Sinne, die Überzeugung eines poetisch-religiösen Glaubens; kein Wunder, dass man dies alles

The victory of Copernicanism rejected and suppressed this demand as an illegitimate interference with the pursuit of science, and established the principle that scientific truth shall take no account of its religious or moral repercussions. But this principle is not incontestable. It is rejected today by the Soviet theory that all science is class science and must be guided by 'partynost', party-spirit. It is contested also by the Catholic Church, as for example in the Encyclical *Humani Generis* of 1950, and similarly by biblical fundamentalists everywhere. My opposition to a universal mechanical interpretation of things, on the ground that it impairs man's moral consciousness, also implies some measure of dissent from the absolute moral neutrality of science. Yet though the issue is not altogether closed, the principle of moral and religious indifference prevails throughout modern science without facing so far any effective rival to its rule, and the outcome of the Copernican controversy continues to form an eminent support for this principle.[1]

Another tenet of modern science which emerged at an early stage from its conflict with the Aristotelian and scholastic tradition is its ideal of empiricism. Though I dissent from this ideal in its absolute form, since I hold that the elimination of personal knowledge from science would destroy science, I acknowledge the decisive achievements of empiricism in opening the way to modern science. Nor do I deny, of course, that science is constantly in danger from the incursion of empty speculations, which must be watchfully resisted and cast out; but I hold that the part played by personal knowledge in science makes it impossible to formulate any precise rule by which such speculations can be distinguished from properly conducted empirical investigations. Empiricism is valid only as a maxim, the application of which itself forms a part of the art of knowing. Some examples of the scientific controversies in which maxims of scientific empiricism have acquired their current meaning will show how controversial and misleading the claims of empiricism have proved in some important instances.

The quixotic attack of the young Hegel on the empirical method of science and his swift defeat at the hands of the scientists was one of the great formative experiences of modern science. In the year 1800 a band of six German astronomers, led by Bode, set out to search for a new planet to fill a gap between Mars and Jupiter in the numerical series of planetary distances, discovered by Titius and known as Bode's Law. The series is obtained by writing down the number 4, followed by the series 3+4,

nicht wollte fahren lassen, dass man sich auf alle Weise einer solchen Lehre entgegensetzte, die denjenigen, der sie annahm, zu einer bisher unbekannten, ja ungeahneten Denkfreiheit und Grossheit der Gesinnungen berechtigte und aufforderte.'

[1] See R. A. Fisher, *Creative Aspects of Natural Law*, Eddington Memorial Lecture, Cambridge, 1950, p. 15: 'We attempt, so far as our powers allow, to understand the world, by reasoning, by experimentation, and again by reasoning. In this process moral or emotional grounds for preferring one conclusion to another are completely out of place.'

$2 \times 3 + 4$, $2^2 \times 3 + 4$, $2^3 \times 3 + 4$... etc. This gives for the first eight places: 4, 7, 10, 16, 28, 52, 100, 196, which can be shown to correspond pretty well to the relative distances of the seven planets known in 1800, provided you leave out the fifth number. Setting the distance of the Earth arbitrarily at 10 you have the table:

BODE'S LAW IN 1800

|         | Predicted | Observed |
|---------|-----------|----------|
| Mercury | 4         | 3·9      |
| Venus   | 7         | 7·2      |
| Earth   | (10)      | (10)     |
| Mars    | 16        | 15·2     |
| ...?    | 28        | ?        |
| Jupiter | 52        | 52       |
| Saturn  | 100       | 95       |
| Uranus  | 196       | 192      |

The young Hegel poured scorn on an enquiry following up a numerical rule which, being meaningless, could only be accidental. Arguing that nature, shaped by immanent reason, must be governed by a rational sequence of numbers, he postulated that the relative spacing of the planets must conform to the Pythagorean series 1, 2, 3, $4(2^2)$, $9(3^2)$, $8(2^3)$, $27(3^3)$ — in which, however, he substituted 16 for 8. This would limit the number of planets to 7 and allow a large gap between the fourth and fifth planet, i.e. Mars and Jupiter. The quest for an eighth planet to fill this gap was therefore chimerical.[1]

However, on January 1st, 1801, Bode's party of astronomers discovered the small planet Ceres in the region in question. Since then over 500 small planets have been found in that neighbourhood,[2] and it may be that these are fragments of a full sized planet that once occupied this place.

Hegel was discomfited and the astronomers triumphed gleefully. This was all to the good, for it confirmed a juster sense of scientific value. But we should realise that it had little else to support it. Whether Bode's Law has any rational foundation or has been fulfilled so far by mere coincidence (as Hegel had thought) is still open to question today; opinions have changed on the subject repeatedly during the last 20 years.[3] Thus Hegel

[1] Hegel, Dissertatio philosophica de Orbitis Planetarum (1801), *Werke*, Berlin, 1834, **16**, p. 28. In his lectures on the Philosophy of Nature, Hegel admitted the presence of Ceres and other asteroids in this gap. He still referred to the numbers of the *Timaeus*, but he now declared that the law of the planetary distances was still unknown, and that one day the scientists would have to turn to the philosophers to find it. Bertrand Beaumont, in discussing Hegel's position (*Mind*, N.S., **63**, 1954, pp. 246–8), suggests that the Platonic series could be extended beyond the original seven; but in terms of Greek mathematics this is impossible.

[2] H. H. Turner, *Astronomical Discovery*, London, 1904, p. 23.

[3] An attempt to interpret Bode's Law rationally by deriving it from a theory of the planetary system was made by C. F. von Weizsäcker in 1943 (*Zs. für Astrophysik*, **22** (1944), p. 319). But from a later paper it appears that the problem is still in flux (see C. F. von Weizsäcker, *Festschrift der Akademie der Wissenschaften in Göttingen*, 1951, p. 120).

may have been right in rejecting the astronomers' grounds for the search for a new planet.

Yet I agree that the astronomers were right and Hegel was wrong. Why so? Because the astronomers' guess lay within a conceivable scientific system, and so it was a kind of guess to which astronomers as scientists are entitled. It was a competent guess, and—if Bode's Law has any truth in it— even a true guess; while Hegel's inference was altogether unscientific, *incompetent*. Fortunately Hegel guessed wrong and the astronomers, though their guess was perhaps unjustified, did hit the mark. But even if Hegel's guess had proved right and the astronomers' wrong, we would still reject Hegel's vision of reality and cleave to that of the astronomers.

The revulsion of scientists against *Naturphilosophie* was violent and lasting. By the middle of the century empiricism ruled unchallenged.[1] But unfortunately the empirical method of enquiry—with its associated conceptions of scientific value and of the nature of reality—is far from unambiguous, and conflicting interpretations of it had therefore ever again to fight each other from either side of a logical gap.

In his doctoral thesis, presented in 1875 to the University of Utrecht, J. H. van't Hoff had put forward the theory that compounds containing an asymmetric carbon atom are optically active. In 1877 there appeared a German translation of this work with a commendatory introduction by Wislicenus, a distinguished German chemist and an authority on optical activity. This publication evoked a furious attack from Kolbe, another leading German chemist, who had recently published an article called 'Signs of the Times',[2] in which he castigated the decline of rigorous scientific training among German chemists; a decline which, he said, had led to a renewed sprouting of

> the weeds of a seemingly learned and brilliant but actually trivial and empty
> Philosophy of Nature, which, after having been replaced some 50 years ago
> by the exact sciences, is now once more dug up by pseudo-scientists from the
> lumber room of human fallacies, and like a trollop, newly attired in elegant
> dress and make-up, is smuggled into respectable company, to which she
> does not belong.

In a second paper,[3] he gave as a further example of this aberration an account of van't Hoff's work which 'he would have ignored like many other efforts of its kind' but for 'the incomprehensible fact' of its warm

---

[1] *Naturphilosophie* lingered on longest in botany, where eminent scientists were ranged on both sides, with Braun and Agassiz standing largely under the influence of Goethe's morphology and of the nature philosophy of Schelling, opposed from the middle of the nineteenth century onward by others, notably Schleiden and Hofmeister, who developed the science of plant morphology on an experimental basis. See K. v. Goebel, *Wilhelm Hofmeister*, London, 1926. We shall see in Part Four that this controversy is still not completely closed today.

[2] A. W. H. Kolbe, *Journ. für praktische Chemie*, **14** (1877), p. 268.

[3] A. W. H. Kolbe, 'Zeichen der Zeit II', *Journ. für prakt. Chem.*, **15** (1877), p. 473. The above summary of the first paper is quoted from the second paper.

recommendation by so distinguished a chemist as Wislicenus. So Kolbe wrote:

> A certain Dr. J. H. van't Hoff, employed by the Veterinary Academy at Utrecht, appears to have no taste for exact chemical research. He found it more convenient to mount Pegasus (borrowed no doubt from the Veterinary Academy) and to proclaim in his *La Chimie dans l'espace* how on his daring flight to the chemical Parnassus the atoms appeared to him disposed in world space.

Kolbe's comment on the introduction given by Wislicenus to van't Hoff's theory reveals even further the principles of his criticism. Wislicenus had written of 'this real and important step in the advancement of the theory of carbon compounds, a step which was organic and internally necessary'. Kolbe asks: What is 'the theory of carbon compounds'? What is meant by saying that 'this step was organic and necessary'? And he goes on: 'Wislicenus has here expelled himself from the ranks of exact scientists and has joined instead the nature philosophers of ominous memory who are separated only by a slender "medium" from the spiritists.'

Scientific opinion eventually repudiated Kolbe's attack on van't Hoff and Wislicenus, but his suspicion of speculative chemistry ('paper chemistry') continues to be shared by most of the leading chemical journals, which refuse up to this day contributions containing no new experimental results. In spite of the fact that chemistry is largely based on the speculations by Dalton, Kekulé and van't Hoff, which were initially unaccompanied by any experimental observations,[1] chemists still remain suspicious of this kind of work. Since they do not sufficiently trust themselves to distinguish true theoretical discoveries from empty speculations, they feel compelled to act on a presumption which may one day cause the rejection of a theoretical paper of supreme importance in favour of comparatively trivial experimental studies. So difficult is it even for the expert in his own field to distinguish, by the criteria of empiricism, scientific merit from incompetent chatter.

Nor does this apply only to purely theoretical discoveries. The great controversy on the nature of alcoholic fermentation which, starting in 1839, went on for almost forty years, showed that the verification of an experimental observation may run into precisely the same difficulties. From 1835 to 1837 no less than four independent observers (Caignard de la Tour, Schwann, Kützing and Turpin) had reported that yeast produced during fermentation was not a chemical precipitate, but consisted of living cellular organisms which multiplied by budding, and had concluded that fermentation was a living function of yeast cells.[2] But this went against the dominant intellectual passion of contemporary scientists. In 1828, Wöhler had synthetized urea from inorganic materials and had triumphantly dis-

---

[1] For the case of John Dalton, see H. E. Roscoe and A. Harden, *A New View of the Origin of Dalton's Atomic Theory*, London, 1896, p. 50.
[2] R. J. Dubos, *Louis Pasteur*, London, 1950, pp. 120–1.

proved thereby the existence of powers hitherto ascribed exclusively to living beings. Liebig had followed suit by laying the foundations of a chemical approach to all living matter, and Berzelius recognized that platinum could speed up reactions occurring in its presence, in the same way in which fermentation was caused by yeast. All three great masters poured scorn on claims which they regarded as a fantastic resurgence of the kind of 'vitalism' they had banned for ever. Wöhler and Liebig published an elaborate skit making fun of these absurd speculations.[1]

In 1857 Pasteur entered the lists on the side of the 'vitalists'. His investigations on yeast and putrefaction involved him at the same time in another fierce controversy of longer standing, the question of 'spontaneous generation'. In this, too, he was on the side considered at the time reactionary (and still so considered, at the time of writing this, in the Soviet Union), which denied that living beings could be produced experimentally from dead matter.[2]

The reason why both these controversies dragged on indefinitely is revealed by a remark of Pasteur concerning his own arguments for regarding fermentation as a function of the living cells of yeast: 'If anyone should say that my conclusions go beyond the established facts (he wrote) I would agree, in the sense that I have taken my stand unreservedly in an order of ideas which, strictly speaking, cannot be irrefutably demonstrated.'[3] This order of ideas was therefore separated by a logical gap from that entertained by Liebig, Wöhler and many other great men of his time. The schism was eventually bridged by a conceptual reform induced by Buchner's discovery, in 1897, of zymase in the liquor squeezed out of yeast cells. The agent of fermentation was proved a dead catalyst of the kind imagined by Liebig and Berzelius, but it also proved a vital organ of yeast cells, as Pasteur and his precursors since Caignard de la Tour had affirmed; the new conception of intracellular enzymes combined these two aspects.[4]

The great scientific controversies which I have just recalled were

---

[1] Wöhler and Liebig. *Annalen der Pharmacie*, **29** (1839), p. 100.

[2] See the violent attacks on Pasteur by Pisarev published in 1865. (A. Coquart, *Dmitri Pisarev*, Paris, 1946, pp. 336 ff.) Experiments acknowledged until recently in the Soviet Union as proofs of the spontaneous generation of cellular organisms were carried out by Lepeshinskaia (see Th. Dobzhansky, *Proceedings of Hamburg Congress on Science and Freedom*, London, 1955, p. 219).

[3] R. J. Dubos, *op. cit.*, p. 128. J. B. Conant (*Pasteur's and Tyndall's Study of Spontaneous Generation*, Harvard Univ. Press, 1953) suggests (p. 15) that the most convincing evidence for the impossibility of spontaneous generation is to be found 'in the whole fabric of the results of the study of *pure* bacterial cultures in the last sixty or seventy years'. The author implies that all the experiments made to decide this question, from the inception of Spallanzani's studies in 1768 up to 1880–90, could be interpreted in terms of either of the opposing systems of thought.

[4] I have illustrated before how an apposite new conception can reconcile two alternative systems of interpretation which hitherto violently opposed each other. Braid's conception of 'hypnosis' acknowledged the reality of the very features of Mesmerism hitherto taken to prove its fraudulence, while rejecting the evidence for 'animal magnetism' which had been advanced as its claim to scientific solidity. (Part Two, ch. 9, p. 108 above.)

conducted in passionate accents, as was inevitable between contestants who shared no common framework within which a more impersonal procedure could be followed. Kolbe could not argue against van't Hoff. He quoted with ironical glee van't Hoff's description of the disposition of atoms in spirals, which to him was sufficient evidence that the new theory was a tissue of fancies. And from his own point of view he was right in refusing to enter into any detailed argument on these lines, since he denied that one could argue rationally in terms of such wild ideas. The ironical caricature by which Wöhler and Liebig replied to the papers of Caignard de la Tour, Schwann and others, who claimed that fermentation is a function of living yeast cells, sprang from the same view that an argument believed to be wholly specious cannot be seriously discussed point by point.[1] A Western scientist challenged to answer Lysenko's biological theories would similarly refuse to discuss them on the Marxist-Leninist grounds on which they were put forward; while, on the other hand, Lysenko refuses to consider the statistical evidence for Mendelism on the grounds that 'in science there is no place for chance'.[2]

We may conclude that empiricism, like the moral neutrality of science, is a principle laid down and interpreted for us by the outcome of past controversies about the scientific value of particular sets of ideas. Our appreciation of scientific value has developed historically from the outcome of such controversies, much as our sense of justice has taken shape from the outcome of judicial decisions through past centuries. Indeed, all our cultural values are the deposits of a similar historic succession of intellectual upheavals. But ultimately, all past mental strife can be interpreted today only in the light of what we ourselves decide to be the true outcome and lesson of this history. And we have to take this decision within the context of contemporary controversies which perhaps challenge these lessons afresh and raise in their turn quite novel questions of principle. The lesson of history is what we ourselves accept as such.

There are serious questions still open today concerning the nature of things. At least, I believe them to be open, though the great majority of scientists are convinced that the view they themselves hold is right and scorn any challenge to it. A notorious example is offered by extra-sensory perception. The evidence for it is ignored today by scientists in the hope that it will one day find some trivial explanation. In this they may be right, but I respect those too who think they may be wrong; and no profitable discussion is possible between the two sides at this stage.

Another example. Neurologists today accept almost without exception

---

[1] Eddington's derivation of the 'fine structure constant' $hc/2\pi e^2 = 137$ was similarly caricatured by a fictitious communication to *Naturwissenschaften*, **19** (1931), p. 39, by G. Beck, H. Bethe and W. Riezler. The authors extended Eddington's argument to the farcical conclusion that the value of $-273°$ C. for the temperature at absolute zero was an integer.

[2] *Pravda*, August 10th, 1948, quoted by Sidney Hook, *Marx and the Marxists*, New York, 1955, p. 235.

the assumption that all conscious mental processes can be interpreted as epiphenomena of a chain of material events occurring in the nervous system. Some writers, like Dr. Mays,[1] myself [2] and Professor R. O. Kapp,[3] have tried to show that this is logically untenable, but to my knowledge only one neurologist, namely Professor J. C. Eccles, has gone so far as to amend the neurological model of the brain, by introducing an influence by which the will intervenes to determine the choice between two possible alternative decisions.[4] This suggestion is scornfully ignored by all other neurologists, and indeed, it is difficult to argue profitably about it from their point of view.

A similar schism is present today between the ruling school of genetics, which explains evolution as a result of a haphazard sequence of mutations, and writers like Graham Cannon in England, Dalcq in Belgium, Vandel and others in France, who consider this explanation inadequate and support the assumption of a harmonious adaptive power controlling the most important innovations in the origin of higher forms of life.

Some people may listen to these illustrations with impatience, for they believe that science provides a procedure for deciding any such issues by systematic and dispassionate empirical investigations. However, if that were clearly the case, there would be no reason to be annoyed with me. My argument would have no persuasive force, and could be ignored without anger.

At any rate, let me make quite clear what I have urged here. I have said that intellectual passions have an affirmative content; in science they affirm the scientific interest and value of certain facts, as against any lack of such interest and value in others. This *selective* function—in the absence of which science could not be defined at all—is closely linked to another function of the same passions in which their cognitive content is supplemented by a conative component. This is their *heuristic* function. The heuristic impulse links our appreciation of scientific value to a vision of reality, which serves as a guide to enquiry. Heuristic passion is also the mainspring of originality—the force which impels us to abandon an accepted framework of interpretation and commit ourselves, by the crossing of a logical gap, to the use of a new framework. Finally, heuristic passion will often turn (and have to turn) into *persuasive* passion, the mainspring of all fundamental controversy.

I am not applauding the outbreak of such passions. I do not like to see

[1] W. Mays, 'Mind-like Behaviour in Artefacts and the Concept of Mind', *Brit. Jl. Phil. Sc.*, **3** (1952-3), p. 191.
[2] Michael Polanyi, 'The Hypothesis of Cybernetics', *Brit. Jl. Phil. Sci.*, **2** (1951-2), p. 312.
[3] R. O. Kapp, 'The Observer, the Interpreter, and the Object Observed', *Methodos*, **7** (1955), pp. 3-12.
[4] J. C. Eccles, *The Neurophysiological Basis of Mind*, Oxford, 1953, ch. VIII, pp. 261 ff.

a scientist trying to bring an opponent into intellectual contempt, or to silence him in order to gain attention for himself; but I acknowledge that such means of controversy may be tragically inevitable.

## 6. The Premisses of Science

At this point we may enquire how far the controversial principles in question may be regarded as the premisses of science. Can science be said to rest on specifiable presuppositions, be it on rules of correct procedure or on substantial beliefs about the nature of things? I shall put the matter first in the light of my own views, based on reflections of the kind made in the course of the foregoing survey. We have seen how perilously poised is the balance between the intrinsic claims of a subject matter, and the passion for exactitude and coherence, e.g. in the scientific study of sentient beings; how the tendency towards a universal mechanistic conception of things may threaten completely to denature our image of man. We have seen other examples too, in the work of Kepler and Einstein, to show how the most powerful visions of scientific truth may reveal later elements of fundamental error. I have mentioned a series of great speculative discoveries which eloquently testified to the veridical powers of intellectual beauty, and have shown at the same time how frequently such discoveries may remain unrecognized by the most expert judges, and that no one— not even their authors—can even remotely discern at first what they imply. The delicate character of the criteria which characterize scientific value was brought home further by the great scientific controversies of the past. We have seen that the outcome of these controversies—together with the result of other upheavals in science—have laid down these criteria for us today, though in the last resort it still remains for us to decide how far we shall accept or modify these interpretations. For in the history of science, as in that of all other human activities, it falls finally to him who tells their story, to endorse or revise all previous assessments of their outcome— while simultaneously responding to contemporary issues unthought of before. Traditions are transmitted to us from the past, but they are our own interpretations of the past, at which we have arrived within the context of our own immediate problems.

The general criteria of scientific value to be derived from the historic instances I have considered, may be tentatively regarded as a fair sample of the premisses of science. Copernicus and his opponents; Kepler and Einstein; Laplace and John Dalton; Hegel and Bode; de Broglie and Dirac; van't Hoff and Kolbe; Liebig and Pasteur; Elliotson and Braid; Freud, Eddington, Rhine and Lysenko; all these and countless other scientists, or persons claiming to be scientists, have held certain allegedly 'scientific' beliefs about the nature of things and the proper method and purpose of scientific enquiry. These beliefs and valuations have indicated to their adherents the kind of questions which seem reasonable and interesting to explore. They have recommended the kind of conceptions and relations that should be upheld as plausible, even when some evidence seemed to contradict them; or which, on the contrary, should be rejected as unlikely, even though there was evidence which seemed to favour them and which could not be readily explained on other grounds.

The rules of scientific procedure which we adopt, and the scientific beliefs and valuations which we hold, are mutually determined. For we proceed according to what we expect to be the case and we shape our anticipations in accordance with the success which our methods of procedure have met with. Beliefs and valuations have accordingly functioned as joint premisses in the pursuit of scientific enquiries. But how exactly should they be defined in this relation? We might be inclined to acknowledge as its premisses the general views and purposes bearing on a future scientific enquiry. But 'premiss' is a logical category: it refers to an affirmation which is logically anterior to that of which it is the premiss. Accordingly, the general views and purposes implicit in the achievement and establishment of a scientific discovery are its premisses, even though these views and purposes may no longer be quite the same as those held *before* the investigation was first seriously thought of. This paradoxical sense seems to be the only one in which we can envisage any premisses of science. But let me first elaborate the same principle briefly in respect to everyday knowledge.

Natural science deals with facts borrowed largely from common experience. The methods by which we establish facts in everyday life are therefore logically anterior to the special premisses of science, and should be included in a full statement of these premisses. The standards of intellectual satisfaction which urge and guide our eyes to gather what there is to see, and which guide our thoughts also to shape our conception of things— the beliefs about the nature of things transmitted by our everyday descriptive language—all these form part of the premisses of science, even though we must allow for the revision of these standards and beliefs within science. On the other hand, the assumption of persistent features in nature is certainly not a sufficient premiss for the establishment of natural science. It gives us grounds for referring to facts and for thinking of the universe as an aggregate of facts. But factuality is not science. Only a comparatively few peculiar facts are scientific facts, while the enormous rest are without scientific interest. Hence principles like that of the Uniformity of Nature (J. S. Mill), or that of Limited Variety (J. M. Keynes), which may account for factuality, cannot account *by themselves* for natural science. Astrology and magic rely as much on the Uniformity of Nature or on its Limited Variety as do the natural sciences, even though science repudiates the facts alleged by astrologers and medicine men.

I have said that the premisses of science are tacitly observed in the practice of scientific pursuits and in the acceptance of their results as true. This is true also of the Uniformity of Nature or of its Limited Variety.

Indeed, it is only by our common acquaintance with facts, which either persist for a while on a single occasion, or keep recurring at different places and times, that we can appreciate what is meant by Uniformity or Limited Variety in Nature. These conceptions would be quite unintelligible to us if we lived in a gaseous universe, in which no circumscribed or recurrent facts could ever be discerned. The logical premisses of factuality are not known to us or believed by us *before* we start establishing facts, but are recognized on the contrary *by reflecting on the way we establish facts.* Our acceptance of facts which make sense of the clues offered by experience to our eyes and ears must be presupposed first, and the premisses underlying this process of making sense must be deduced from this afterwards. Since the process of discovering the logical antecedent from an analysis of its logical derivate cannot fail to introduce a measure of uncertainty, the knowledge of this antecedent will always be less certain than that of its consequent. We do not believe in the existence of facts because of our anterior and securer belief in any explicit logical presuppositions of such a belief; but on the contrary, we believe in certain explicit presuppositions of factuality only because we have discovered that they are implied in our belief in the existence of facts.

The same peculiar logical structure will be seen to apply to the more specific premisses of science and indeed, far beyond these, to the logical antecedents of all informal mental processes, some of which enter into every rational act of man. The simplest illustration of this structure is given in the practice of skills like swimming, cycling or playing the piano; to recall my analysis of them will strip the foregoing formulation of its paradoxical features. Swimming may be said to presuppose the principle of keeping afloat by retaining an excessive residue of air in the lungs; and we can state certain operational principles of cycling and piano playing which can likewise be regarded as the premisses underlying these performances. But we have seen that we achieve and practise these skills without any antecedent focal knowledge of their premisses. Indeed, the premisses of a skill cannot be discovered focally prior to its performance, nor even understood if explicitly stated by others, before we ourselves have experienced its performance, whether by watching it or by engaging in it ourselves. In performing a skill we are therefore acting on certain premisses of which we are focally ignorant, but which we know subsidiarily as part of our mastery of that skill, and which we may get to know focally by analysing the way we achieve success (or what we believe to be success) in the skill in question. The rules of success which we thus derive can help us to improve our skill and to teach it to others—but only if these principles are first re-integrated into the art of which they are the maxims. For though no art can be exercised according to its explicit rules, such rules can be of great assistance to an art if observed subsidiarily within the context of its skilful performance.[1]

---

[1] See p. 31 above and p. 49.

We can accordingly amend our previous formulation as follows. The logical antecedents of an informal mental process like fact finding, or more particularly, the finding of a fact of science, come to be known subsidiarily in the very act of their application; but they can become known focally only later, from an analysis of their application, and, once focally known, they can be applied by re-integration to guide subsidiarily improved performances of the process.

The first step towards establishing the premisses of a mental achievement like science—or music, or law, etc.—is, therefore, to acknowledge its authentic instances. We do not accept every allegation of a fact as true, nor can we acknowledge every intended contribution to science, or music, or law, as part of true science, or music, or law. The question, which 'facts' are facts, what 'science' is science, what 'music' is music, what 'law' is law, may indeed be highly controversial. In order to elucidate the assumptions underlying the establishment of facts, and particularly of the facts of science, we must take a stand on such doubtful issues. For we must reflect on facts and parts of science which we acknowledge as valid, or at least on facts and parts of science which we regard as competently alleged, even if not validly established. I would accept, for example, the statement that according to the facts as known to Kepler the number of planets was six, though this is not the fact according to our own knowledge; and I acknowledge Pythagorean speculations as part of Kepler's scientific work, even though I do not believe them to be correct; while at the same time I reject Kepler's horoscopes and all his astrology as incompetent, both as statements of fact and as scientific work.

Any attempt to define the body of science more closely comes up against the fact that the knowledge comprised by science is not known to any single person. Indeed, nobody knows more than a tiny fragment of science well enough to judge its validity and value at first hand. For the rest he has to rely on views accepted at second hand on the authority of a community of people accredited as scientists. But this accrediting depends in its turn on a complex organization. For each member of the community can judge at first hand only a small number of his fellow members, and yet eventually each is accredited by all. What happens is that each recognizes as scientists a number of others by whom he is recognized as such in return, and these relations form chains which transmit these mutual recognitions at second hand through the whole community. This is how each member becomes directly or indirectly accredited by all. The system extends into the past. Its members recognize the same set of persons as their masters and derive from this allegiance a common tradition, of which each carries on a particular strand.

This analysis of the scientific consensus will be carried further in the next chapter, on Conviviality. Suffice it here to say that anyone who speaks of science in the current sense and with the usual approval, accepts this organized consensus as determining what is 'scientific' and what

163

'unscientific'. Every great scientific controversy tends therefore to turn into a dispute between the established authorities and a pretender (Elliotson, Kützing, Rhine, Freud, van't Hoff, Lysenko, etc.) who is as yet denied the status of a scientist, at least with respect to the work under discussion.

These pretenders do not deny the authority of scientific opinion in general, but merely appeal against its authority in a particular detail and seek to modify its teachings in respect of that detail. Indeed, every thoughtful submission to authority is qualified by some, however slight, opposition to it. The position is similar—and closely linked—to that already stated in respect to tradition. When I speak of science I acknowledge both its tradition and its organized authority, and I deny that anyone who wholly rejects these can be said to be a scientist, or have any proper understanding and appreciation of science. Consequently, nothing that I—who accept the traditions and authority of science—may say about science can mean anything to such a person, and this holds also in reverse. Yet I do not enter this commitment unconditionally, as shown by the fact that I refuse to follow both the tradition and authority of science in its pursuit of the objectivist ideal in psychology and sociology. I accept the existing scientific opinion as a *competent* authority, but not as a *supreme* authority, for identifying the subject matter called 'science'.

This distinction is implicit in the remarks I have just made about Kepler. It is indispensable to any survey of the historic progress of science. For to limit the term science to propositions which we regard as valid, and the premises of science to what we consider to be its true premises, is to mutilate our subject matter. A reasonable conception of science must include conflicting views within science and admit of changes in the fundamental beliefs and values of scientists. To acknowledge a person as a scientist—and even as a very great scientist—is merely to acknowledge him as competent in science, which admits the possibility that he was, or is, in many ways mistaken.

I may observe then, as I have done already, that the modern physical sciences went through three stages, each of which had its own scientific values and its corresponding vision of ultimate reality. Scientists of the first period believed in a system of numbers and geometrical figures, the next in one of mechanically constrained masses, the last in systems of mathematical invariances. In attaching themselves to the pursuit of these successive fundamental guesses about the nature of things, the intellectual passions of scientists underwent profound changes—changes similar in extent, and perhaps even not unrelated, to those which the appreciation of visual arts underwent from the Byzantine mosaics to the works of the Impressionists and from these to Surrealism. But there was a similar transcendence of an enduring passion in both cases. Granted that many of the arguments of Copernicus, Galileo and Kepler, and even of Newton, Lavoisier and Dalton, seem misguided today, and that their presuppositions have led to conclusions which we now consider to be false; and granted also that these giants of the past, if they returned today, might not readily accept relativity and quantum mechanics as satisfying systems of science; yet so much of earlier science has remained true, and has even revealed as true ever more of its deeper implications, that the pioneers of science have kept growing through the centuries in our respect. In this sense then science embraces a consistent pursuit of gradually changing, and —I believe—on the whole, ever more enlightened and elevated intellectual aspirations.

Such is the general framework within which the pursuit of science can be defined and the assumptions underlying its achievements identified. This perspective will have to be considerably widened—as I shall show in Part Four—to include the biological sciences; the inclusion of psychology and sociology would raise further, sharply controversial questions which I shall touch upon only in passing.

So much for the breadth of the field. I can only hint at the kind of detail into which any substantial investigation of the premises underlying scientific discovery and verification would have to go. Perhaps it would have to examine above all the great discoveries made—particularly in this century—by scientists speculatively pursuing some specific guesses at a rational interpretation of nature; it would have to consider how obscure and controversial some of these speculations appeared at first; how many similar speculations were in fact empty or mistaken; and yet how astoundingly true and deeply prophetic some of them proved in several famous instances. Exceptional sensibility would be needed in order to discover what general ideas about the nature of things have guided these remarkable guesses. But even so, such an account would only reveal the premises of past scientific achievements. The actual premises of science, at the moment of writing, are present only in the yet unformed discoveries maturing in the minds of scientific investigators intent on their work. The visitor to the research school of a great master—whose intuitive apprehensions can filter through, even though imperfectly, to an inner circle of collaborators— may dimly discern the premises of prospective discoveries by the way the collaborators talk about their work. We can get no closer to the current premises of science.

We have now before us a true image of science in the process of emerging from initial vagueness and conjecture to greater precision and certainty. It is here, in the course of discovery and verification, that the premises of science exercise their guidance over the judgment of scientists. It is obvious that no formulation of these premises ever proposed (or yet to be proposed) would have enabled a person lacking the special gifts and training of a scientist, competently to decide any of the serious uncertainties that have arisen in the various controversial or doubtful issues I have mentioned. Indeed, when we try to apply any of these formulations for deciding a great question in science, we find that they prove ambiguous precisely to the extent of allowing both alternatives to be equally arguable.

Take Mach's principle of 'mental economy', according to which science is the simplest description or the most convenient summary of the facts. Imagine the puzzled examiners of de Broglie's doctoral thesis having recourse to this criterion as to the scientific value of the work. How could they? Most of the facts which the theory eventually was found to describe were yet undiscovered. They would have had to limit themselves to the facts known to be described by the theory. Should they have arranged a competition to determine whether the new theory was simpler, in the sense that it would make it easier to memorize these facts or to teach them in schools; or that the theory could be written down in a smaller space or in a more familiar vocabulary? The idea is farcical. What they had to decide first and last was whether this work was a substantial discovery or a mere jeu-d'esprit. This would decide also whether it was the scientifically simplest description of the facts. For as a mere conceit it was a fancifully far-fetched way of looking at things; while as a true statement, it was an amazingly simple short-cut to vast new perspectives.

Or apply the concept of 'simplicity' to the controversy about Rhine's experiments on card guessing. Extra-sensory perception is of course the simplest explanation for them, if you are prepared to believe in extra-sensory perception. Yet most scientists today would prefer some other explanation, however complicated, if only it lay within the scope of hitherto known physical interactions. To them it appears more 'economical' not to introduce a new principle if we can possibly manage with those already accepted; and they are even prepared to disregard Rhine's observations until such time as these can be fitted into the existing framework of natural laws. Again, the question of simplicity of description, in the ordinary sense of the word 'simple', plays, and can play, no part whatsoever in the controversy. On the contrary, in whatever way the issue is finally decided, this will determine what will be the simpler explanation in the scientific sense.

This ambiguity of the term 'simple' within Mach's formula follows from the fact which I stated before,[1] namely that it functions here as a pseudo-substitute for 'true'. In consequence, the answer to the question what is simple in a given case must always be exactly as doubtful as the answer to the question what is true in the same case. The same ambiguity will adhere to other pseudo-substitutes for truth, like the pragmatist criterion that a theory 'works'. This ambiguity is revealed in the same way as that of 'simplicity', if we apply practicality in place of simplicity as a criterion of such doubtful or controversial issues as de Broglie's thesis or Rhine's experiments. The same test will yield similar results in respect to the criterion of fruitfulness, of which I have already shown how grotesquely it mimics the functions of truth.

I have said before (p. 161) that the premisses of science determine the methods of its pursuit and vice versa. Yet the enquiry into scientific pro-

[1] P. 16.

cedure has in fact been conducted on separate and much more systematic lines. Its aim has been the discovery of some strictly formulated rule by which valid general propositions can be derived from available empirical observations. One such scheme, based on the process of collecting evidence which will increase the probability of an empirical proposition to the point of practical certainty, I have fully discussed and criticized in ch. 2. The additional material collected in the present chapter should enable me to re-formulate that criticism, and apply it also to all attempts made to formulate the process of induction on the lines of J. S. Mill's principles of Agreement and Difference.

Specific rules of empirical inference claim (*a*) to proceed by a prescribed operation from clues to discovery or at least (*b*) to show how to verify, or at the very least (*c*) how to falsify, an empirical proposition according to some such rules. Claim (*a*) must be rejected in view of the demonstrable fact that discovery is separated by a logical gap from the grounds on which it is made. It is, as I said before, a travesty of the scientific method to conceive of it as an automatic process depending on the speed of piling up evidence for hypotheses chosen at random (Part One, ch. 2, p. 30). The history of the great scientific controversies teaches us now that claims (*b*) and (*c*) are equally unfounded.

The reasons are similar to those I used for criticizing terms like 'simple' as a substitute for 'true'. All formal rules of scientific procedure must prove ambiguous, for they will be interpreted quite differently, according to the particular conceptions about the nature of things by which the scientist is guided. And his chances for reaching true and important conclusions will depend decisively on the correctness and penetration of these conceptions. We have seen that there is a type of empirical discovery that is achieved without any process of induction. De Broglie's wave theory, the Copernican system and the theory of Relativity, were all found by pure speculation guided by criteria of internal rationality. The triumph of the Michelson-Morley experiment, despite its giving the wrong result, the tragic sacrifice of D. C. Miller's professional life to the pursuit of purely empirical tests of a great theoretical vision, are sardonic comments on the supposed supremacy of experiment over theory. Admittedly, other controversies, like those of fermentation, hypnotism and extra-sensory perception, seem to centre altogether on questions of factual evidence. But looking at these disputes more closely it appears that the two sides do not accept the same 'facts' as facts, and still less the same 'evidence' as evidence. These terms are ambiguous precisely to the extent to which the two opposing opinions differ. For within two different conceptual frameworks the same range of experience takes the shape of different facts and different evidence. Indeed, one side may disregard some of the evidence altogether in the confident expectation that it will somehow turn out to be false. I shall give further illustrations of this power of scientific theory over scientific facts in a later chapter (Part Three, ch. 9, 'A Critique of Doubt').

We should also remember that the rules of induction have lent their support throughout the ages to beliefs that are contrary to those of science. Astrology has been sustained for 3000 years by empirical evidence confirming the predictions of horoscopes. This represents the longest chain of historically known empirical generalizations. For many prehistoric centuries the theories embodied in magic and witchcraft appeared to be strikingly confirmed by events in the eyes of those who believed in magic and witchcraft. Lecky[1] rightly points out that the destruction of belief in witchcraft during the sixteenth and seventeenth centuries was achieved in the face of an overwhelming, and still rapidly growing, body of evidence for its reality. Those who denied that witches existed did not attempt to explain this evidence at all, but successfully urged that it be disregarded. Glanvill, who was one of the founders of the Royal Society, not unreasonably denounced this method as unscientific, on the ground of the professed empiricism of contemporary science. Some of the unexplained evidence for witchcraft was indeed buried for good, and only struggled painfully to light two centuries later when it was eventually recognized as the manifestation of hypnotic powers.

Moreover, a whole realm of more familiar facts has been overlooked by philosophers who strove to justify science by ascribing a unique reliability to the inductive method.[2] Constant conjunction would lead to absurd predictions all over the enormous range of processes the course of which is determined by decay or the satiation of appetites. Our expectation of life does not increase with the number of days we have survived. On the contrary, the experience of living through the next 24 hours is much less likely to recur after it has happened 30,000 consecutive times than after only 1000 times. Attempts to train a horse to do without food will break down precisely after the longest series of successes; and the certainty of amusing an audience by one's favourite joke does not increase indefinitely with the number of its successful repetitions. It is true that in conditioning experiments animals tend to expect that an event which is repeatedly preceded by a sign will recur after the sign is given again; but when children are asked to guess in a random sequence of red and green lights appearing in two parallel rows which of the two will come out next, they will expect that it will be the one which has so far occurred fewer times.[3] We can easily imagine a universe in which all recurrences would be limited in number, so that new recurrences would invariably become steadily *less* likely with the number of their previous occurrences.

The decisive reason why such obviously inadequate formulations of the principles of science were accepted by men of great intellectual distinction

---

[1] Lecky, *Rationalism in Europe*, London, 1893, **1**, pp. 116–7.

[2] As when R. B. Braithwaite in *Scientific Explanation*, Cambridge, 1953, p. 272, says, 'The non-inductive policies are not starters'.

[3] Up to 90 per cent of the subjects may predict that the hitherto non-preponderant alternative will appear next (J. Cohen and C. E. M. Hansel, *Risk and Gambling*, London, New York, Toronto, 1956, pp. 10–36).

lies in a desperate craving to represent scientific knowledge as impersonal. We have seen that this is achieved by two alternative recipes: (1) by describing science in terms of some secondary feature (simplicity, economy, practicality, fruitfulness, etc.), and (2) by setting up some formal model in terms of probabilities or constant conjunctions. In both cases the scientist would be left uncommitted; in the first because he would say nothing more than a telephone directory, in the second because he would have a machine to speak for him, impersonally. Since the latter solution still leaves over the personal act of accrediting the machine, this act may be played down on the lines of recipe 1 by describing it as a mere 'policy'. But to justify a scientific procedure by its practical advantage as a policy, is to conceal the fact that this advantage is expected to accrue only because we hold certain beliefs about the nature of things which make this expectation reasonable.

I shall presently (p. 191) have more to say on the curious logical dilemma in which any formal axiomatization of science (or mathematics) leads itself ad absurdum. At the moment I only wish to explain how the paramount desire for impersonal knowledge could succeed in rendering plausible such flagrantly inadequate formulations of science as given either by recipe 1 or 2. We owe this immense power for self-deception to the operation of the ubiquitous tacit coefficient by which alone we can apply any articulate terms to a subject matter described by them. These powers enable us to evoke our conception of a complex ineffable subject matter with which we are familiar, by even the roughest sketch of any of its specifiable features. A scientist can accept, therefore, the most inadequate and misleading formulation of his own scientific principles without ever realizing what is being said, because he automatically supplements it by his tacit knowledge of what science really is, and thus makes the formulation ring true.

Since this process is essential to the mechanism of pseudo-substitution, to which I ascribe some importance as an instrument of a misguided critical philosophy, I shall digress on it a little further. A most dramatic instance of self-deception, caused by the intervention of the inarticulate powers of the observer, occurred in the case of Clever Hans: the horse which could tap out with his hoofs the answer to all kinds of mathematical problems, written out on a blackboard in front of him. Incredulous experts from all relevant branches of knowledge came and tested him severely, only to confirm again and again his unfailing intellectual powers. But at last Mr. Oskar Pfungst had the idea of asking the horse a question to which he, Pfungst, did not know the answer. This time the horse went on tapping and tapping indefinitely, without rhyme or reason. It turned out that all the severely sceptical experts had involuntarily and unknowingly signalled to the horse to stop tapping at the point where they—knowing the right answer—expected him to stop.[1] This is how they made the answers invariably come out right; and this is exactly also how philosophers make

[1] Oskar Pfungst, *Das Pferd des Herrn von Osten (Der kluge Hans)*, Leipzig, 1907.

their descriptions of science, or their formalized procedures of scientific inference, come out right. They never use them to decide any open scientific problem, whether past or present, but apply them to scientific generalizations which they regard as indubitably established.[1] This belief eliminates all the ambiguities which the formal procedures of constant conjunction—or of the progressive confirmation of hypotheses according to their increasing probability—leave open, and thus makes either process give invariably the right result. And again, you can successfully conceal from yourself the unaccountable fact that you are absolutely convinced of (say) the law of gravitation, by calling it a mere working hypothesis, or a shorthand description of the facts, etc. For a belief which can be touched by no shadow of a doubt remains unaffected by such understatements. So these formulae can be safely uttered to appease a strictly empiricist conscience. It is only when we are confronted with the anxious dilemma of a live scientific issue, that the ambiguity of the formal processes and of the various attenuated criteria of scientific truth becomes apparent, and leaves us without effective guidance.[2]

These formal criteria can of course function legitimately as *maxims* of scientific value and scientific procedure. To every change in scientific value, from Kepler to Laplace and from Laplace to Einstein, there has corresponded a change in scientific method, which can be formulated in changing maxims of procedure. We have seen such legislation occurring in the past as the outcome of great controversies and upheavals in science. They have formed the scientific tradition which it falls to us ultimately to interpret within the context of our own controversial issues.

Such is, in effect, the legitimate purpose and meaning of exploring the logical antecedents of science. But this meaning is obscured by any attempt at formulating these antecedents as the axiomatic presuppositions of empirical inference. What such postulates can say is not in itself convincing, nor indeed clearly comprehensible. They derive their meaning and convincing power from our anterior belief in a body of natural sciences which appears to imply their validity, and only as we become imbued with the knowledge of natural science and learn to apply its methods to new problems, can we learn to appreciate these postulates as guiding principles on which we rely.

[1] Morris R. Cohen concludes a critique of the traditional 'canons of induction' by saying: 'If the true cause is not included in our major premise the "canons of induction" will not enable us to discover it. If anyone thinks that I have understated the case for these canons of induction as methods of discovery, let him discover by their means the cause of cancer or of disorders of internal secretion' (*A Preface to Logic*, London and New York, 1944, p. 21).

[2] There is a variant to the Clever Hans fallacy in what may be called the illusion of 'You can't miss it'. Persons very familiar with a district are the worst at giving directions to a stranger. They tell you 'just to keep going straight on', forgetting the forks at which you will have to decide which way to go. They cannot realize that their indications are altogether ambiguous, because to them they are not. So they say confidently, 'You can't miss it'.

If we fail to realize that the logical antecedents of science are internal to science, they will inevitably appear as propositions accepted prior to the pursuit of science. If we then reflect on them and find that they are not logically inescapable, we are faced with the insoluble problem of finding a justification for them. The problem is insoluble, for it seeks an explanation for a non-existent state of affairs. Nobody has ever affirmed the presuppositions of science by themselves. The discoveries of science have been achieved by the passionately sustained efforts of succeeding generations of great men, who overwhelmed the whole of modern humanity by the power of their convictions. Thus has our scientific outlook been moulded, of which these logical rules give a highly attenuated summary. If we ask why we accept this summary, the answer lies in the body of knowledge of which they are the summary. We must reply by recalling the way each of us has come to accept that knowledge and the reasons for which we continue to do so. Science will appear then as a vast system of beliefs, deeply rooted in our history and cultivated today by a specially organized part of our society. We shall see that science is not established by the acceptance of a formula, but is part of our mental life, shared out for cultivation among many thousands of specialized scientists throughout the world, and shared receptively, at second hand, by many millions. And we shall realize that any sincere account of the reasons for which we too share in this mental life must necessarily be given as part of this life.

Science is a system of beliefs to which we are committed. Such a system cannot be accounted for either from experience as seen within a different system, or by reason without any experience. Yet this does not signify that we are free to take it or leave it, but simply reflects the fact that it *is* a system of beliefs to which we are committed and which therefore cannot be represented in non-committal terms. In leading up to this position, the logical analysis of science decisively reveals its own limitations and points beyond itself in the direction of a fiduciary formulation of science, to which I propose to move on at a later stage of this enquiry.

## 7. PASSIONS, PRIVATE AND PUBLIC

I have described before the passionate preoccupation with a problem which alone can elicit discovery, and the protracted struggles against doubts of its significance and validity by which its announcement is often followed. Such a struggle, in which the ardour of discovery is transformed into a craving to convince, is clearly a process of verification in which the act of making sure of one's own claims is coupled with the effort of getting them accepted by others.

Yet as we pursue scientific discoveries through their consecutive publication on their way to the textbooks, which eventually assures their reception as part of established knowledge by successive generations of students, and through these by the general public, we observe that the intellectual passions aroused by them appear gradually toned down to a faint echo of their discoverer's first excitement at the moment of Illumination. A theory like that of relativity continues to attract the interest of ever new students and laymen by intimations of its beauty yet hidden to their understanding: a beauty which is rediscovered every time a new mind apprehends the theory. And it is still for the sake of this remote and inaccessible beauty, and not for its few useful formulae (which could be memorized in a minute), that relativity continues to be valued as an intellectual triumph and accepted as a great truth. All true appreciation of science by the public continues to depend on the appreciation of such beauty—even though sensed only at second hand; it offers an indirect tribute to the values that the multitude have been taught to entrust to a group of men whose cultural guidance they have accepted. Though the torrent widening towards the ocean no longer breaks new paths, the intellectual passions which had urged on the discoverer still pulsate in the common valuation of science.

A transition takes place here from a heuristic act to the routine teaching and learning of its results, and eventually to the mere holding of these as known and true, in the course of which the personal participation of the knower is altogether transformed. The impulse which in the original heuristic act was a violent irreversible self-conversion of the investigator and may have been followed by an almost equally tempestuous process of converting others, is first repeated as a milder version of itself in the eventual acceptance of the discovery by the public, and will thus assume finally a form in which all dynamic quality is lost. Personal participation changes from an impetuous pouring out of oneself into channels of untried assumptions, into a confident holding of certain conclusions as part of one's interpretative framework. The driving power of originality is reduced to a static personal polarization of knowledge; the intellectual effort which led to discovery and guided its verification is transformed into the force of a conviction which holds it to be true—in exactly the same way as the effort of acquiring a skill is transformed into a sense of its mastery. This kind of emotional cascade could be detected on parallel lines over many different domains of knowledge which are originally shaped by pioneers and subsequently held by their successors. But I shall postpone this analysis of apprenticeship, and turn now to a comparison of the affirmative functions of our bodily emotions with those of our intellectual passions.

Not all emotions have a sufficiently pointed bearing on something outside the person moved by them to imply an affirmation. Languor, vivacity or restlessness, anxiety (as distinct from fear), drunken hilarity are all pervasive changes of personality that imply no affirmation on the part of the person affected by them about anything outside himself. But we find the same pointed character that intellectual passions invariably possess in the force of drives, the lure of lust, the grip of fear. We have indeed acknowledged such drives before as the most primitive manifestations of the active principle by which we grasp knowledge and hold it.

Yes, hunger, sex and fear are the motives of quests pursued with passion, and these quests seek to discover the means to satisfy their motives by consummatory acts like eating, copulation or flight; and hence it follows also that the gratification of appetites is a manner of verification —the proof of the pudding which is in the eating. Yet we must allow for the possibility of the pudding's being poisoned, and not consider that everything that an animal swallows is proper food for it. While we shall hold that animals are competent to choose their food, we shall not deem their choice infallible.

The parallel with intellectual passions is clear, and so is the contrast to them. As the pursuit of our drives implies the supposition that there exist objects which we have reason to desire or to fear, so similarly, all passions animating and shaping discovery imply a belief in the possibility of a knowledge of which these passions declare the value; and again, in accrediting these passions with the power to recognize the truth, we do not assume their infallibility—since no rule of scientific procedure is certain of finding the truth and avoiding error—but we accept their competence. Our intellectual passions, however, differ essentially from the cravings and emotions which we share with the animals. The satisfaction of these terminates the situation which evoked them. Discovery likewise terminates the problem from which it started, but it leaves behind knowledge, which gratifies a passion similar to that which sustained the craving for discovery. Thus intellectual passions perpetuate themselves by their fulfilment.

This distinctive quality of intellectual passions is largely due to the fact that they are attached to an articulate framework. A scientist seeks to discover a satisfying theory, and when he has found it, he can enjoy its excellence permanently. The intellectual passion which animates the student to grapple with the difficulties of mathematical physics is gratified when he feels at last that he understands it, but it is the resulting sense of mastery that gives him permanent intellectual satisfaction. While the purely intellectual joy of the animal who has contrived a trick already shows the same enduring quality, the articulate powers of man can extend the range of such joys to whole systems of cultural gratification.

This wider perspective brings us back to the fact that scientific value must be justified as part of a human culture extending over the arts, laws and religions of man, all contrived likewise by the use of language. For this great articulate edifice of passionate thought has been reared by the force of the passions to which its erection offered creative scope, and its lasting fabric continues to foster and gratify these same passions. Young men and women brought up in this culture accept it by pouring their minds into its fabric, and so live the emotions which it teaches them to feel. They transmit these emotions in their turn to succeeding generations, on whose responding fervour the edifice relies for its continued existence.

By contrast to the satisfaction of appetites, the enjoyment of culture creates no scarcity in the objects of offering gratification, but secures and ever widens their availability to others. Those who obtain such goods increase their universal supply and teach others to enjoy them by practising what they were taught. The pupil submits to what he acquires and improves himself by its standards.

Accordingly, the social lore which satisfied our intellectual passions is not merely desired as a source of gratification; it is listened to as a voice which commands respect. Yielding to our intellectual passions, we desire to become more satisfying to ourselves, and accept an obligation to educate ourselves by the standards which our passions have set to ourselves. In this sense these passions are public, not private: they delight in cherishing something external to us, for its own sake. Here is indeed the fundamental difference between appetites and mental interests. We must admit that both are sustained by passions and must ultimately rely on standards which we set to ourselves. For even though intellectual standards are acquired by education, while our appetitive tastes are predominantly innate, both may deviate from current custom; and even when they conform to it, they must both ultimately be accredited by ourselves. But while appetites are guided by standards of private satisfaction, a passion for mental excellence believes itself to be fulfilling universal obligations.

This distinction is vital to the existence of culture. If it is repudiated, all cultural life becomes subordinated in principle to the demand of our appetites and of the public authorities responsible for the advancement of material welfare. I shall deal with this situation further when defining the relation of pure science to technology.

## 8. SCIENCE AND TECHNOLOGY

In the list of three kinds of learning of which animals are capable I have placed trick learning before sign learning, since motoricity is fully developed in the lower animals before they achieve the capacity for recording complex perceptions. Yet the capacity to perform a useful action presupposes some purely intellectual control over the circumstances in which the action is to take place. Technology always involves the application of some empirical knowledge and this knowledge may be part of natural science. Our contriving always makes use of some anterior observing.

Putting it this way, we become aware of the incommensurability of the two things combined in a technical performance. Suppose you hammer in a nail. Before starting, you look at the hammer, the nail and the board into which you will drive it; the result is knowledge which you can put into words. Then you hammer in the nail. The result is a deed: something is now firmly nailed on. Of this you can have knowledge, but it is not itself knowledge. It is a material change which counts as an achievement. Knowledge can be true or false, while action can only be successful or unsuccessful, right or wrong.

It follows that an observing which prepares a contriving must seek knowledge that is not merely true, but also useful as a guide to a practical performance. It must strive for applicable knowledge.

The conceptual framework of applicable knowledge is different from that of pure knowledge. It is determined primarily in terms of the successful performances to which such knowledge is relevant. Take hammering again. This performance implies the conception of a hammer, which defines a class of objects that are (actual or potential) hammers. It will include, apart from the usual tools of this kind, rifle butts, shoe heels and fat dictionaries, and establish at the same time a grading of these tools according to suitability. The suitability of an object to serve as a hammer is an observable property, but it can be observed only within the framework defined by the performance it is supposed to serve.

There are three kinds of observable things which can be defined by their participation in practical performances: (1) materials, (2) tools, including all manner of installations, and (3) processes. Timber, textiles, fuels, are technical materials; hammers, engines, houses, railways, are tools or installations; fermentation, cooking, smelting, are technical processes. Many of these technical conceptions comprise a variety of otherwise disparate objects (for example, different kinds of textiles, from cotton and wool to nylon and glass fibres, and different means of lighting, from candles to discharge lamps), but all these objects are specially prepared, shaped or otherwise so contrived as to make them technically suitable. To this extent these classes of objects or processes are known, and the individual objects or processes themselves are intelligible only within the framework of a useful performance which they successfully serve. Pure knowledge, lacking this framework, and pure science in particular, ignore these classes and cannot understand these contrivances. We cannot eliminate instrumentality from technical knowledge, any more than we can represent natural science in terms of practical procedure.

A gap is opening up here between two kinds of knowledge, both of which refer to material things: one derived from an acknowledged purpose, the other unrelated to any such purpose. The disparity of science and technology which I am examining here will prove relevant later to the relation between the science of inanimate things, in which no purpose is apparent, and that of living beings which can be understood only in teleological terms. We should keep this prospect in mind while proceeding to elucidate further the characteristic logical structure of technology.

Primitive technology may be regarded as a mere extension of bodily skills employed for the satisfaction of bodily appetites. And even in highly complex and predominantly articulate branches of technology, like the manufacture of cloth or the production of steel, there is involved a measure of unspecifiable know-how which is essential to the efficiency of labour and the quality of its product. Manufacturing experience remains a valuable qualification to a technician, and its possession by the aggregate of a country's technicians is a great national asset. But even though the teachings of technical science can become effective only by their skilful execution, the foundation of modern man's technical mastery lies in the explicit exposition of technology by textbooks, journals, patents, etc.

Technology teaches action. This is made plain when it speaks in imperatives, as it often does in cookery books or directions for the use of machinery. The symbol at the head of a prescription is an imperative prefacing an order to make up a medicine; crafts like weaving or welding are taught in imperatives. All technology is equivalent to a conditional command, for it is not possible to define a technology without acknowledging, at least at second hand, the advantages which technical operations might reasonably pursue. It is true, of course, that anything a man does or can conceivably do, could be described as the pursuit of an advantage if we imputed to him the purpose of achieving the consequences of this action; but a technology which would teach all such imputable purposes would be as meaningless as a science which would give a list of all observable facts. A technology must therefore declare itself in favour of a definite set of advantages, and tell people what to do in order to secure them.

Technology teaches only actions to be undertaken for *material* advantages by the use of *implements* according to (more or less) *specifiable rules*.[1] Such a rule is an operational principle. As implements are defined and understood in terms of an action which they serve, they are defined and understood likewise in terms of the operational principle which tells how to perform such an action.[2]

I have spoken before of the operational principles which we observe subsidiarily in the performance of a skill, and also of the operational principles applied—again for the most part subsidiarily—in achieving scientific knowledge. I have shown symbolic operations carried out according to certain explicit rules and have noted that such operations require that symbols should be manageable, just as tools have to be serviceable. Modern electronic devices used for the automatic control of technical processes show that some highly formalized operational principles of technology can be readily affiliated to mathematical operations. The meaning of technical implements resembles that of mathematical symbols, in so far as they are both intended for use in a certain range of operations,

[1] 'Material advantages' should exclude *inter alia* the achievement of symbolic expression or of human interactions. Thus the construction of churches and prisons or the manufacture of handcuffs are tasks of technology, but the ultimate uses of these objects are not part of technology. The word 'implement' is meant to designate all three classes of useful things: materials, devices and processes. Action according to 'specifiable rules' excludes artistic performances.

[2] Operational principles will be taken to include here the constructional principles which tell us the way technical devices, like machines or houses, are to be built.

in the service of which they can be replaced by a whole class of equally serviceable, though otherwise disparate, entities. This kinship can be pursued through the whole subsequent analysis of operational principles.

The difference between scientific knowledge and an operational principle of technology is recognized by patent law, which draws a sharp distinction between a *discovery*, which makes an addition to our knowledge of nature, and an *invention*, which establishes a new operational principle serving some acknowledged advantage. New inventions rely as a rule on known facts of experience, but it may happen that a new invention involves a new discovery. Yet the distinction between the two will still hold: only the invention will be granted protection by a patent, and not the discovery as such.

The reason is obvious. A patent has two functions, namely, publicly to disclose its subject matter, and to grant a monopoly in respect to its use. If applied to new knowledge its first function would preclude the second: once such knowledge is publicly disclosed it can no longer be anyone's monopoly. But the patent can grant and enforce a monopoly for the practice of any new operational principle; it can restrain unauthorized persons from using the new invention which it makes generally known.[1]

Invention has it in common with discovery that it can claim to be what it is only if it is surprising. It must be separated from its antecedents by a considerable logical gap. I have mentioned already that in case of doubt the courts undertake to assess whether this logical gap is wide enough to warrant the acknowledgment of an invention. This width measures the ingenuity of the invention.

But a new operational principle may be acknowledged by patent law, and yet not be an invention in the technological sense. A new ingenious process for extracting tap-water from champagne may be an invention in the sense of patent law, but it would not be acknowledged as such by technology. For in addition to the disclosure of a new operational principle, technology requires that an invention should be economic and thus achieve a material advantage.

Hence any invention can be rendered worthless, and indeed farcical, by a radical change in the values of the means used up and the ends produced by it. If the price of all fuels went up one hundred-fold, all steam engines, gas turbines, motor-cars and aeroplanes, would have to be thrown on the junk heap. A brilliant invention is often rendered non-sensical overnight by a better invention: tram-cars are as absurd today as the horse-drawn buses which they once displaced. By contrast to this, the validity of a scientific observation cannot be affected by changes in the value of goods. If diamonds became as cheap as salt is today, and salt as

---

[1] The law could try to grant a monopoly for the future practical applications of a new discovery; but no patent law does this, for it is impracticable. The law endorses thereby once more the sharp distinction between a knowledge of the facts of nature (achieved by discovery) and the knowledge of an operational principle (achieved by invention).

precious as diamonds are now, this would not invalidate any part of the physics and chemistry of diamonds or of salt. If either of the two minerals became so rare as to be practically inaccessible, this might affect the interest attached to their study, but it would leave unimpaired the validity of its results. Nor is there any true parallel in science to the extinction of an invention by the emergence of a more profitable way of achieving the same advantage.

The beauty of an invention differs accordingly from the beauty of a scientific discovery. Originality is appreciated in both, but in science originality lies in the power of seeing more deeply than others into the nature of things, while in technology it consists in the ingenuity of the artificer in turning known facts to a surprising advantage. The heuristic passion of the technician centres therefore on his own distinctive focus. He follows the intimations, not of a natural order, but of a possibility for making things work in a new way for an acceptable purpose, and cheaply enough to show a profit. In feeling his way towards new problems, in collecting clues and pondering perspectives, the technologist must keep in mind a whole panorama of advantages and disadvantages which the scientist ignores. He must be keenly susceptible to people's wants and able to assess the price at which they would be prepared to satisfy them. A passionate interest in such momentary constellations is foreign to the scientist, whose eye is fixed on the inner law of nature.

Hence there arises a conflict of values which makes it difficult to mix the two occupations. From his experience of developing atomic weapons in Los Alamos during the Second World War, J. R. Oppenheimer wrote: 'The scientist is irritated by the practical preoccupations of the man concerned with development, and the man concerned with development thinks that the scientist is lazy and of no account and is not doing a real job anyway. Therefore the laboratory very soon gets to be all one thing or all the other.' [1]

This sharp division between science and technology is entirely compatible with the existence of domains which in one respect or another form a transition between them. The older crafts which still form the

---

[1] J. R. Oppenheimer, 'Functions of the International Agency in Research and Development', *Atomic Scientific Bulletin*, 1947, p. 173. See also V. B. Wigglesworth, 'The Contribution of Pure Science to Applied Biology', *The Annals of Applied Biology*, **42** (1955), pp. 34–44. Speaking of pure scientists working on practical war-time problems, Wigglesworth writes: 'In the pure science to which they were accustomed, if they were unable to solve problem A they could turn to problem B, and while studying this with perhaps small prospect of success they might suddenly come across a clue to the solution of problem C. But now they must find a solution to problem A, and problem A alone, and there was no escape. Furthermore, there proved to be tiresome and unexpected rules which made the game unnecessarily difficult: some solutions were barred because there was not enough of the raw material available: others were barred because the materials required were too costly; and yet others were excluded because they might constitute a danger to human life or health. In short, they made the discovery that applied biology is not "biology for the less intelligent", it is a totally different subject requiring a totally different attitude of mind' (p. 34).

majority of modern industries were invented by mere trial and error, without the aid of science. By contrast, electrotechnics and much of chemical technology are derived from the application of pure science to practical problems. Hence there is the following interrelation between science and technology. To the extent to which a technical process is an application of scientific knowledge it contributes nothing to science, while empirical technology, which is itself unscientific, may well offer—for this very reason—important material for scientific study.[1]

We have, correspondingly, two forms of enquiry that lie between science and technology. Technologies founded on an application of science may form a scientific system of their own. Electrotechnics and the theory of aerodynamics are examples of *systematic technology* which *can be cultivated in the same way as pure science.* Yet their technological character is apparent in the fact that they might lose all interest and fall into oblivion, if a radical change of economic relationships were to destroy their practical usefulness. On the other hand, it may happen that some parts of pure science offer such exceptionally ample sources of technically useful information that they are thought worth cultivation for this reason, though they would otherwise lack sufficient interest. The scientific study of coal, metals, wool, cotton, etc., are branches of such *technically justified science.*

Systematic technology and technically justified science are two fields of study lying between pure science and pure technology. But the two fields may overlap completely. The discovery of insulin as a cure for diabetes was an important contribution to science, owing to the intrinsic interest of its subject matter; it was also the invention of an operational principle serving to cure diabetes. The same quality applies over large parts of pharmacology. It holds, indeed, wherever a process inherent in nature is interesting to science owing to the importance of its outcome, while at the same time it can also be operated at will for achieving this desirable outcome. Such coincidences between science and technology are fully accounted for by the same principles which define them in general as completely disparate domains.[2]

---

[1] On the range of undisclosed knowledge buried in empirical technology, see p. 52 above.

[2] In the address by Wigglesworth just cited (p. 178, n. 1), the author describes the varying relationships which obtain between pure and applied science in the biological field. These two 'totally different subjects' may contribute to each other's good in a number of ways. E.g. for the pure scientist 'one of the most efficient correctives to the dangers of over-specialization is provided by the stimulus of contact with practice' (p. 36). On the other hand, applied biology may turn to pure science for the systematic explanation of its practical discoveries (p. 38); and of course the applied biologist 'in thinking about any practical problem . . . is continually making use of the whole range of scientific knowledge that exists about all its component parts' (p. 40). Yet the authorities are warned that this mutual advantage depends in the last analysis on the independence of pure biology from the narrower demands of the applied subject: '. . . the D.S.I.R. . . . makes grants for any research proposals which are of exceptional "timeliness and promise". The difficulty is that the most original ideas are at the outset both unpromising and untimely. Only research which is totally unfettered can advance into the most unpromising fields. . . . I very much doubt whether it would have been reasonable for the

Nothing could have appeared more obvious until recently than this difference between pure science and technology. It is unquestioningly embodied in the general framework of higher education, as shown by its division into universities and colleges of technology; it is expressed in the current distinctions between pure and applied chemistry, pure and applied physics, pure and applied mathematics, etc., in the description of university chairs, journals and international congresses; it determines the conditions of employment of scientists in universities on the one hand and industrial laboratories on the other; it underlies the operation of the patent law.

This framework survives practically unchanged in the countries not subject to Marxism and has not been altogether abandoned in the Soviet Union either. But since the rise around 1930 of the Neo-Marxian theory of science, which became within the subsequent decade the official doctrine of the U.S.S.R. and gained widespread influence outside it, the distinction between science and technology, even where still upheld in practice by the continued operation of these institutions, is violently challenged in principle.

This is part of the drive, described earlier on, for subordinating cultural values to a radically utilitarian conception of the public good: a materialistic outlook paradoxically imbued by inordinate moral aspirations. Such an attack is of course double-edged. It denies the effectiveness of pure intellectual passions in guiding scientific discovery, by affirming that every important step in the progress of science occurs in response to a specific practical interest; while it also denounces the pursuit of science for its own sake as irresponsible, selfish, immoral. Taken literally, the two attacks are mutually incompatible, for something that does not really happen cannot be denounced as morally wrong. But the materialistic interpretation of culture is a disguised imperative: it both declares that culture really is, and decrees that it ought to be, the servant of welfare. This is part of the Laplacean system in which morality must seek the sanction of science by representing itself in terms of scientific predictions.[1]

I am not much concerned here with the question how serious this menace to science may prove in practice. While the official repudiation by Stalinist orthodoxy of science pursued for its own sake led to the persecution and death in 1942 of Russia's most distinguished biologist, N. I. Vavilov, and had resulted by 1948 in the suppression or serious distortion of various branches of biology, it seems otherwise to have imposed on natural scientists little more restraint than the obligation falsely to declare their work

---

A.R.C. to have supported, for example, Darwin's experiments on the curving of bean shoots or the early experiments of the Wents on the growth of the oat coleoptile—because no one could have foreseen the impact that these observations were going to have on the agriculture of the future. . . . But at least the Research Councils should take great care not to impede the advance of pure science. . . . Knowledge is a delicate plant, and . . . it is an undesirable practice to keep pulling plants up to see how the roots are getting on' (pp. 42-3).

[1] The mechanism of this transformation will be examined in the next chapter.

to be guided by its practical usefulness. And this may be all. People may perhaps continue indefinitely to cultivate pure science, while professing a theory of science which exposes this occupation as a pretence or condemns it as an abuse. Yet the spread of this doctrine among scientists in countries where they are not compelled to subscribe to it, does raise the question which is relevant for us here, whether the distinctive passions which animate the cultivation of science may be superseded one day by other passions, or may even simply fade away for lack of response to them.

I have answered the last question in the positive sense, when warning that science may be once more discredited, as it was by St. Augustine, if it cannot avoid denaturing our conception of man.[1] The appreciation of natural science is of recent origin and its tradition is rooted in a limited area. It is a single shoot of one civilization among many others of equal antiquity and richness. The Greeks never developed a systematic natural science, nor did Byzantium or China, despite their technological achievements.[2] Today we can speak confidently of sixteenth- and seventeenth-century science only because, with modern hindsight, we can easily separate the genuine works of science from unscientific admixtures. Kepler's *Harmonics*, published in 1619, was imbued with astrology, and it is typical in this respect of much subsequent writing among scientists for the following two or three generations. I have mentioned already that Glanvill, one of the founders of the Royal Society in 1660, argued persistently for the acknowledgment of witchcraft. Another founding fellow, John Aubrey, published little else than a treatise on occult phenomena.[3] The Cartesian spirit dominating France at that time was *a-prioristic* rather than experimental. Newton himself still occasionally used religious arguments in science; for example when he suggested that God gave the world an atomic structure, as most conducive to his purpose. The great controversies of the nineteenth and twentieth centuries show that the struggle against intrusion of extraneous points of view into science have never ceased and that grave differences continue to persist in respect to these issues between a dominant majority and various dubiously established minorities of scientists. Yet we may acknowledge that by the time Newton's influence became prevalent, and particularly through his *Optics*, the method of observational science became effectively consolidated. Since then, in spite of such uncertainties and vagaries as I have described in the section on Scientific Controversies, we may recognize a coherent body of men, standing in the same scientific tradition, moved by the proper temper and true appreciation of science. Arago acclaiming Leverrier's discovery of Neptune in 1846 as 'one of the noblest titles of his country to the gratitude and

---

[1] P. 141 above.
[2] Stephen Runciman, *Byzantine Civilization*, London, 1936, ch. IX, and Joseph Needham, *Science and Civilization in China*, **2**, Cambridge, 1956, pp. 26-9, 84.
[3] Lytton Strachey, *Portraits in Miniature*, London, 1931, p. 23.

admiration of posterity'[1] expressed this in clear accents. No contribution to knowledge could be more useless than was the discovery of this remote new planet.

Actually, up to that time natural science had made no major contribution to technology. The industrial revolution had been achieved without scientific aid. Except for the Morse telegraph, the great London Exhibition of 1851 contained no important industrial devices or products based on the scientific progress of the previous fifty years. The appreciation of science was still almost free from utilitarian motives.

But these sentiments were held within a very small area and were shared at no time by more than a minority of the local population. The migration of science overseas and into Asian and African countries occurred slowly at a later period, when the medical, industrial and military value of science had greatly increased and could serve to recommend the reception of science to industrially less developed countries. These auspices did not favour a true appreciation of science. In all parts of the world where science is just beginning to be cultivated, it suffers from a lack of response to its true values. Consequently, the authorities grant insufficient time for research; politics play havoc with appointments; businessmen deflect interest from science by subsidizing only practical projects. However rich the fund of local genius may be, such an environment will fail to bring it to fruition. In the early phase in question, New Zealand loses its Rutherford, Australia its Alexander and its Bragg, and such losses retard further the growth of science in a new country. Rarely, if ever, was the final acclimatization of science outside Europe achieved, until the government of a country succeeded in inducing a few scientists from some traditional centre to settle down in their territory and to develop there a new home for scientific iife, moulded on their own traditional standards.[2]

Encircled today between the crude utilitarianism of the philistine and the ideological utilitarianism of the modern revolutionary movement, the love of pure science may falter and die. And if this sentiment were lost, the cultivation of science would lose the only driving force which can guide it towards the achievement of true scientific value. The opinion is widespread that the cultivation of science would always be continued for the sake of its practical advantages. It was expected, for example, that Lysenko's theories, if false, would be soon abandoned by the Soviet Government because they could produce no useful results. This expectation overlooked the fact that such questions cannot be decided in practice. Lysenko's theories are actually the theoretical conclusions which Michurin in Russia and

[1] See W. M. Smart, 'John Couch Adams and the Discovery of Neptune', *Nature*, **158**, (1946), pp. 648–52. Or listen to Ball commenting on the fact that Lalande would have discovered Neptune in 1795 if only he had believed what he saw on the 8th and 10th of May in that year. 'But had he done so, how lamentable would have been the loss to science. The discovery of Neptune would then merely have been an accidental reward to a laborious worker, instead of being one of the most glorious achievements in the loftiest department of human reason' (Sir R. S. Ball, *The Story of the Heavens*, London, 1891, p. 288).          [2] On tradition, see also p. 53 above.

Burbank in the U.S. derived from their substantial successes as plant-breeders.[1] Almost every major systematic error which has deluded men for thousands of years relied on practical experience. Horoscopes, incantations, oracles, magic, witchcraft, the cures of witch doctors and of medical practitioners before the advent of modern medicine, were all firmly established through the centuries in the eyes of the public by their supposed practical successes. The scientific method was devised precisely for the purpose of elucidating the nature of things under more carefully controlled conditions and by more rigorous criteria than are present in the situations created by practical problems. These conditions and criteria can be discovered only by taking a purely scientific interest in the matter, which again can exist only in minds educated in the appreciation of scientific value. Such sensibility cannot be switched on at will for purposes alien to its inherent passion. No important discovery can be made in science by anyone who does not believe that science is important—indeed supremely important—in itself.[2]

In saying this, I have acknowledged that values which I deem to be transcendent may be known only transiently to a small minority of mankind. There is no contradiction in this: it correctly reflects the fact that universal validity is not an observed fact. When we say that a statement is generally accepted or that no sane person would deny it, etc., we are saying something about the attitude of people towards the statement, which accredits the statement only if we accredit those people's judgment of it. But there is no general warrant to do this: the maxim 'quod semper, ubique, ab omnibus' has often proved erroneous. The standards by which we observe or appraise can never be derived from statistical surveys.

Indeed, we cannot look at our standards in the process of using them, for we cannot attend focally to elements that are used subsidiarily for the purpose of shaping the present focus of our attention. We attribute absoluteness to our standards, because by using them as part of ourselves we rely on them in the ultimate resort, even while recognizing that they are actually neither part of ourselves nor made by ourselves, but external to ourselves. Yet this reliance can take place only in some momentary circumstance, at some particular place and time, and our standards will

---

[1] See T. Dobzhansky, 'The Fate of Biological Science in Russia', *Proceedings of the Hamburg Congress on Science and Freedom*, London, 1955, p. 216. The attempt to define science in terms of its practical success has already been shown to be logically untenable (see p. 169 above).

[2] Some parallels from remoter fields may throw light on the principle involved here. Suppose it were decided by psychiatrists that a general increase in psychoneurotic ailments could only be checked by a restoration of religious faith; this would not make us all believe in God. In fact no ulterior advantage can make us believe in God, while if we do believe in God no consequent disadvantage can make us lose our faith. Or suppose that the people of the United States came clearly to the conclusion from a study of British experience that they would live together more intimately if their common affections were attached to a King and a Royal Family. This would not in itself produce such affections, or establish a monarchy in the U.S. No genuine affections can ever be produced by ulterior motives; they must discover and uphold their satisfaction in themselves.

be granted absoluteness within this historical context. So I could properly profess that the scientific values upheld by the tradition of modern science are eternal, even though I feared that they might soon be lost for ever. This duality will be stabilized later within the concept of commitment.

### 9. MATHEMATICS

Natural science is an expansion of observing; technology, of contriving; mathematics, of understanding. I have illustrated inarticulate understanding in animals by the way they know their way about a complex topography. Another illustration of it, taken from human experience, was that of an engineer grasping how the parts of a machine fit together and function jointly. A process of understanding is carried out articulately by operations which transform a given set of formulae into another set of formulae, implied in the first, or by a construction which transforms a geometrical figure into another, determined by the first. The result can be expressed either as a law, such as the laws of number theory and the theorems of geometry, or as a rule of procedure, such as we have for solving equations or for constructing geometrical figures from given elements. In the first case mathematics appears as a set of declaratory sentences resembling natural science, in the second as a set of recipes resembling technology. But these declaratory sentences record no observations concerning specific objects of nature, and the recipes disclose no operational principles for achieving any specific material advantage. Both the affirmations of mathematical formulae and the recipes given in mathematical proofs deal with conceptions which may have no specific bearing on experience. Valid formulae acknowledge the identity of two alternative aspects of the same conception, while proofs induce the identification of two such alternatives. The first can be said to be true or false like the statements of natural science; the second to be successful or unsuccessful (right or wrong) like the operational principles of technology. But both are merely articulate means of reorganizing the conceptions of which they speak: one stating the result of the reorganization, the other prescribing the procedure for achieving it.

Accordingly, mathematics can be equally well affiliated either to natural science or technology. Physics and mathematics coincide when mechanics is transposed into a four-dimensional non-Euclidean geometry, and when three-dimensional geometry is regarded as comprising the metric relations of rigid solids. The conception of integers is part of physics in so far as it affirms the existence of permanently discrete objects; while on the other hand, mathematical operations can form part of automatic technical processes and a strictly formalized technology may be regarded as part of mathematics.[1] Mathematical symbolism and mathematical operations thus

---

[1] Remember in this connection that mathematical exercises in problem-solving are of two types, one of which ('prove that . . .') is a contriving, as in technology, while the other ('find an *X* such that . . .') is a discovering, as in science.

show themselves appropriate to the exercise of intellectual control both over things and manipulations, but the instances to which they apply are so varied as to leave extremely little of any particular experience attached to the mathematical framework controlling them. Even elementary mathematics denotes conceptions and operations of great generality, and these conceptions are further attenuated by mathematical inventiveness, which keeps extending the conceptual framework of mathematics ever further beyond its original contacts with experience.

This process is principally guided by two closely related cravings. The first strives for ever greater generality. Descartes' triumph in discovering that the theorems of analytical geometry were but an illustration of algebra elevated the human mind into a region where numbers and diagrams were merged in a common harmonious understanding. Since then mathematics has made innumerable further strides towards generalization. Moreover, aspirations for greater generality frequently entail demands for greater rigorousness—the second craving that guides mathematical invention. Euclid had not hesitated to construct an equilateral triangle by connecting the point at which two circles intersect with the centres of the two circles. But once lines were defined, by a generalization of arithmetic as point-sets, it seemed no longer obvious that two circles crossing each other had a point of intersection. Contrary to traditional common sense, curves could now be conceived as discontinuous at every point. Modern set theory has thus raised new critical scruples in geometry, and by satisfying these scruples it has established more exacting standards of geometrical proof.[1]

---

[1] The scruple in question was raised already by Leibniz and was eliminated only by the addition of a new axiom by Dedekind (1872) to those of Euclid (Weyl, *op. cit.*, p. 40). The process of successive conceptual development and accompanying increase of rigour has been meticulously traced by Daval and Guilbaud in *Le Raisonnement Mathématique* (Presses Universitaires de France, 1945), for the series of creative acts leading up to Bolzano's theorem. The starting point lay in the elementary process of successive approximation by which we can determine to any degree of accuracy the solution of an equation such as $x^3 = 4$. This method was known by the end of the sixteenth century. But since no approximation gave *the* solution, this method left open a problem—a problem which became clearer through the generalization and geometric expression of the equation. If the equation $y^3 = 4$, or more generally, $f(x) = c$, is expressed in the form:

$$y = f(x)$$
$$y = c,$$

the solution will lie at the intersection of the curve $y = f(x)$ and the line $y = c$. But what guarantees that the two will meet? Cauchy, in 1821, proved the theorem (since called Bolzano's): if $f(x)$ is continuous in the interval $x = a$, $x = b$, and if $c$ is a number lying between $f(a)$ and $f(b)$, the equation $f(x) = c$ has always at least one solution in the interval $(a, b)$. But what is meant by 'continuity'? With the help of the conception of convergence, Cauchy defined continuity: a function $(x)$ is continuous at point $a$ if

$$\lim_{x \to a} f(x) = f(a).$$

Thus the all-important idea of continuity came to light, say Daval and Guilbaud, 'grâce à un regard jeté sure une opération mentale déjà effectuée, mais qui se cristallise du fait que l'esprit regarde ce qu'il a fait au lieu de continuer à faire' (p. 117).

185

We have seen the importance of conceptual reforms in the natural sciences, and they play their part also in technology. But in mathematics they take on a new power: they create a universe of discourse which is interesting in itself. This is (as I have said before) like inventing a game by the creation of entirely new concepts, the symbols of which denote nothing but that they are the proper subjects for certain operations.

The imaginative acts by which mathematics creates the object of its own discourse are acceptable only if something interesting can be said about these objects that is not immediately apparent from their original definition.[1] I have said this before in extending the concept of reality to mathematics and recalled there how Lobatschesvky's assumption that a multiplicity of parallels to one straight line can be drawn through one point outside it, eventually carried conviction to mathematicians because it could be shown to involve a whole system of noteworthy implications. In algebra we have a striking instance of this process in the imaginary roots of negative numbers, first defined as such in the sixteenth century (Cardan, Bombelli), the justification of which remained questionable until the discovery of their far-reaching functions in the calculus of complex numbers (the sums of real and imaginary numbers) by Gauss in the nineteenth century. Some further examples will be mentioned later.

No sharp distinction can be drawn between mathematical theories which apply to external objects, and mathematical inventions which are interesting only in themselves, for there is always a possibility that a mathematical theorem may prove applicable to experience some time. Yet the fact that this is not necessarily true, and indeed appears very unlikely for the far greater part of mathematics, is a distinctive feature of this science.[2] Not being primarily concerned with foretelling what is going to happen, or with contriving what anyone wished to happen, but merely with understanding exactly how alternative aspects of a certain set of conceptions are logically connected, mathematics can extend its subject matter indefinitely by conceiving new problems of this sort, without any reference to experience. New conceptions are thus consolidated, as their wider implications and more extensive operability come into view, and this pursuit perpetuates itself by throwing up ever new opportunities for further conceptual innovations.

It now appears that the logical structure of this process is not quite that of inventing a game, but rather that of the continued invention of a game in the very course of playing the game. This kind of game-inventing is akin to the writing of a novel, and the parallel is indeed quite close up to a point. There never was a person called Sherlock Holmes, nor even a person like Sherlock Holmes. Yet this character was well defined by the description

[1] Cf. Émile Borel, *L'Imaginaire et le Réel en Mathématiques et en Physique*, Paris, 1952, p. 100: '. . . le but généralement poursuivi par les mathématiciens, c'est de trouver pour chacune des êtres mathématiques qu'ils ont défini, une propriété distincte de leur définition.'

[2] G. H. Hardy, *A Mathematician's Apology*, Cambridge, 1940, pp. 71–83.

of his consistent behaviour in a series of fictitious situations. Once Conan Doyle had composed a few good stories with Sherlock Holmes as their hero, the image of the detective—however absurd in itself—was clearly fixed for the purposes of any further such stories. The main difference between a fictitious mathematical entity, like a complex number, and a fantastic character like Sherlock Holmes, lies in the greater hold which the latter has on our imagination. It is due to the far richer sensuous elements entering into the conception of Sherlock Holmes. That is why we acquire an image and not merely a conception of the detective.

## 10. The Affirmation of Mathematics

We have seen that a statement is of value to *natural science* if it (1) corresponds to the facts, (2) is relevant to the system of science and (3) bears on a subject matter which is not without intrinsic interest; and that a statement is of value in technology (1) if it reveals an effective and ingenious operational principle which (2) achieves, in existing circumstances, a substantial material advantage. Mathematics is a much freer creation than either natural science or technology. While its early primitive conceptions and operations were no doubt originally suggested by experience and have served to control the manipulation of material things, these empirical and practical contacts do not enter effectively into its present appreciation.

What, then, is mathematics? The 'grim and inflexible resolve' of objectivism has given strange answers to this question. For while we can attempt to achieve impersonality for natural science by basing it on the supposed commands of empirical fact, and technology can be grounded in the requirements of practical life, the only impersonal (or at least apparently impersonal) justification that is left to mathematical statements is their freedom from self-contradiction. Accordingly, mathematics has been described as a set of tautologies.

To this it must be objected in the first place that it is false. Tautologies are necessarily true, but mathematics is not. We cannot tell whether the axioms of arithmetic are consistent; and if they are not, any particular theorem of arithmetic may be false. Therefore these theorems are not tautologies. They are and must always remain tentative, while a tautology is an incontrovertible truism.

But even supposing mathematics were wholly consistent, the criterion of consistency, which the 'tautology' doctrine is intended to support, would still be ludicrously inadequate for defining mathematics. One might as well regard a machine which goes on printing letters and typographical signs at random as producing the text of all future scientific discoveries, poems, laws, speeches, editorials, etc. For just as only a tiny fraction of true statements about matters of fact constitute science and only a tiny fraction of conceivable operational principles constitute technology, so also only a tiny fraction of statements believed to be consistent constitute mathematics. Mathematics cannot be properly defined without appeal to the principle which distinguishes this tiny fraction from the overwhelmingly predominant aggregate of other non-self-contradictory statements.

We may try to supply this criterion by defining mathematics as the totality of theorems derived from a certain set of axioms according to certain operations which will assure their self-consistency, provided the axioms themselves are mutually consistent. But this is still inadequate. First, because it leaves completely unaccounted for the choice of axioms, which hence must appear arbitrary—which it is not; second, because not all mathematics considered to be well established has ever been completely formalized according to strict procedure; and third, because—as K. R. Popper has pointed out—among the propositions that can be derived from some accepted set of axioms there are still, for every single one that represents a significant mathematical theorem, an infinite number that are trivial.[1]

All these difficulties are but consequences of our refusal to see that mathematics cannot be defined without acknowledging its most obvious feature: namely, that it is interesting. Nowhere is intellectual beauty so deeply felt and fastidiously appreciated in its various grades and qualities as in mathematics, and only the informal appreciation of mathematical value can distinguish what is mathematics from a welter of formally similar, yet altogether trivial statements and operations. And we shall see that this emotional colour of mathematics also justifies its acceptance as true. It is by satisfying his intellectual passions that mathematics fascinates the mathematician and compels him to pursue it in his thoughts and give it his assent.

I have said earlier on[2] that we can understand mathematics only by our tacit contribution to its formalism. I have shown how all the proofs and theorems of mathematics have been originally discovered by relying on their intuitive anticipation; how the established results of such discoveries are properly taught, understood, remembered in the form of their intuitively grasped outline; how these results are effectively reapplied and developed further by pondering their intuitive content; and that they can therefore gain our legitimate assent only in terms of our intuitive approval. I have indeed shown that all articulation depends on a tacit component of the same kind for conveying a meaning accredited by the person uttering it. And also that this comprehension-cum-affirmation is continuous with the active principle of animal life by which we both shape and accept our knowledge at all its levels, down to that of drives, motoricity and perception, with which as animals we are equipped by nature.

[1] K. R. Popper, *British Journal for the Philosophy of Science*, **1** (1950-1), p. 194. Poincaré seems to point out the same condition of significance in the words: 'Discovery consists precisely in not constructing useless combinations, but in constructing those that are useful, which are an infinitely small minority' (*Science and Method*, London, 1914, p. 51).          [2] Part Two, ch. 5, pp. 117 ff.

The inarticulate coefficient by which we understand and assent to mathematics is an active principle of this kind; it is a passion for intellectual beauty. It is on account of its intellectual beauty, which his own passion proclaims as revealing a universal truth, that the mathematician feels compelled to accept mathematics as true, even though he is today deprived of the belief in its logical necessity and doomed to admit forever the conceivable possibility that its whole fabric may suddenly collapse by revealing a decisive self-contradiction. And it is the same urge to see sense and make sense that supports his tacit bridging of the logical gaps internal to every formal proof.

There is in fact ample evidence that such intellectual passions are intrinsic to the affirmation of mathematics. Modern mathematics has emerged from a long series of conceptual reforms tending towards greater generality and rigour, as well as from more radical conceptual inventions opening up altogether new perspectives. The acceptance of such conceptual innovations is a self-modifying mental act in search of a truer intellectual life. It has been authoritatively stated that 'the moments of greatest creative advancement in science frequently coincide with the introduction of new notions by means of a definition'.[1] This can be true only because the acceptance of a new conception, even when it is specified by a definition, is ultimately an informal act: a transformation of the framework on which we rely in the process of formal reasoning. It is the crossing of a logical gap to another shore, where we shall never again see things as we did before. To the extent, therefore, to which mathematics is the accumulated product of past conceptual innovations, our affirmation of mathematics is likewise an irreversible, informal act.

Such an act can be said to be rational if it satisfies our standards of excellence, and the intellectual beauty of mathematics, upheld by the passionate connoisseurship of mathematicians, is such a standard. Accordingly, fundamental progress in mathematics, involving conceptual reform, is found to be guided by a search for beauty.[2] The position appears essentially the same as in mathematical physics; intellectual beauty is recognized as a token of a hidden reality. But while in the natural sciences the feeling of making contact with reality is an augury of as yet undreamed future empirical confirmations of an imminent discovery, in mathematics it betokens an indeterminate range of future germinations within mathematics itself.

Since the convincing power of a mathematical proof operates through our tacit understanding of it, the acceptance of a proof may also involve radical conceptual innovations. 'There are beautiful theorems in the "theory of aggregates" (*Mengenlehre*) such as Cantor's theorem of the

[1] A. Tarski, 'The Semantic Conception of Truth and the Foundations of Semantics', *Philosophy and Phenomenological Research*, **4** (1944), p. 359.

[2] See the case histories collected by J. Hadamard, *The Psychology of Invention in the Mathematical Field*, Princeton, 1945, pp. 126–33.

"non enumerability" of the continuum', writes G. H. Hardy, 'the proof [of which] is easy enough once the language has been mastered, but considerable explanation is necessary before the meaning of the theorem becomes clear.'[1] Cantor's proofs traversed a logical gap across which only those willing to enter into their meaning and capable of grasping it could follow him. Reluctance or incapacity to do so caused divisions among mathematicians, similar to those which arose between van't Hoff and Kolbe on the subject of the asymmetric carbon atom, or between Pasteur and Liebig on that of fermentation as a vital function of yeast. Hadamard describes how he and the great Lebesgue, finding themselves on opposite sides of this dispute, were compelled to recognize the impossibility of understanding each other. 'We could not avoid the conclusion that what is evident—the very starting point of certitude in every domain of thought —had not the same meaning for him and for me.'[2] The fundamental conceptual changes involved in Cantor's work were so repulsive to Kroneker, who dominated German mathematics in the 1880's, that he barred Cantor from promotion in all German universities and even from having his papers published in any German mathematical journal.[3] Hadamard confesses that in another field of great modern discoveries, the theory of groups, 'though being eventually able to use it for simple applications, he met with insuperable difficulty in mastering more than a rather elementary and superficial knowledge of it'.[4] Some important innovations have, admittedly, been established by proofs requiring no such far-reaching conceptual adaptation. But even so, their intellectual excellence has contributed to the consolidation of the fundamental conceptions on which their success was based.

## 11. AXIOMATIZATION OF MATHEMATICS

Once more we are brought up against the image of a living science, groping its way towards the satisfaction of the intellectual passions upholding its values. We see it originating thousands of venturesome guesses which had long obsessed their authors until they laboriously brought them to the test of completion, and often battled for them against protracted objections, until they finally gained their established places in the textbooks. And again we see the curious contrast between this image and the ideal of casting the result of this heuristic process—and by implication any

[1] G. H. Hardy, *op. cit.*, p. 38.

[2] J. Hadamard, *op. cit.*, p. 92 (I have taken the liberty of revising the text slightly in the sense which I believe to have been intended).

[3] *loc. cit.* See also *ibid.*, p. 119, where Hadamard writes of the discoveries of Galois (1811–1831), which were appreciated only after his death: 'All these profound ideas were at first forgotten and it was only after fifteen years that, with admiration, scientists became aware of the memoir which the Academy had rejected. It signifies a total transformation of higher algebra, projecting a full light on what had been only glimpsed thus far by the greatest mathematicians, and, at the same time, connecting that algebraic problem with others in quite different branches of science'.

[4] *ibid.*, p. 115.

further continuation of it—into a strictly formalized system of axioms and symbolic operations. Indeed, for mathematics this ideal was not left to be pursued by philosophers, as was the case in the natural sciences, but was included in the endeavour towards an ever greater generality and rigour inherent in the pursuit of mathematics itself.[1]

As such it does not concern us here. But we must ask—as we did in the case of the natural sciences—what the logical position of this system of formal premisses is and, in particular, on what grounds we accept it as valid. The answer to both questions follows closely what has been said about the axiomatization of the natural sciences. When certain undefined terms, axioms and symbolic operations are established as the logical antecedents of mathematics, these are based on the prior assumption that mathematics is true. Our acceptance of what is logically anterior is based on our prior acceptance of what is logically derivative, as being implied in our acceptance of the latter.

Axiomatization has indeed proved a powerful method in the pursuit of greater generality and rigour in all deductive sciences. But it has not supplied a formalized organon for the process of future discovery. Nor has it become the supreme arbiter in deciding controversial issues in mathematics. It is not necessary to show in this case—as we did for natural science—why this is so, for it has been demonstrably established that apart from fairly elementary problems which in a sense are trivial, there cannot exist a method which would always lead in a finite number of steps to the solution of a problem, nor is there any formal procedure that can tell us what problem would be thus decidable.[2] This conclusion could have been partly anticipated from the fact that major steps in mathematics often involve conceptual decisions which can by their very nature never be rigorously proved right.

We can now turn to the paradox of a mathematics based on a system of axioms which are not regarded as self-evident and indeed cannot be known to be mutually consistent. To apply the utmost ingenuity and the most rigorous care to prove the theorems of logic or mathematics, while the premisses of these inferences are cheerfully accepted, without any grounds being given for doing so, as 'unproven asserted formulae', might seem altogether absurd. It reminds one of the clown who solemnly sets up in the middle of the arena two gateposts with a securely locked gate between them, pulls out a large bunch of keys, and laboriously selects one which opens the lock, then passes through the gate and carefully locks it after himself—while all the while the whole arena lies open on either side of the gateposts where he could go round them unhindered. A fully axiomatized deductive system is like a carefully locked gate in the midst of an

[1] Nobody has yet tried to formalize the premisses of technology. The analysis of teleological systems in Part Four may be regarded as a contribution to this task.

[2] Comp. e.g. A. M. Turing, 'Solvable and Unsolvable Problems', *Science News*, **31** (1954), pp. 7–23. Cf. Part Two, ch. 5, p. 126.

infinite empty area. If the acceptance of any proof requires the acceptance without proof of some presuppositions from which the proof is ultimately derived, it follows that the principle of rejecting any unproven statement in mathematics implies also the rejection of all proven statements and therefore of all mathematics.

The solution lies in rejecting the rule which denies acceptance to unproven statements, by admitting that our belief in logically anterior maxims of mathematical procedure is based on our previous acceptance of this procedure as valid. And let us remember once more that logical antecedents derived from the prior acceptance of their consequents are necessarily less certain than the consequents. It is clearly unreasonable, therefore, to regard these antecedents as the grounds on which we accept their consequents.

We should declare instead candidly that we dwell on mathematics and affirm its statements for the sake of its intellectual beauty, which betokens the reality of its conceptions and the truth of its assertions. For if this passion were extinct, we would cease to understand mathematics; its conceptions would dissolve and its proofs carry no conviction. Mathematics would become pointless and would lose itself in a welter of insignificant tautologies and of Heath Robinson operations, from which it could no longer be distinguished.

Mathematics has once already fallen into oblivion by becoming incomprehensible. After the death of Apollonius in 205 B.C. there occurred a break in the oral tradition which alone made the mathematical texts of the Greeks intelligible to students.[1] This was probably due in part to a growing distrust of mathematics, owing to its conflict with the conception of number at the point where it led to magnitudes like $\sqrt{2}$ which could not be expressed in terms of integers. In our own time Gödel's theorem of uncertainty might conceivably erode confidence, likewise, in our own mathematics. And other influences might deepen this distrust. Ideological utilitarianism censures Archimedes today for speaking lightly of his own practical inventions and his passion for intellectual beauty, which he expressed by desiring his grave to be marked by his most brilliant geometrical theorem, is dismissed as an aberration. This movement would discredit the core of mathematics, which is its intellectual beauty. The transmission of mathematics has today been rendered more precarious than ever by the fact that no single mathematician can fully understand any longer more than a tiny fraction of mathematics. Modern mathematics can be kept alive only by a large number of mathematicians cultivating different parts of the same system of values: a community which can be kept coherent only by the passionate vigilance of universities, journals and meetings, fostering these values and imposing the same respect for them on all mathematicians.[2] Such a far-flung structure is

[1] Van der Waerden, *op. cit.*, pp. 266–7.
[2] The structure of this system of mutual surveillance is described in the next chapter.

highly vulnerable and, once broken, impossible to restore. Its ruins would bury modern mathematics in an oblivion more complete and lasting than that which enveloped Greek mathematics twenty-two centuries ago.

## 12. THE ABSTRACT ARTS

Our acknowledgment of intellectual passions in science will gain support by extending our perspective to other emotions that are kindred to it. This kinship is manifest in the most ancient achievement of scientific theory due to Pythagoras, which derives the pleasing effect of a succession of musical notes from the integer ratio between the lengths of the chords struck in producing them. Sustained by this striking fact, the Pythagorean tradition maintained for centuries a musical appreciation of the mathematical laws controlling the celestial order. These were extravagances; but they stemmed from the existing kinship between different kinds of order and beauty, whether discovered in nature, conceived in mathematics, or imaginatively created by art. The relation is closest between pure mathematics and the abstract arts, such as music and abstract painting.

Both visual and musical compositions are appreciated for the beauty of a set of complex relations embodied in them. And as in pure mathematics, so also in the abstract arts, these interesting relationships are discovered, or created, within structures composed of utterances denoting no tangible object. Among the abstract arts music stands out by its precise and complex articulation, subject to a grammar of its own. In profundity and scope it may compare with pure mathematics. Moreover, both of these testify to the same paradox: namely that man can hold important discourse about nothing. For they both speak to us. We do not merely hear music but listen to it and enjoy it by understanding it, even as we enjoy mathematics. Like mathematics, music articulates a vast range of rational relationships for the mere pleasure of understanding them.

Abstract painting creates pleasing visible relationships. That is why we not merely see a canvas, but look at it and try to understand it. Its design bears the same kinship to geometry as music does to arithmetic. Witness the theories of cubism, or the attempts made ever since Vitruvius, to formulate geometrical rules for the appreciation of harmonious pictorial and architectural composition.

It is true that mathematics differs radically from the abstract arts by its practice of symbolic operations; a mathematical symbol signifies the manner in which it functions within such operations. But while the elementary utterances of abstract art can have no such meaning, they can rely instead on their sensuous content. A patch of colour, a musical note are so substantial in themselves, that they can speak their part in articulating a relationship with other patches of colour, or other musical notes, without pointing beyond themselves. Instead of denoting something—

whether an external object or their own use—they emphatically present their own striking sensuous presence.

The decisive part which intellectual passions have been shown to play in the several domains of natural science, engineering, and mathematics, demonstrates the ubiquity of such participation. In each of these domains it is the relevant intellectual passion which affirms the distinctive intellectual values by which any particular performance may qualify for admittance to the domain. The arts appear then no longer as contrasted but as immediately continuous with science, only that in them the thinker participates more deeply in the object of his thought.

The emotional life engendered by an articulate culture is, of course, primordially rooted in the emotions of inarticulate creatures. We have seen that the exhilaration shown by apes and babies when solving a problem prefigures the intellectual joys of science. A game of chess creates its own pleasures, but could not do so if babies could not play with rattles; though a joke is not an expression of hilarity, it can create hilarity because men can laugh. Laments for the dead and songs of love are likewise formulations of earlier shapeless emotions, which are refashioned and amplified into something new by words and music. The originally experienced sentiments are not expressed but alluded to, just as objects are alluded to rather than represented in a painting. Such allusions may be as remote as the allusions to the existence of solids in the theory of numbers, or as close as the allusion to observed crystals in geometrical crystallography; and all art lies between two limits of this kind. However abstract, it will echo some experience, and would be as meaningless to someone lacking any such experience, as arithmetic would be to a person living in a gaseous universe. And again, however meticulously descriptive and plainly expressive a work of art may be, it must never come any closer in referring to experience than crystallography does to crystals; no closer than a representation of a conceivable experience, framed in its own harmonious terms, can come to actual experience. Precise statements of fact or exact expressions of sentiment contained in a work of art tend to flatten it out to a map, a report or a personal communication.[1]

---

[1] The statements made in a poem are, therefore, not 'pseudo-statements', as I. A. Richards called them, at least not any more than those of geometry. (See I. A. Richards, *Science and Poetry*, London, 1926, ch. VI, 'Poetry and Beliefs', pp. 55 ff.) The elucidation of obscure hints in a work of art may help us to enjoy it: Picasso's visual explanation of one of his cubistic pictures in the Museum of Modern Art in New York guides the viewer's eye to a better understanding of it. On the other hand, A. E. Housman, in *The Name and Nature of Poetry* (Cambridge, 1933), demonstrates the disastrous effects of tracing the precise meaning of the symbolism in Poe's 'Haunted Palace'. Some of my favourite poems gave me more genuine pleasure as a child, when I did not understand them.

On allusiveness in architecture, see Geoffrey Scott, *The Architecture of Humanism*, 2nd edn., New York, 1954, p. 95: 'Architecture . . . is an art of spaces and of solids, a felt relation between ponderable things, an adjustment to one another of evident forces, a grouping of material bodies subject *like ourselves* to certain elementary laws. Weight and resistance, burden and effort, weakness and power, are elements in our own

An intelligence which dwells wholly within an articulate structure of its own creation accentuates by doing so a paradox that is inherent in the exercise of all intellectual passions. The practice of the visual and musical arts which releases, formulates and disciplines our faculties for harmonious experience, exerts to the utmost the artist's powers of invention and discrimination merely for the purpose of satisfying the standards of appreciation which the artist has set for himself. A symphony is obviously something new achieved by the human mind; but in calling it a symphony its composer demands recognition for it as something inherently excellent. The natural scientist and the engineer are not so free to satisfy themselves; no scientific theory is beautiful if it is false and no invention is truly ingenious if it is impracticable. Yet this merely modifies the conditions of a process of self-satisfaction. The standards of scientific value and of inventive ingenuity must still be satisfied, and these standards are set by the scientist's and the engineer's own intellectual passions.

To this extent, then, whether thought operates indwellingly within a universe of its own creation, or interprets and controls nature as given to it from outside, the same paradoxical structure prevails throughout the articulate systems so far surveyed. There is present a personal component, inarticulate and passionate, which declares our standards of values, drives us to fulfil them and judges our performance by these self-set standards.

## 13. DWELLING IN AND BREAKING OUT

A valid articulate framework may be a theory, or a mathematical discovery, or a symphony. Whichever it is, it will be used by dwelling in it, and this indwelling can be consciously experienced. Astronomic observations are made by dwelling in astronomic theory, and it is this internal enjoyment of astronomy which makes the astronomer interested in the stars. This is how scientific value is contemplated from within. But awareness of this joy is dimmed when the formulae of astronomy are used in a routine manner. It is only when he reflects on its theoretic vision, or consciously experiences its intellectual powers, that the astronomer may be said to contemplate astronomy. Similarly for mathematics. Between the practice of hackneyed exercises on the one hand and the heuristic visions of the lonely discoverer on the other, lies the major domain of established mathematics on which the mathematician consciously dwells by losing himself in the contemplation of its greatness. A true understanding of science and mathematics includes the capacity

experience, and inseparable in that experience from feelings of ease, exultation, or distress. But weight and resistance, weakness and power, are manifest elements also in architecture, which enacts through their means a kind of human drama. Through them the mechanical solutions of mechanical problems achieve an aesthetic interest and an ideal value.'

for a contemplative experience of them, and the teaching of these sciences must aim at imparting this capacity to the pupil. The task of inducing an intelligent contemplation of music and dramatic art aims likewise at enabling a person to surrender himself to works of art. This is neither to observe nor to handle them, but to live in them. Thus the satisfaction of gaining intellectual control over the external world is linked to a satisfaction of gaining control over ourselves.

The urge towards this dual satisfaction is persistent; yet it operates by phases of self-destruction. The construction of a framework which will handle experience on our behalf begins in the infant and culminates in the scientist. This endeavour must occasionally operate by demolishing a hitherto accepted structure, or parts of it, in order to establish an even more rigorous and comprehensive one in its place. Scientific discovery, which leads from one such framework to its successor, bursts the bounds of disciplined thought in an intense if transient moment of heuristic vision. And while it is thus breaking out, the mind is for the moment directly experiencing its content rather than controlling it by the use of any pre-established modes of interpretation: it is overwhelmed by its own passionate activity.

The scientist's urge to ponder new problems and break new paths in seeking to solve them, presents us with the essential restlessness of the human mind, which calls ever again in question any satisfaction that it may have previously achieved. We may trace this back primordially to the level of the animal. It is true that when provoked into action by a problematical situation the animal tends to establish a new habit which meets the situation and renders further intelligent effort unnecessary; but in higher animals this general trend is occasionally opposed and overcome by playfulness. Animals at play seek excitement, and even when they have outgrown the playful stage they need activity. Human beings develop this desire for tension in a variety of forms. Man is one of the few animals who continue to play throughout adult life. Men have also at all times gone out in search of adventure and enjoyed tales of adventure. We all appreciate feats of craftiness, or the solving of puzzles, and enjoy in innumerable ways the sudden relaxation of a tension in which we have become involved, whether by actual participation or merely in imagination. Our gigantic modern amusement industry betokens the popular forms of this desire; but our craving for mental dissatisfaction enters also into the highest forms of man's spontaneous originality.

The most radical manifestation of this urge to break through all fixed conceptual frameworks is the act of ecstatic vision. When we abandon ourselves to the contemplation of the stars we attend to them in a way which is not an astronomical observation. We look at them with great interest but without thinking about them. For if we did, our awareness of the stars would pale into that of mere instances of apposite conceptions: the focus of our interest being shifted beyond them, our awareness of them would become subsidiary to this focus and their vivid impact on the eye and mind would be lost.[1]

As observers or manipulators of experience we are guided *by* experience and pass *through* experience without experiencing it *in itself*. The conceptual framework by which we observe and manipulate things being present as a screen between ourselves and these things, their sights and sounds, and the smell and touch of them transpire but tenuously through this screen, which keeps us aloof from them. Contemplation dissolves the screen, stops our movement through experience and pours us straight into experience; we cease to handle things and become immersed in them. Contemplation has no ulterior intention or ulterior meaning; in it we cease to deal with things and become absorbed in the inherent quality of our experience, for its own sake. And as we lose ourselves in contemplation, we take on an impersonal life in the objects of our contemplation; while these objects themselves are suffused by a visionary gleam which lends them a new vivid and yet dreamlike reality. It is dreamlike, for it is timeless and without definite spatial location.[2] It is not an objective reality; for it is not the focus of an intelligent perception anticipating future confirmation by tangible things, but resides merely in the coloured patches of various shapes which the things present to the eye. Correspondingly, the impersonality of intense contemplation consists in a complete participation of the person in that which he contemplates and not in his complete detachment from it, as would be the case in an ideally objective observation. Since the impersonality of contemplation is a self-abandonment, it can be described either as egocentric or as selfless, depending on whether one refers to the contemplator's visionary act or to the submergence of his person.

The religious mystic achieves contemplative communion as a result of an elaborate effort of thought, supported by ritual. By concentrating on the presence of God, who is beyond all physical appearances, the mystic seeks to relax the intellectual control which his powers of perception instinctively exercise over the scene confronting them. His fixed gaze no longer scans each object in its turn and his mind ceases to identify their particulars. The whole framework of intelligent understanding, by which he normally appraises his impressions, sinks into abeyance and uncovers a world experienced uncomprehendingly as a divine miracle. The process is known in Christian mysticism as the *via negativa* and the tradition which prescribes it as the only perfect path to God stems from the *Mystic Theology* of the Pseudo-Dionysius. It invites us, through a succession of

---

[1] If one looks at a landscape with one's head bent sideways the intensity of its colours is increased. The loss of meaning caused by the unusual posture is compensated for by increased sensory vividness.

[2] Aldous Huxley, *The Doors of Perception* (London, 1954, p. 14), writes of visual experience under mescaline '. . . along with indifference to space there went an even completer indifference to time'. Cf. also W. Mayer-Gross, 'Experimental psychoses and other mental abnormalities produced by drugs', *Brit. Med. Journ.* (1951), **2**, p. 317.

'detachments', to seek in absolute ignorance union with Him who is beyond all being and all knowledge.[1] We see things then not focally, but as part of a cosmos, as features of God.

The Christian mystic's communion with the world seeks a reconciliation which is part of the technique of redemption. It is man's surrender to the love of God, in the hope of gaining his forgiveness and admission to His presence. The radical anti-intellectualism of the *via negativa* expresses the effort to break out of our normal conceptual framework and 'become like little children'. It is akin to the reliance on the 'foolishness of God', that short-cut to the understanding of Christianity, of which St. Augustine said enviously that it was free to the simple-minded but impassable to the learned. The Christian faith in everyday action is just such a sustained effort at breaking out, sustained by the love and desire for God, a God who can be loved but not observed. Proximity to God is not an observation, for it overwhelms and pervades the worshipper. An observer must be relatively detached from that which he observes, and religious experience transforms the worshipper. It stands in this respect closer to sensual abandon than to exact observation. Mystics speak of religious ecstasy in erotic terms, describing communion with God or with Christ as the union of bride and bridegroom. In the orgiastic rituals of fertility cults religion and sensual fervour are openly blended. But religious ecstasy is an articulate passion and resembles sensual abandon only in the surrender achieved by it.

This surrender corresponds to the degree to which the worshipper dwells within the fabric of the religious ritual, which is potentially the highest degree of indwelling that is conceivable. For ritual comprises a sequence of things to be said and gestures to be made which involve the whole body and alert our whole existence. Anyone sincerely saying and doing these things in a place of worship could not fail to be completely absorbed in them. He would be partaking devoutly in religious life.

But the dwelling of the Christian worshipper within the ritual of divine service differs from any other dwelling within a framework of inherent excellence, by the fact that this indwelling is not enjoyed. The confession of guilt, the surrender to God's mercy, the prayer for grace, the praise of God, bring about mounting tension. By these ritual acts the worshipper accepts the obligation to achieve what he knows to be beyond his own unaided powers and strives towards it in the hope of a merciful visitation from above. The ritual of worship is expressly designed to induce and sustain this state of anguish, surrender and hope. The moment a man were to claim that he had arrived and could now happily contemplate his own perfection, he would be thrown back into spiritual emptiness.

The indwelling of the Christian worshipper is therefore a continued attempt at breaking out, at casting off the condition of man, even while humbly acknowledging its inescapability. Such indwelling is fulfilled most

[1] V. Lossky, *Essai sur la Theologie mystique de l'Église d'Orient*, Paris, 1944, p. 25.

completely when it increases this effort to the utmost. It resembles not the dwelling within a great theory of which we enjoy the complete understanding, nor an immersion in the pattern of a musical masterpiece, but the heuristic upsurge which strives to break through the accepted frameworks of thought, guided by the intimations of discoveries still beyond our horizon. Christian worship sustains, as it were, an eternal, never to be consummated hunch: a heuristic vision which is accepted for the sake of its unresolvable tension. It is like an obsession with a problem known to be insoluble, which yet follows, against reason, unswervingly, the heuristic command: 'Look at the unknown!' Christianity sedulously fosters, and in a sense permanently satisfies, man's craving for mental dissatisfaction by offering him the comfort of a crucified God.

Music, poetry, painting: the arts—whether abstract or representative —are a dwelling in and a breaking out which lie somewhere between science and worship. Mathematics has been compared with poetry: 'The true spirit of delight, the exaltation, the sense of being more than Man, which is the touchstone of the highest excellence, is to be found in mathematics as surely as in poetry', writes Bertrand Russell.[1] Yet there is a great difference in the range of these delights. Owing to its sensuous content a work of art can affect us far more comprehensively than a mathematical theorem. Moreover, artistic creation and enjoyment are contemplative experiences more akin than mathematics to religious communion. Art, like mysticism, breaks through the screen of objectivity and draws on our pre-conceptual capacities of contemplative vision. Poetry 'purges from our inward sight the film of familiarity which obscures from us the wonder of our being', it breaks into 'a world to which the familiar world is chaos' (Shelley).

The mechanism by which a negative theology opens access to the presence of God is applicable here to a process of artistic creation. But the negation of familiar meaning may go beyond this. It may usher us into the presence of nothingness. Sartre's 'Nausée' contains the classic description of this process. It is a generalization of the technique for rendering a word incomprehensible by repeating it a number of times. You say 'table, table, table . . .' until the word becomes a mere meaningless sound. You can destroy meaning wholesale by reducing everything to its uninterpreted particulars. By paralysing our urge to subordinate one thing to another, we can eliminate all subsidiary awareness of things in terms of others and create an atomized, totally depersonalized universe. In it the pebble in your hand, the saliva in your mouth and the word in your ear all become external, absurd and hostile items. This universe is the counterpart of the cosmic vision, with despair taking the place of hope. It is the logical outcome of utterly distrusting our participation in holding our beliefs. Left strictly to itself, this is what the world is like.

---

[1] B. Russell, *Mysticism and Logic* (London, 1918), p. 62.

Modern art has moved along with existentialist philosophy towards the exploration of increasingly radical negations. Surrealism distrusts all meaning and so does modern poetry. It regards easiness as vulgar and intelligibility as dishonest. Fragmentation alone can then be trusted; only an aggregate of fragments can carry a meaning that is wholly ineffable and protected thereby against self doubt.

I have said that the visionary powers of the scientist which lead him to new discoveries subside, once discovery is achieved, into a peaceful contemplation of the result—while religious practices culminate in an endeavour which they seek ever again to achieve. The arts are in an intermediate position. As in science, the heuristic passion of the originator far exceeds in intensity the sentiments induced by his finished product. But the work of art is more akin to an act of religious devotion in remaining, even in its finished form, an instrument of more active and comprehensive contemplation. Though the artist cannot make the public re-live his creative hours, he does make them enter a wide world of sights, sounds and emotions which they had never seen, heard or felt before. 'To achieve this,' writes Marcel Proust,

> the creative painter, the creative writer proceed like the eye specialist. The treatment—with the help of their paintings, their writings—is not always pleasant. When the treatment is concluded they tell us: you can look now. And thus the world which hasn't been created only once, but is recreated every time a new artist emerges, appears to us perfectly comprehensible— so very different from the old. We now adore the women of Renoir and Giraudoux, whereas before the treatment we refused to recognize them as women. And we would love to go for a stroll in those woods which previously seemed to represent anything but woods, for example a tapestry woven of thousands of shades with just the colourings of a forest missing. Such is the passing and new universe created by the artist, which survives only until a new artist arises.[1]

Proust speaks too gently here of the unpleasantness incurred in the treatment of our eyes by new works of art. We are shocked by the offer of an unfamiliar system purporting to be meaningful. When the public is pressed to enter the new framework so as to discover its meaning, their bewilderment turns into indignation. They are outraged by the respect paid to what seems to them deserving of contempt and angry at the implied contempt for their own standards of excellence. There were scenes of violence around the exhibitions of the early Impressionists in Paris. There was fighting in the Parisian audiences of Stravinsky in 1913 [2] and similar disturbances had occurred in various countries at the first performance of some of Wagner's operas. In such conflicts the two sides are actually fighting for their lives, or at least part of their lives. For in the existence of each there is an area which can be kept in being only by denying reality to an area in the existence of the other. And such a denial is a

---

[1] Marcel Proust, Preface to P. Morand, *Tendres Stocks*, Paris, 1921.
[2] E. W. White, *Stravinsky*, London, 1947, p. 42.

shock to the conviction of the other and an attack against his being, to the extent to which he lives in this conviction.

We have seen that the major revolutions in science—both in mathematics and the natural sciences—have also evoked such existential conflicts. The religious wars of the past and the ideological wars of the present will be spoken of later on similar lines. But here I must yet recall that in dealing with scientific controversies of the past we must inevitably judge their outcome ourselves, in the present. All our cultural values are the deposit of a succession of past upheavals, but it is ultimately for us to say what these upheavals signify: whether a triumph or a disaster. It might be thought that artistic innovations are not so comprehensive, so that the new achievements of our time may be added on to the unchanged appreciation of our previous possession. But this is not so. New movements of art include a re-appreciation of their ancestry and a corresponding shift in the valuation of all other artistic achievements of the past. And this necessity re-evokes the paradox which had already presented itself when I contrasted our belief in the eternal value of scientific beauty with our fears for its continued cultivation. For we must admit that truth and beauty may not prevail, or may not prevail for long. We know how monstrous the judgment of posterity can be. Special kilns were used in Medieval Rome, on the site of the Forum and the Campus Martius, to reduce ancient works of art to lime,[1] and at the moment of writing this some of the greatest art treasures of Soviet Russia—paintings by Matisse, Cézanne, Picasso, Renoir, etc.—are condemned as degenerate and stowed away in an attic in Moscow.[2] Yet all the same, the acceptance of a novel work must presume the approval of posterity. Artistic beauty is a token of artistic reality, in the same sense in which mathematical beauty is a token of mathematical reality. Its appreciation has universal intent, and bears witness beyond that to the presence of an inexhaustible fund of meaning in it which future centuries may yet elicit. Such is our commitment to indwelling.

A personal knowledge accepted by indwelling may appear merely subjective. It cannot be fully defended here against this suspicion. But we can already distinguish between the accrediting of an articulate framework, be it a theory, a religious ritual or a work of art, and the accrediting of an experience, whether within such a framework or as visionary contemplation. It might appear questionable whether there can be anything to accredit where nothing seems to be asserted. We see what we see, we smell what we smell and feel what we feel, and there seems no more to it. Experiences that make no claim whatever would be truly incorrigible. But we must allow in the first place for the fact that what we see or feel depends very much on the way we make sense of it, and in this respect it is corrigible.

[1] H. Jordan, *Topographie der Stadt Rom im Alterthum*, Berlin, 1878, **1**, p. 65.
[2] Hélène and Pierre Lazareff, in *L'URSS a L'Heure Malenkov* (Paris, 1954), reproduce a photograph of these pictures stored in an attic.

A white patch may turn black when we take in the fact that it is part of a black cloth bathed in sunshine. A child may feel hungry and not know that it wants to eat, until it is offered food. But even apart from this, any deliberate existential use of the mind may be said to succeed or fail in achieving a desired experience. The worshipper strenuously concentrates on his prayer for the sake of achieving devotion to God; he may succeed or fail. Monks and nuns afflicted by 'acidia' are tormented by failure to pray wholeheartedly. Experiences can be compared in *depth*, and the more deeply they affect us, the more *genuine* they may be said to be. Besides, *reports* of experience can be doubted even if they are correct. A person may correctly report the colours of a large number of objects, yet not really see any difference between green and red; so that when it is eventually found that he is red-green colour blind, we shall conclude that his previous reports, though correct, were not authentic.

The acceptance of different kinds of articulate systems as mental dwelling places is arrived at by a process of gradual appreciation, and all these acceptances depend to some extent on the content of relevant experiences; but the bearing of natural science on facts of experience is much more specific than that of mathematics, religion or the various arts. It is justifiable, therefore, to speak of the verification of science by experience in a sense which would not apply to other articulate systems. The process by which other systems than science are tested and finally accepted may be called, by contrast, a process of *validation*.

Our personal participation is in general greater in a validation than in a verification. The emotional coefficient of assertion is intensified as we pass from the sciences to the neighbouring domains of thought. But both *verification* and *validation* are everywhere an acknowledgment of a commitment: they claim the presence of something real and external to the speaker. As distinct from both of these, *subjective* experiences can only be said to be *authentic*, and authenticity does not involve a commitment in the sense in which both verification and validation do.

# 7

# CONVIVIALITY

## 1. Introduction

ARTICULATE systems which foster and satisfy an intellectual passion can survive only with the support of a society which respects the values affirmed by these passions, and a society has a cultural life only to the extent to which it acknowledges and fulfils the obligation to lend its support to the cultivation of these passions. Since the advancement and dissemination of knowledge by the pursuit of science, or technology and mathematics forms part of cultural life, the tacit coefficients by which these articulate systems are understood and accredited, and which uphold quite generally our shaping and affirmation of factual truth, are also coefficients of a cultural life shared by a community.

I propose now to show first that this tacit sharing of knowing underlies every single act of articulate communication. I shall then take in the whole network of tacit interactions on which the sharing of cultural life depends, and so lead on to a point at which our adherence to the truth can be seen to imply our adherence to a society which respects the truth, and which we trust to respect it. Love of truth and of intellectual values in general will now reappear as the love of the kind of society which fosters these values, and submission to intellectual standards will be seen to imply participation in a society which accepts the cultural obligation to serve these standards.

Once we fully recognize these civic coefficients of our intellectual passions, we shall be confronted once again, and even more dangerously, with the realization that we hold with universal intent a set of convictions acquired by our particular upbringing. For if we believe that we hold these convictions merely because we were taught them, they may appear to be external to us; while to the extent to which we acknowledge that we have actively decided to accept them, they will tend to appear arbitrary. Moreover, these unsettling reflections will now challenge also the framework of society. At all points where men in authority are seen to impose

203

on others intellectual values which on reflection may come to appear adventitious, the justification of this authority may be called in question. The exercise of authority will tend to appear as bigoted or as hypocritical, if it asserts as universal what is actually parochial.

Thus the disturbance of our own convictions, caused by the sight of our own ubiquitous participation in the shaping of truth, will expand into a civic predicament, and the struggle to regain our mental balance in this philosophic situation will gain a new significance. We shall realize that on its success depends the possibility of upholding the intellectual and moral culture of our society.

Unfortunately, while the realization of the civic usefulness of our philosophic aim will sharpen our interest in it, it will also further complicate our task, for it will extend to a deeper level our suspicion that in holding our convictions as valid in themselves we are acting in bad faith. This doubt will have to be carried over to the next chapter, in the hope of dispelling it there within the proposed reform of the conception of truth.

## 2. COMMUNICATION

In my chapter on Articulation I have restricted myself to the intellectual advantages which a solitary individual may conceivably derive from the use of language. This restriction will now be abandoned and with it also the restriction to the declaratory mode and descriptive use of language.

My argument has, of course, repeatedly overflowed already from the descriptive to the interactive and expressive uses of language. The affirmation of a scientific theory was seen to convey an appreciation of its beauty and all the statements of mathematics were seen to carry a whole gamut of delicate aesthetic appreciations. And again, the operational principles of technology and the formal demonstrations of mathematics were seen to be rules of a successful action which could be most appositely cast in the imperative form, even though they were being considered so far only in their solitary use.

The expressive and imperative components of descriptive language become more marked when declarations of fact are used for purposes of interpersonal communication. Communication is a form of address, calling someone's attention to its message and to its speaker. Yet the possibility of communicating information to others is already foreshadowed in the mere descriptive powers of language. A small set of consistently used symbols which, owing to their peculiar manageability, enable us to think about their subject matter more swiftly in terms of its symbolic representation, can be used to carry information to other people if they can use this representation as we do. This can happen only if speakers and listeners have heard the terms used in similar circumstances, and have derived from these experiences the same relation between the symbols and the recurrent features (or functions) which they represent. Both speakers and listeners must also have found the symbols in question manageable, as otherwise they could not have acquired any fluency in their use.

I believe that even though people may conceivably misunderstand any particular words addressed to them, they can, as a rule, convey information to each other reliably enough by speech. For I think that the tacit judgments involved in the process of denotation do tend to coincide between different people and that different people also tend to find the same set of symbols manageable for the purpose of skilfully reorganizing their knowledge.[1] Let me develop this belief now in a wider context.

The interpersonal coincidence of tacit judgments is primordially continuous with the mute interaction of powerful emotions. The sexual embrace wordlessly communicates an intense mutual satisfaction. Animals which rear their young establish between parent and offspring a mutual satisfaction, coloured by dominance and submission. A baby smiles back at a smiling adult and is frightened into crying by a frowning countenance, without any practical experience of their corresponding dispositions.[2] Judging by Piaget's observation, the companionship of children in play is so close, that they insufficiently realize the distinction between themselves and their playmates. They react in an 'autistic' manner, which may appear selfless or egocentric, depending on whether one regards them as losing themselves or as appropriating the person of the others. The conviviality of gregarious animals, of which I shall speak later, seems akin to this.

Diffuse emotional conviviality merges imperceptibly into the transmission of specific experiences in the kind of physical sympathy which overcomes the onlooker at the sight of another's sharp suffering. One has specially to train oneself in order to stand the sight of a surgical operation. Even experienced doctors may faint or get sick at the sight of a deep incision in the eye of a patient. Sadism is the transmutation of transmitted pangs into pleasurable excitement; it is a masochistic sharing of another man's torment and is known to be associated with masochism in the subject. Even the most determined criminals are liable to be effected by physical compassion. It is on record that when the head of the Gestapo, Himmler, desiring to test the technique of extermination at first hand, ordered the killing of a hundred Jews in his presence, he came near to fainting at the sight. In spite of deliberate training to merciless cruelty, upheld by a firm conviction of its rightness, the horrible sight of their

[1] Bees can communicate with each other by symbols without being able to use these for the purpose of discursive thought. So the connection between the solitary and the social use of symbols affirmed in the text does not hold in reverse.

[2] See Katz, who comments further as follows: 'The understanding of the mental life of another person must be something quite primitive, even though it is perhaps modified and refined now and then by individual experience' (my translation from D. Katz, *Gestaltpsychologie*, Basel, 1944, p. 80).

deeds proved a serious difficulty to the persons charged with mass exterminations and it was in order to reduce this 'seelische Belastung', that the gas chamber method was eventually adopted.[1]

Knowledge (as distinct from a single experience) is transmitted on a primordial level from one generation of animals to the next by an imitative process which students of animal behaviour call *mimesis*.[2] But communication at this level is not readily distinguishable from the actions determined by the inheritance of instincts. A true transmission of knowledge stemming from conviviality takes place when an animal shares in the intelligent effort which another animal is making in its presence. There are telling photographs by W. Köhler of chimpanzees watching a fellow animal's attempt to perform a difficult feat and revealing by their gestures that they participate in another's efforts. Such interpersonal transmission seems at work whenever animals learn something by example, which they obviously do when a trick invented by a more intelligent chimpanzee is immediately taken up by another, who would never have been able to think of it on his own. Köhler, giving instances of this process, convincingly asserts that it is no blind parrot-like imitation, but a genuine transmission of an intellectual performance from one animal to another: a real communication of knowledge on the inarticulate level.[3]

All arts are learned by intelligently imitating the way they are practised by other persons in whom the learner places his confidence. To know a language is an art, carried on by tacit judgments and the practice of unspecifiable skills. The child's way of learning to speak from his adult guardians is therefore akin to the young mammal's and young bird's mimetic responses to its nurturing, protecting and guiding seniors. The tacit coefficients of speech are transmitted by inarticulate communications, passing from an authoritative person to a trusting pupil, and the power of speech to convey communication depends on the effectiveness of this mimetic transmission.

Spoken communication is the successful application by two persons of the linguistic knowledge and skill acquired by such apprenticeship, one person wishing to transmit, the other to receive, information. Relying on what each has learnt, the speaker confidently utters words and the listener confidently interprets them, while they mutually rely on each other's correct use and understanding of these words. A true communication will take place if, and only if, these combined assumptions of authority and trust are in fact justified.

We become aware of the precariousness of these conditions when they are grossly unfulfilled, as in the conversation of children, who, as Piaget says, 'fail to understand one another . . . because they think that they do

[1] Edward Crankshaw, *Gestapo*, London, 1956, p. 30, p. 169.

[2] E. A. Armstrong, 'The Nature and Function of Animal Mimesis', *Bull. of Animal Behaviour*, No. 9, 1951, p. 46.

[3] Köhler, *Mentality of Apes, op. cit.*, ch. VII, pp. 185 ff. Piaget, *Psychology of Intelligence*, pp. 125-8, also confirms the role of imitation in the development of thought.

understand one another';[1] while at the same time 'the words spoken are not thought of from the point of view of the person spoken to, and the latter . . . selects them according to his own interest, and distorts them in favour of previously formed conceptions'.[2] As writers, speakers and listeners, we know the perils of such vagaries and are constantly on the alert against them. Speaking and writing is an ever renewed struggle to be both apposite and intelligible, and every word that is finally uttered is a confession of our incapacity to do better; but each time we have finished saying something and let it stand, we tacitly imply also that this says what we mean and should mean it therefore also to the listener or reader. Though these ubiquitous tacit endorsements of our words may always turn out to be mistaken, we must accept this risk if we are ever to say anything.

### 3. Transmission of Social Lore

The combined action of authority and trust which underlies both the learning of language and its use for carrying messages, is a simplified instance of a process which enters into the whole transmission of culture to succeeding generations.

Our modern culture is highly articulate. If another Flood came over us, the largest liner afloat would not suffice to carry the millions of volumes, the many thousands of paintings and hundreds of different instruments, musical, scientific and technical, together with the host of specialists uniquely qualified to use these means of articulation, by which we might transmit to post-diluvian humanity even the crudest remains of our civilization. The current transmission of this immense aggregate of intellectual artefacts from one generation to another takes place by a process of communication which flows from adults to young people. This kind of communication can be received only when one person places an exceptional degree of confidence in another, the apprentice in the master, the student in the teacher, and popular audiences in distinguished speakers or famous writers. This assimilation of great systems of articulate lore by novices of various grades is made possible only by a *previous act of affiliation*, by which the novice accepts apprenticeship to a community which cultivates this lore, appreciates its values and strives to act by its standards. This affiliation begins with the fact that a child submits to education within a community, and it is confirmed throughout life to the extent to which the adult continues to place exceptional confidence in the intellectual leaders of the same community. Just as children learn to speak by assuming that the words used in their presence mean something, so throughout the whole range of cultural apprenticeship the intellectual junior's craving to understand the doings and sayings of his intellectual superiors assumes

[1] Piaget, *Language and Thought of the Child*, London, 1932, p. 101.
[2] *ibid.*, p. 98.

that what they are doing and saying has a hidden meaning which, when discovered, will be found satisfying to some extent.

I have spoken before of heuristic intimations in problem-solving and shown their kinship with the learner's anticipation that what he tries to understand is in fact reasonable. The learner, like the discoverer, must believe before he can know. But while the problem-solver's foreknowledge expresses confidence in himself, the intimations followed by the learner are based predominantly on his confidence in others; and this is an acceptance of authority.

Such granting of one's personal allegiance is—like an act of heuristic conjecture—a passionate pouring of oneself into untried forms of existence. The continued transmission of articulate systems, which lends public and enduring quality to our intellectual gratifications, depends throughout on these acts of submission.[1]

These self-modifying processes are inherently informal, irreversible and to this extent a-critical. Admittedly, once discovery is achieved or the learner has mastered his subject, the conjectural tension is reduced: the discoverer can then demonstrate his result and the learner can justify the knowledge he has acquired. But the amount of knowledge which we can justify from evidence directly available to us can never be large. The overwhelming proportion of our factual beliefs continue therefore to be held at second hand through trusting others, and in the great majority of cases our trust is placed in the authority of comparatively few people of widely acknowledged standing.

Moreover, what is true of the acquisition of knowledge applies likewise to all other intellectual satisfactions. The current cultivation of thought in society depends throughout on the same kind of personal confidence which secures the transmission of social lore from one generation to the next. I shall go into this presently in detail when describing the administration of culture.

Meanwhile, I have yet to add an essential qualification to the principle of authority. Every acceptance of authority is qualified by some measure of reaction to it or even against it. Submission to a consensus is always accompanied to some extent by the imposition of one's views on the consensus to which we submit. Every time we use a word in speaking and writing we both comply with usage and at the same time somewhat modify the existing usage; every time I select a programme on the radio I modify a little the balance of current cultural valuations; even when I make my purchase at current prices I slightly modify the whole price system. Indeed, whenever I submit to a current consensus, I inevitably modify its teaching; for I submit to what I myself think it teaches and by joining the consensus on these terms I affect its content. On the other hand, even the sharpest dissent still operates by partial submission to an existing consensus: for the revolutionary must speak in terms that people can under-

[1] Cf. p. 173.

stand. Moreover, every dissenter is a teacher. The figures of Antigone and of the Socrates of the *Apology* are monuments of the dissenter as law-giver. So are also the prophets of the Old Testament—and so is a Luther, or a Calvin. All modern revolutionaries since the Jacobins demonstrate likewise that dissent does not seek to abolish public authority, but to claim it for itself.

Admittedly, submission to authority is in general less deliberately assertive than is an act of dissent. But not always. St. Augustine's struggle for belief in revelation was much more dynamic and original than is the rejection of revelation by a religiously brought up young man today. In any case, at every step of the process by which we are brought up and continue to participate in an established consensus, we exercise *some* measure of choice between different degrees of conformity and dissent, and *either* of these choices may mean a more passive or a more assertive reaction.

We should realize at the same time how inevitable, and how unceasing and comprehensive are such accreditive decisions. I cannot speak of a scientific fact, of a word, of a poem or a boxing champion; of last week's murder or the Queen of England; of money or music or the fashion in hats, of what is just or unjust, trivial, amusing, boring or scandalous, without implying a reference to a consensus by which these matters are acknowledged—or denied to be—what I declare them to be. I must continually endorse the existing consensus or dissent from it to some degree, and in either case I express what I believe the consensus ought to be in respect to whatever I speak of. The present text, in which I have described in my own way the interaction of every utterance with the public consensus, is no exception to what I have said in the text about utterances of this kind. Throughout this book I am affirming my own beliefs, and more particularly so when I insist, as I do here, that such personal affirmations and choices are inescapable, and, when I argue, as I shall do, that this is all that can be required of me.

### 4. PURE CONVIVIALITY

The sentiments of trust and the persuasive passions by which the transmission of our articulate heritage is kept flowing, bring us back once more to the primitive sentiments of fellowship that exist previous to articulation among all groups of men and even among animals. Evidence of the primordial character of such conviviality and of the lively emotions engendered and gratified by its interplay is supplied by the experience both of animals and men.

A newly hatched chicken soon learns to join the flock around its mother and to seek protection under her wings. This educational process goes on so swiftly that it normally escapes notice, but it is clearly revealed by the experiment of letting a chicken grow up in isolation. When the chicken brought up in solitude is released after a fortnight, and brought together with its sisters and brothers, which have formed a flock around their mother, it behaves in a frantic manner. It pecks wildly at its fellow chicks and runs around terror-stricken.[1] We may say therefore that the earliest inter-personal interaction between chickens affects their emotions towards each other. They usually succeed in developing a rationally balanced emotional life, which would be stunted and deranged by artificial isolation.

The emotional comfort which chickens seem to enjoy when brought up in a flock is not unrelated to the bodily satisfaction of shared warmth and protection, but it is yet distinct from the pleasure of mere drive-satisfaction which an animal enjoys in finding food and shelter. A hungry dog will jump about and bark when its meal is approaching and this excitement has emotional colouring, but the companionship afforded by a dog to a man, by which it may vitally participate in his existence, is rooted in richer and more disinterested passions. Indeed, a dog will attach its affection to a master who plays with it, goes for walks with it and generally shows interest in it, rather than to the person who feeds it.[2] The comprehensive scope of convivial relations has been expressed by Köhler in the aphorism that a solitary chimpanzee is not a chimpanzee. All its physical needs are satisfied, yet it languishes through emotional starvation. It lacks that sharing and interplay of life between fellow animals, the manifold forms of which are reflected in a whole gamut of varied sentiments.

Companionship among men is often sustained and enjoyed in silence. Mr. Utterson, in Stevenson's *Dr. Jekyll and Mr. Hyde*, puts off any business, however important, to take his regular walk with his friend Mr. Richard Enfield, during which neither of them pronounces a single word. But conviviality is usually made effective by a more deliberate sharing of experience, and most commonly by conversation. The exchange of greetings and of conventional remarks is an articulation of companionship, and every articulate address of one person to another makes some contribution to their conviviality, in the sense of their reaching out to each other and sharing each other's lives. Pure conviviality, that is, the cultivation of good fellowship, predominates in many acts of communication; indeed, the main reason for which people talk to each other is a desire for company.[3] The torment of solitary confinement is that it deprives one not of information but of conversation, however uninformative.

The fostering of good fellowship within small groups of people living together, be it as families, as school-fellows, as shipmates, as fellow members of a congregation or of a workshop or office team, is a direct

---

[1] D. Katz, *Animals and Men*, London, 1937, p. 216.

[2] *ibid.*, p. 40.

[3] This, it would seem, underlies the integrating effect of gossip as described by M. Gluckman in a broadcast on 'The Sociology of Gossip', September 30th, 1956.

contribution to the fulfilment of man's purpose and duty as a social being. But the process is also of practical use in making the joint activities of the group more effective. Naval commanders know that the crew of a happy ship will fight well. Industrial psychologists have observed that the output of a workshop increases when its operatives find pleasure in each other's company.[1] Many are the instances in which the improvement of conviviality is deliberately advanced for the sake of such advantageous results; and this offers further confirmation of the substantial character that we have ascribed to the feelings of companionship.

It also forms a transition to a second kind of pure conviviality: from the sharing of experience to a participation in joint activities. Such co-operation is usually incidental to a purpose jointly aimed at, but it becomes purely convivial in the joint performance of a ritual. By fully participating in a ritual, the members of a group affirm the community of their existence, and at the same time identify the life of their group with that of antecedent groups, from whom the ritual has descended to them. Every ritual act of a group is to this extent a reconciliation within the group and a re-establishment of continuity with its own history as a group. It affirms the convivial existence of the group as transcending the individual, both in the present and through times past. The occasions for these emotional reaffirmations are anniversary dates or the recurrent changes by which the group undergoes reconstitution. Its coherence is renewed ritually to the annual rhythm of the seasons, or else when the occurrence of death, birth, marriage, or other alterations of status, are solemnly consecrated in traditional terms.[2]

Since rituals are a celebration of convivial existence they incur the hostility of individualism, which denies value to group life as a form of being, not accessible to the isolated individual. Ritual is deprecated also both by the utilitarian on the grounds of serving no tangible purpose, and by the romantic (the emotionalist brother of the utilitarian) for suppressing people's genuine spontaneous feelings, in favour of the standardized public emotions which they are forced insincerely to pretend to share. Traditionalism is even more fundamentally discredited by reflecting on the fact that the solemnity to which we surrender in performing a ritual is of man's own making. We seem both to generate it and submit to it as something external to us, and in doing so we seem to dupe ourselves and deceive our fellows. We see reappear here the internal insecurity of self-set standards, in the fuller range of their social setting.

[1] See e.g. W. J. H. Sprott, *Science and Social Action*, London, 1954, ch. IV, 'The Small Group', pp. 64 ff. Cf. the account of the Hawthorne experiment of the Western Electric Company by F. J. Roethlisberger and W. J. Dickson (*Management and the Worker*, Cambridge, Mass., 1939), and the use made of this material by G. C. Homans, *The Human Group*, London, 1951.

[2] See Arnold van Gennep, *Les Rites de Passage*, Paris, 1909; M. Fortes, 'Ritual Festivals and Social Cohesion in the Hinterland of the Gold Coast', *American Anthropologist*, N.S., **38** (1936), pp. 590 ff.; M. Gluckman, *Rituals of Rebellion in South-East Africa*, The Frazer Lecture, 1952, Manchester, 1954.

## 5. The Organization of Society

The picture of society which I have so far outlined is like a newly launched ship—a frame without an engine. I have traced the tacit personal interactions which make possible the flow of communications, the transmission of social lore from one generation to the other and the maintenance of an articulate consensus I have shown also how the same interactions gratify a desire for companionship, a pure conviviality to which a participation in common rituals gives the firmest expression. These features of group life suffice for the formation of a fellowship, but not of an organized society. We can understand the latter only when we recognize the framework of interpersonal obligations imposed by the social lore of the group.

Yet the mere sharing of intellectual passions directed towards no other persons establishes already a wide range of common values, which are continuous with the interpersonal appreciations laid down by morality, custom and law. Moreover, such sharing constitutes an orthodoxy upholding certain intellectual and artistic standards, and an undertaking to engage in the pursuits guided by them, which amounts in effect to a recognition of cultural obligations. Finally, since the passions expressed in a ritual affirm the value of group life, they declare that the group has a claim to the conformity of its members, and that the interests of group life may legitimately rival and sometimes overrule those of the individual. This acknowledges a *common good* for the sake of which deviation may be suppressed and individuals be required to make sacrifices for defending the group against subversion and destruction from outside.

At this stage, the framework of cultural and ritual fellowship reveals primordially the four coefficients of societal organization which jointly compose all specific systems of fixed social relations. Two of these coefficients recall the two ways of satisfying intellectual passions on an articulate level, namely by affirmation or indwelling: the first is the *sharing of convictions*, the second the *sharing of a fellowship*. The third coefficient is *co-operation*; the fourth the exercise of *authority or coercion*.

These four titles refer to four aspects of society which must always be seen in conjunction with each other, for only together can they form stable features in the shape of social institutions. Yet in modern society, based on elaborate articulate systems and on a high degree of specialization, we find certain institutions which predominantly embody each of these four coefficients in turn. (1) Universities, churches, theatres and picture galleries, serve the sharing of convictions, in the wide sense of the term which I am applying here. They are institutions of *culture*. (2) Social intercourse, group rituals, common defence, are predominantly convivial institutions. They foster and demand *group loyalty*. (3) Co-operation for a joint material advantage is the predominant feature of society as *an economic system*. (4) Authority and coercion supply the *public power* which shelters and controls the cultural, convivial and economic institutions of society.

Primitive illiterate peoples cannot operate such distinctive institutions and present therefore throughout an intimate amalgam of all four social coefficients. At this stage no fundamental tension can exist between power and thought in society. Nor did such a tension arise even after power and thought had been embodied in separate institutions, so long as society accepted its own structure as permanently established, as it did during the greater part of recorded history. For in spite of many great reforms—like those of Solon and Cleisthenes, of Gregory the Great or Luther, of Richelieu or Peter the Great—made during the first 2300 years of European history, a hierarchical social structure was for the most part regarded as essential to the very existence of the body politic. Only after the American and French revolutions did the conviction gradually spread over the world that society could be improved indefinitely by the exercise of political will of the people, and that the people should therefore be sovereign, both in theory and fact.

This movement gave rise to modern dynamic societies, of which there are two kinds. When a society is resolved on a sudden complete renewal of itself, its dynamism is revolutionary; if it aims at a more gradual approach to perfection, its dynamism is reformist. In the rest of this chapter I shall deal at length with the status of scientific truth and of other intellectual values in these two kinds of societies, elaborating thereby the distinction, at which I have hinted already, between a totalitarianism which tries to fulfil the Laplacean programme by subordinating all thought to welfare, and a free society which accepts in principle the obligation to cultivate thought according to its inherent standards. But I must make clear first, however sketchily, how both types of modern dynamic society differ, with respect to their relation to thought, from the static societies from which they have emerged.

For this purpose we must acknowledge the difference between freedom of thought and the recognition of thought as a real force. No static society ever denied the intrinsic power and worth of thought: religion, morality, law and all the arts were respected in their own right. Though their enterprises were restricted by a set of specific beliefs which they were forbidden to challenge, cultural pursuits flourished within these limits. Moreover, the established orthodoxy was imposed by rulers who accepted it also for their own guidance. Though the quest for truth was restricted by the acceptance of certain teachings as indubitably true, the obligatory respect for the authority of these teachings implied a deep respect for truth.[1]

The intellectual control exercised by modern revolutionary governments differs from this in principle. Its rulers propose to re-shape society,

---

[1] Bertrand de Jouvenel in *Sovereignty*, Cambridge, 1957, p. 290, says of the dogmatic authorities of this period: 'For them Truth was the all important value.' I have found support for many of my views in this book.

including its thought, in the service of its welfare. They deny thereby any independent status or free activity to thought, even though they may in fact often admit its authority as a tacit concession to common sense.

This is totalitarianism. By contrast both to it and to a static society, a free society accords both independent status and a theoretically unrestricted range to thought, though in practice it fosters a particular cultural tradition, and imposes a public education and a code of laws which uphold existing political and economic institutions.

In principle, the free society claims the right of self-determination for the purpose of self-perfection as absolutely as the modern revolutionary regimes. Indeed, these aspirations form part of the original forces that created the free societies; they stem from the unfettered thoughts and generous feelings which overthrew the static authoritarianism of the Middle Ages. Yet at the same time they have set up a menacing contradiction in the free society that they produced. The great movement for independent thought instilled in the modern mind a desperate refusal of all knowledge that is not absolutely impersonal, and this implied in its turn a mechanical conception of man which was bound to deny man's capacity for independent thought. Such objectivism must represent the public good in terms of welfare and power and set in motion thereby the self-destruction of freedom. For when open professions of the great moral passions animating a free society are discredited as specious or utopian, its dynamism will tend to be transformed into the hidden driving force of a political machine, which is then proclaimed as inherently right and granted absolute dominion over thought.

The seriousness of this civic predicament—resulting in the last resort from the inherent instability of our convictions—will be presently explored in some detail. But I must expand my perspective first by introducing explicitly the moral aspirations of man as an extension of his more specifically intellectual passions.

## 6. TWO KINDS OF CULTURE

The principal purpose of this book is to achieve a frame of mind in which I may hold firmly to what I believe to be true, even though I know that it might conceivably be false. The cultivation of thought in general is only examined as the context in which truth may be upheld. Yet I shall now have to include explicitly the domains of morality, custom and law within the system of culture.

Moral judgments are appraisals and as such are akin to intellectual valuations. The thirst for righteousness has the same capacity for satisfying itself by enriching the world that is proper to intellectual passions. And like the artist and scientist, moral man strives to satisfy his own standards, to which he attributes universal validity.

But moral judgments cut much deeper than intellectual valuations. A

214

man may be consumed by an intellectual passion; he may be a man of genius, yet be also sycophantic, vain, envious and spiteful. Though a prince of letters, he would be a despicable person. For men are valued as men according to their moral force; and the outcome of our moral striving is assessed, not as the success or failure of any external performance of ours, but by its effect on our whole person. Accordingly, moral rules control our whole selves rather than the exercise of our faculties, and to comply with a code of morality, custom and law, is to live by it in a far more comprehensive sense than is involved in observing certain scientific and artistic standards.

Moral rules are therefore an instrument of civic power in the hands of those who administer moral culture, and morality is allied to custom and law. Men form a society to the extent to which their lives are ordered by the same morality, custom and law, which jointly constitute the *mores* of their society.

We recognize here an important division in the administration of social lore. For we see that while some systems of social lore are cultivated for the sake of our intellectual life as individuals, others are cultivated by the act of ordering our lives socially in accordance with them. The first is a social fostering of essentially *individual* thought, the second an administration of society in accordance with essentially *civic* thought.

All thought is valid by its own standards and its progress is everywhere prompted by its own passions. If thought is to be cultivated socially, these standards and passions must be shared by a group of people. To secure this sharing, society must establish appropriate sets of rights and duties which constitute its *cultural institutions.* This will make the life of thought in society dependent at second hand on the *civic institutions* of society, that is on group loyalty, property and power. But this dependence will enter differently into the two types of thought in society, since *civic culture* itself sustains the civic institutions of society, while *individual culture* is, on the contrary, itself sustained by these institutions.

I shall not try to assign here either logical or historical priority to any of the three closely interwoven civic institutions. It may be, for example, that men establish a social order in the first place for the sake of making a living, and that the defence of property is thus the key to group loyalty and to the exercise of power both within the group, and in defence of it. But whatever the connection between the three civic institutions, loyalty is parochial, property appetitive and public authority violent. Thus the civic pole relies ultimately on coefficients that are essentially at variance with the universal intent of intellectual or moral standards.

On the other hand, no order of society is thoughtless: it embodies the civic sense and moral convictions of those who believe in it and live by it. To a happy people its civic culture is its civic home; and to this extent the intellectual passions sustaining this culture are in fact esoteric. But again, in a critical age, this intertwining of civic exigencies with the ideals of morality will remain precarious. The genuineness of moral standards will be rendered suspect when it is realized that they are upheld by force, based on property and imbued with local loyalty. Indeed, such conflicts may call in question altogether the intrinsic force of civic thought, and if in this conflict thought is the loser, thought will be denied here—and here in the first place—its essential autonomy. Morality will then be reduced to a mere ideology, and this depreciation of thought will tend to spread and to bring about eventually the subjection of all thought to local patriotism, economic interest and the power of the state. Let me now develop this pattern of conflicting tendencies.

## 7. ADMINISTRATION OF INDIVIDUAL CULTURE

We shall start with the condition of thought in a free society, and let the advancement of science serve as our principal example for the cultivation of individual thought to which independent status is accorded by society.

The organization of the scientific process is determined, in the first place, by the fact that modern science is so vast that any single person can properly understand only a small section of it. The Royal Society has eight sub-committees for the election of Fellows, each of which has a separate field of research allotted to it. One of these fields, for example, is mathematics; but individual mathematicians are further specialized and are competent to deal only with a small part of mathematics. It is a rare mathematician—we are told—who fully understands more than half a dozen out of fifty papers presented to a mathematical congress. 'The very language in which most of the other forty-four are presented goes clear over the head of the man who follows the six reports nearest his own speciality.' [1] Adding to this evidence my own experience in chemistry and physics, it seems to me that the situation may be similar for all major scientific provinces, so that any single scientist may be competent to judge at first hand only about one hundredth of the total current output of science.

Yet this group of persons—the scientists—administer jointly the advancement and dissemination of science. They do so though the control of university premises, academic appointments, research grants, scientific journals and the awarding of academic degrees which qualify their recipients as teachers, technical or medical practitioners, and opens to them the possibility of academic appointment. Moreover, by controlling the advancement and dissemination of science, this same group of persons, the scientists, actually establish the current meaning of the term 'science', determine what should be accepted as science, and establish also the current meaning of the term 'scientist' and decide that they themselves and those designated by themselves as their successors should be recognized

---

[1] E. T. Bell, *Mathematics, Queen and Servant of Science*, London, 1952, p. 7

as such. The cultivation of science by society relies on the public acceptance of these decisions as to what science is and who are scientists.

We are accustomed to take this consensus for granted when we are partners to it. It is commonly regarded as the obvious outcome of the fact that you could repeat and confirm any observation recorded by science. But the affirmation of this supposed fact is actually but another manner of expressing our adherence to the consensus in question. For we never do repeat any appreciable part of the observations of science. And besides, we know perfectly well that if we tried to do this and failed (as we mostly would), we would quite rightly ascribe our failure to our lack of skill. We should keep in mind also that even if we could reliably repeat the facts recorded by science, this would still not justify our acceptance of the generalizations which science bases on these facts, and would still less warrant the anterior selection of these particular facts as subjects of scientific observation. We should also take into account that even the truth of a generalization does not establish it as a part of science, since the reliability of an affirmation is only one of the three coefficients composing the scientific value of a statement. The consensus which accepts as science what it declares to be science endorses the scientific value of it as graded on the threefold scale of reliability, systematic interest, and intrinsic interest.

We see, therefore, that the consensus of scientific opinion goes far beyond an agreement concerning a common experience. It is a joint appraisal of an intellectual domain, of which each consenting participant can properly understand and judge only a very small fraction. One may well wonder how such an agreement can ever be reasonably established. I think the underlying principle is this. Each scientist watches over an area comprising his own field and some adjoining strips of territory, over which neighbouring specialists can also form reliable first-hand judgments. Suppose now that work done on the speciality of B can be reliably judged by A and C; that of C by B and D; that of D by C and E; and so on. If then each of these groups of neighbours agrees in respect to their standards, then the standards on which A, B and C agree will be the same on which B, C and D agree, and on which also C, D and E agree, and so on, throughout the whole realm of science. This mutual adjustment of standards occurs of course along a whole network of lines which offers a multitude of cross-checks for the adjustments made along each separate line; and the system is amply supplemented also by somewhat less certain judgments made by scientists directly on professionally more distant achievements of exceptional merit. Yet its operation continues to be based essentially on the 'transitiveness' of neighbouring appraisals—much as a marching column is kept in step by each man's keeping in step with those next to him.

By this consensus scientists form a continuous line—or rather a continuous network—of critics, whose scrutiny upholds the same minimum level of scientific value in all publications accredited by scientists. More than that: by a similar reliance of each on his immediate neighbour they even make sure that the distinction of scientific work above this minimum level, and right up to the highest degrees of excellence, is measured by equivalent standards throughout the various branches of science. The rightness of these comparative appreciations is vital to science, for they guide the distribution of men and subsidies between the different lines of study, and they determine, in particular, the crucial decisions by which recognition and assistance are granted to new departures in science or else withheld from them. Though it is admittedly easy to find instances in which this appreciation has proved mistaken, or at least sadly belated, we should acknowledge that we can speak of 'science' as a definite and on the whole authoritative body of systematic knowledge only to the extent to which we believe that these decisions are predominantly correct. Otherwise, scientific institutions would no longer serve the advancement of science but bring about its progressive mutilation. The title of 'scientist' (mutually granted to each other by persons thereby called 'scientists') would then gradually cease to carry its true meaning, and so would the word 'science' as used by these persons in description of their own pursuits.

Let me expand this. Suppose for a moment that all scientists were charlatans, as some certainly are; or, to make the assumption more plausible, that they were all self-deluded like Lysenko, or else either dishonest or forced to conform to the views of people who are themselves either dishonest or self-deluded, as Lysenko's followers mostly were. Or suppose that standards of scientific reliability and significance were generally so debased as they are even now in some parts of the world; or going a step beyond this, that the natural sciences were replaced altogether by the occult sciences based on cabalistic methods. There might still exist a consensus between the various specialists acknowledging each other as scientists, and mutually acknowledging also the validity and significance of their respective domains of pseudo-science, and the public might be deluded by their joint assurances to accept what they call 'science' as science. But clearly, if I knew what lay behind such a consensus, I should regard it as a consensus of rogues and fools, deceiving both each other and their public—the result of an accident or a conspiracy, and in either case devoid of any true significance.

Of course, even a consensus on a single common experience might be illusory, and if we accredit it, we do so on certain assumptions about the nature of this shared experience. But the consensus which mutually recognized scientific specialists achieve by relying on each other, and the further consensus of the public with the agreed judgment of this body of specialists, imply assumptions of a more far-reaching kind. Scientists must assume that the various domains of science are so coherent that the scientific value of work done by the several specialists within a multitude of separate provinces is in fact assessable by essentially similar standards; and that they (the scientists) can *and will* actually so supervise each other's assessments across the boundaries of their particular specialities, that they may continue safely to trust each other—even through the passage of succeeding generations—to apply these similar standards everywhere. Moreover, the public must share these assumptions if it is to grant confidence—as it does—to the whole body of science, of which it hardly knows anything at all, and to accept unseen—as we may hope it will—the future pronouncements not only of living scientists, but even of those of their successors to be accredited one day as scientists by the scientific opinion of the day.

We have here the assumptions of a cultural ideal: the ideal of a highly differentiated intellectual life pursued collectively; or more precisely, of a cultural élite actively conducting such an intellectual life within a society responsive to the intellectual passions of this élite. The acceptance of these assumptions seals a pact of mutual confidence within the community of scientists and seals the dedication of society as a whole to the support of their scientific pursuits. This dedication takes effect in the establishment of scientific institutions, set up for the advancement of science and for its dissemination throughout society under the authority of scientific opinion. Anyone who integrally belongs to the society in question will share thereby its cultural dedication and the assumptions which underlie this dedication.

The contrast in which I have set this scientific consensus with the specious coherence of a company of frauds or fools has shown that I share the assumptions underlying this consensus. My unqualified use of accreditive terms like 'scientific standards' and 'scientific values' in formulating these assumptions had already implied that I subscribed to what I was describing. I believe that tacit endorsements of this kind are unavoidable in referring to beliefs and valuations which one shares, and I shall return to this fact later on. Meanwhile, let me add that in subscribing to the assumptions and passions shared by a society in the cultivation of science I also lend—to this extent—my support to such a society. Any sociology which accredits the beliefs on which a society is founded forms a justification of this society. And if the writer is a member of the society in question, his sociology is a declaration of loyalty to it. Indeed, consistency requires that in the affirmation of socially shared values our declarations should agree with our participation in any social activities based on the assumed validity of these affirmations.[1] Yet it is precisely this consistency which renders the universal intent of such declarations suspect, since it shows that they lend support to established powers, after having been instilled in us by the very society which they vindicate. This dilemma will reappear—for reasons adumbrated in the previous section—even more acutely in the realm of civic thought.

Let me generalize meanwhile what I have said of the cultivation of science to the cultivation of other kinds of individual thought, though, conforming

---

[1] This is a simplified variant of the consistency requirement which I called 'self-confirmatory progression' and formulated as follows in Part Two, ch. 6, p. 142: 'Only by accrediting the exercise of our intellectual passions in the act of describing man and society can we form conceptions of man and society which both endorse this accrediting and uphold the freedom of culture in society.'

to the plan of this book, I can again cast only a cursory glance at these provinces of thought.

The administration of the humanities, the arts, or of the practice of various religions are all entrusted, like that of science, to a chain of authoritative specialists. The position and power of these may be institutionally established, as it is in the churches, or it may depend entirely on the respect in which they are held by their admirers and followers, as is the case with poets or painters. In all these domains there is much greater divergence of views than there is between scientists. Most countries of the Western type include different religious bodies. Moreover, apart from religion (but not excluding theology), the culture of our time is predominantly a cult of innovation. The arts, like the sciences, are most alive in the process of renewing themselves; fame is earned in the arts, as in science, by creativity. But artistic originality involves as a rule more comprehensive changes of outlook than does originality in science, and tends to produce therefore more pronounced divisions of opinion between the innovator seeking to establish his authority, and the leaders of previously established art. Accordingly, rival schools of thought, which in science are infrequent and transitory, are essential to a vigorous cultivation of modern art. And of course, even apart from this, the arts are not, and never can become, systematically coherent after the fashion of sciences. There can, therefore, exist no such clear division of labour between different kinds of artists, nor such firm consensus of opinion among them, as we have within the community of scientific specialists.

A cultural élite may be publicly subsidized or dependent on private earnings. Until the beginning of the nineteenth century scholarship and literary pursuits were largely the occupation of wealthy people, living on their private incomes. Today, however, few members of the cultural élite belong to that class and intellectual life depends correspondingly to a greater extent on the material support given to a creative minority by the mass of non-creative citizens. This raises the question whether, in paying for cultural pursuits, society is fulfilling an obligation to enlarge its intellectual possessions by the standards of its creative leaders, or is merely hiring these persons for the purpose of serving either its own amusement or some civic interest, like the moral and political edification of the people.

We can reply to this by recalling our answer to the equivalent question in respect to the advancement of science (p. 183). No important discovery can be made in science by anyone who does not believe that science is important in itself, and likewise no society which has no sense for scientific values can cultivate science successfully. The same applies to all cultural life: a society may be said to have a cultural life only to the extent to which it respects cultural excellence. As in science, this appreciation can rarely be the expression of a first-hand judgment. The humanities, the arts, the various religions, are all extensive and highly differentiated aggregates of which no one can fully understand and judge more than a tiny fraction. Yet each of us respects very much larger areas of these cultural domains. I know for example that Dante's *Divine Comedy* is a great poem though I have read very little of it, and I respect Beethoven's genius though I am almost deaf to music. These are genuine second-hand appreciations, formed in the same way in which scientists appreciate the whole of science and in which the public follows suit. Indirect appreciations of this kind are, again, the roots through which society as a whole nurtures cultural life. By following their chosen intellectual leaders, the non-experts can even participate up to a point in the works of these leaders and beyond this in the whole range of culture accredited by them.[1]

The folklore of primitive societies does not run into millions of volumes and is not subject to continuous innovations. Hence their cultural life does not require a legion of experts to administer it and much of it can be shared at first hand by everybody. Popular art and religious life is shared also in modern societies, but this is a small part of modern culture. Therefore, a modern society which does not accept cultural guidance from a set of authoritative individuals cuts itself off from any culture living within its borders. Its philistinism, deaf to original thought, renders its intellectuals homeless in their own country.

In the Western type of modern society the authority of science is firmly established through the educational system, but all other cultural authorities have to fight for public response and also contest their position against strong rivals. Members of the public may shift their allegiance from one leader to his opponent; they may change from the camp of an academician to that of some innovator, be converted to religion or abandon their faith, drop out of any particular movement and join another. Sanity forbids such shifts to be very frequent, and even so their scope is limited to choices between potential leaders. They still leave the guidance of thought to a small number of individuals, popularly accepted and rewarded as leaders of certain recognized cultural domains. Our society may be said to possess a single culture to the extent to which our cultural leaders harmoniously supplement each other; and to this extent these leaders may be said to uphold the common intellectual standards of our society: both in their own work and by guiding the public appreciation of culture and enjoining society to fulfil its cultural obligations.

Owing to the clash between different philosophies, religious or artistic movements, adherents of one persuasion may refuse to recognize any intellectual merit in those of a rival persuasion, calling them cranks, frauds or fools. People will differ accordingly also in their use of such professional descriptions as 'composer', 'poet', 'painter', 'priest' and in that of accreditive terms like 'expert', 'reputable' or 'distinguished', applied to persons claiming to be composers, poets, etc. Yet the fact that most rival leaders

---

[1] Considering that *Who's Who* contains about 15,000 names of scientists, artists, writers, etc., we may estimate that the 250 million English speaking people rely on about 20–30,000 intellectual leaders; i.e. on one in ten thousand.

share the same status in a pluralistic society demonstrates a measure of consensus in according *some* intellectual merit to *most* of them. This implies also the acknowledgment of a process of thought underlying all these rival affirmations: of a process guided throughout by standards which, though manifestly disparate, are descended from a common inheritance of values and beliefs. This belief in an autonomous process of coherent thought is (as in science) the fundamental condition for the social cultivation of thought, guided by its own standards and prompted by its own passions.

## 8. ADMINISTRATION OF CIVIC CULTURE

Such are the cultural institutions which sustain the freedom of individual thought in a free society. From these institutions we can pass on to the ideal of popular government by extending their principles to the cultivation of civic thought.

The machinery of self-government equips civic opinion with coercive powers to enforce, if necessary, any reforms of the existing mores that it holds to be right. If, therefore, opinion concerning civic matters is allowed to take shape by the same principles which effectively sustain the freedom of individual thought, civic thought will also grow freely and the power wielded by it will be the power of free thought. This is what would happen in an ideal free society. The shaping and dissemination of moral convictions should take place in it under the guidance of intellectual leaders, spread out over thousands of special domains and competing at every point with their rivals for the assent of the public.[1]

To describe the institutional framework within which moral, legal and political opinions are thus continuously re-moulded in a free society would lead us too far. Suffice it to give some of the results of this process, which has radically changed life in the free countries since the principles of social reform gained wider acceptance some 130 years ago. There has taken place a far reaching humanization of the criminal law and of the prison system, and similarly of discipline in the army and the navy, while the same changes have gone on in the schools, asylums, hospitals and within the family itself; the Factory Laws have enforced more humane conditions of employment in an immense variety of ways; new welfare institutions have been set up to provide for the sick and the aged, for the disabled, the unemployed and the slum dwellers; free education has greatly widened the prospects of poorer people's children; the legal disabilities of women, of Catholics, Jews and of the colonial peoples have been removed or at least greatly reduced; the extension of the franchise and the recognition of Trade Unions have shifted the balance of power in favour of hitherto subordinate classes. All these were moral improvements of society which

---

[1] The function of authoritative individuals is generally recognized for the interpretation of the Constitution itself in Britain.

in England's history, for example, can be traced back to a series of specific movements appealing to the public conscience; movements which had usually been evoked in the first place by persuasive individuals devoted to the advocacy of one particular reform. Such is the dynamism of the modern free society. It consists in the moral progress of civic thought, which transmits its conclusions, through the machinery of self-government, into acts of social reform. It is the practical outcome of an intellectual process, moved by its own passions and guided by its own standards.

The constitution of a free society expresses its acknowledgment of these passions and standards. Its government bows in advance to the moral consensus freely arrived at by its citizens, not because they so decide, but because they are deemed competent to decide *rightly*, as the authentic spokesman of the social conscience. I know that this runs counter to current legal positivism, which refuses to qualify in any way the ultimate authority of the 'basic norm' of a given legal structure.[1] Let me add, therefore, that reforms of law are in fact merely components of social reform. The laying down of new coercive rules proceeds within a medium of voluntary informal changes: changes in manners of intercourse, in family customs, in moral rules. Moreover, the law itself is being changed informally through new judicial interpretations; great new institutions are founded privately and the whole network of existing contractual relations renewed voluntarily, in a thousand ways.[2] Legislative reforms are embedded in these broader voluntary, private, informal modifications of society, which the new laws serve to consolidate and to provide with a new framework for ever new departures. There can be no doubt that these broader changes of civic culture, which form the dominant matrix of legislative reform, are determined by a process of thought guided by its own standards and prompted by its own passions.[3]

It might be objected that the passing of new laws is rarely unanimous, and also that in society at large civic values are not universally shared in the way in which scientific values or even artistic values are. But the difference is only superficial: the clash of contending opinions is perhaps more marked in civic matters, but even so it is restricted to contemporary affairs. Few of the innumerable social reforms carried out in Britain during the past 150 years would be repudiated today by any important minority. If a nation could not agree to this extent on its past civic achievements it would be in a state of latent civil war and could not be held to legislate freely for itself. Its self-government would be the coercive rule of a majority. The ruling class might still follow the guidance of a persistent moral impulse, as absolute rulers and dictators have sometimes done too, but the image of a society continuously re-shaping its own life in the pursuit of civic virtues freely fostered in its midst would no longer apply. We may

---

[1] H. Kelsen, *General Theory of Law and the State*, Cambridge, Mass., 1947, pp. 115–16.
[2] C. K. Allen, *Law in the Making*, Oxford, 1939, pp. 39–40.
[3] Cf. A. V. Dicey, *Law and Public Opinion in England*, London, 1905, repr. 1948.

take it, then, that in an ideal free society civic life would be continuously improved solely by the cultivation of moral principles.

## 9. NAKED POWER

But let us remember the facts of power and material ends. Though men be harmoniously guided by their agreed convictions, they must yet form a government to enforce their purpose. Civic culture can flourish only thanks to physical coercion. It is sown in corruption. We must expose now the instability of our moral beliefs in face of this fact.

It may be that strictly speaking nobody can be forced to do anything. During the past wars and revolutions many prisoners have endured tortures of utmost cruelty, steadfastly refusing to betray secrets entrusted to them or to give false evidence against innocent persons. When some yielded to torture combined with 'brainwashing', this may have meant an enforced change of personality, such as is achieved by drugs, by brain surgery or by treatments which induce a neurosis or psychosis—a change which it is not in the nature of man's will to resist. Yet we have to admit all the same that most men *can* be induced to bend their will and reluctantly to obey commands given under sufficiently serious threats: a yielding which may properly be said to be compelled by force.

Indeed, all commands issued with some kind of threat behind them are to this extent coercive, and laws must be effectively coercive, since otherwise they create injustices by rewarding the law-breakers at the expense of the law-abiders. While it is not inconceivable that laws may be enforced by mere moral disapproval, we need not consider such a remote possibility, particularly since it would hardly alter our conclusion that coercion is both possible and indispensable in human society.

It is commonly assumed that power cannot be exercised without some voluntary support, as for example by a faithful praetorian guard.[1] I do not think this is true, for it seems that some dictators were feared by everybody; for example, towards the end of his rule everyone feared Stalin. It is, in fact, easy to see that a single individual might well exercise command over a multitude of men without appreciable voluntary support on the part of any of them. If in a group of men each believes that all the others will obey the commands of a person claiming to be their common superior, all will obey this person as their superior. For each will fear that if he disobeyed him, the others would punish his disobedience at the superior's command, and so all are forced to obey by the mere supposition of the others' continued obedience, without any voluntary support being given to the superior by any member of the group. Each member of the group would even feel compelled to report any signs of dissatisfaction among his comrades, for he would fear that any complaint made in his presence

[1] Cf. Hume, 'Of the First Principles of Government', *Essays*, I, Essay IV (Green & Grosse edn., p. 110) and Dicey, *op. cit.*, p. 2.

might be a test applied to him by an *agent provocateur* and that he would be punished if he failed to report such subversive utterances. Thus the members of the group might be kept so distrustful of each other, that they would express even in private only sentiments of loyalty towards a superior whom they all hated in secret. The stability of such naked power increases with the size of the group under its control, for a disaffected nucleus which might be formed locally by a lucky crystallization of mutual trust among a small number of personal associates, would be overawed and paralysed by the vast surrounding masses of people whom they would assume to be still loyal to the dictator. Hence it is easier to keep control by force of a vast country than of the crew of a single ship in mid-ocean. And hence also it is the standard tactics of an insurrection to spread the rumour that insurrections have broken out already in other places.

This principle of naked power seems indubitably real and effective. It is difficult to imagine any exercise of power that is quite free from a coefficient of this kind, and a regime of terror may well rely preponderantly on this principle. At the same time, no continued exercise of supreme power is likely to consist merely in coercion. For no ruler (this side of sanity) could go on commanding his subjects without some public purpose in mind, nor his subjects go on living by his orders without accepting this purpose to some extent, and no dictator (unless mad) would forgo the measure of popularity which such inclination towards rational conduct would gain for his regime. We may expect, in fact, that no dictator will fail to use his coercive powers for inculcating loyalty to himself in his subjects. For if everybody can be convinced to some extent that it is right to obey and wrong to resist his power, incipient disaffection will be discouraged by a sense of being wrong; and if it is manifested in spite of this, its voice may be silenced by the mere weight of social disapprobation. A claim to legitimacy is a most formidable instrument of power. Even men like Hitler and Stalin, who had perfected to the utmost the machinery of naked power, have never ceased to supplement it by a flow of public self-justification.[1]

Attempts at self-justification will involve the acceptance of a measure of consistency in the wielding of power, according to rules and policies which might be regarded as reasonable by the governed. The more reasonable the rules appear to be, the more assured will be the government which imposed them, but also the more restricted will become in consequence the range of its decisions. Indeed, any argument—however

---

[1] The Bolsheviks fought hard to obtain the support of the Congress of Soviets, after they had captured sufficient armed power to dissolve this assembly. (Leonard Schapiro, *The Origin of the Communist Autocracy*, London, 1955, p. 68.) Hitler had been Chancellor for a month when he set in motion a series of manœuvres by which he eventually forced the Reichstag to invest him with absolute powers. Both Stalin and Hitler used their coercive powers regularly to compel expressions of popular support for themselves and continued to address assemblies of men elected at their command in order to earn their unanimous applause. Napoleon struggled throughout his career to strengthen the legality of his rule and his fall was due to the fact that he never fully achieved this aim.

mendacious and absurd—that naked power might invoke in its own support, necessarily accredits some widely acceptable principles on which the argument rests. The conversations of Stalin and Hitler reveal that, in spite of their cynicism, they were convinced of the rightness of their despotic rule and that, except while engaged in some specific piece of treachery, they interpreted the world in terms not very different from those used in their own propaganda.[1]

People under totalitarian dictatorship may bitterly dislike their rulers. But so long as these effectively prevent the formation of an independent intellectual leadership, even a universal repudiation of the official orthodoxy will produce no alternative movement of thought. In consequence, official ideologies will frequently be used automatically by people for the current interpretation of events, even though they do not support these ideologies. Totalitarianism has clearly demonstrated that no modern culture—whether individual or civic—can survive, except by the operation of authoritative institutions.

## 10. Power Politics

We have seen that even though a public power were originally based on terror, it could not fail to supplement its coercive force by persuasion, and that the thoughts cultivated for the purpose of controlling their people would inevitably gain ascendancy also to some degree over the rulers' own behaviour. Thus the abuse of a moral appeal for immoral purposes seems to confirm by its outcome the intrinsic liberating power of morality.

But the restraint which power incurs as the price of employing morality for its own coercive purposes proves only that morality is an indispensable, though self-willed, ally to power. It does not demonstrate that morality can ever control power according to its own principles; civic culture still remains dependent on force and material ends, and remains therefore suspect. Nor does the history of free societies dispel this suspicion. We see, on the contrary, how every new moral issue has evoked a clash of interests; how often moral progress had to be forced upon the privileged by the pressure of the oppressed; how the existing distribution of privilege has always granted its beneficiaries considerable powers to resist reforms that curtail their advantage, and how they have perpetuated injustice by force. It may indeed be argued (and I shall return to this point later) that since any single reform of detail must rely on the existing social structure as on its matrix, this structure and any iniquities inherent in it can never be

[1] Stalin must have believed the extorted confessions of the Kremlin doctors accused of attempting to assassinate the Soviet leaders; for he could have had no other reason for ordering the execution of these politically insignificant men who were rendering him valuable professional services. Hitler's secret extermination of the Jews, as well as his obstinate wooing of England whose attitude he could not understand, were both determined by his belief in the racial theories used in his propaganda.

Berkley, California

Minotti + The medium

fundamentally improved by any series of piecemeal reforms. We may still doubt, therefore, whether the rulers of any society, however freely self-governed, will ever observe the claims of morality beyond what is needed in order to delude their subjects (and their allies abroad) to trust their professions of morality.

This doubt goes back to antiquity; in modern times it was first revived by Machiavelli. Friedrich Meinecke, writing at the close of the First World War, traced from Machiavelli—through a series of great thinkers—the steadily growing acceptance by Continental political theory of the necessary immorality of public power, both in ruling at home and in the conduct of foreign affairs.

Meinecke interprets the ideological conflict between Germany and her opponents in these terms. He thinks that Germany was accused of immorality only because she frankly declared that Might was Right, while the Anglo-Saxon powers, who acted no less unscrupulously, continued to pay lip-service to morality. They gained an unfair moral advantage for themselves by pointing at Germany's honest professions of the power-political principles which they themselves covertly followed. Meinecke traces the origin of this situation to the realization by German thought of the inevitable sinfulness of power and the bold attempt of German philosophy to overcome this antinomy by conceiving of morality as immanent in the rise of an intrinsically superior power. He admits that the Germans were misguided by a brutalization of this philosophy to which it is liable, but believes that the Anglo-Saxons avoided a similar outcome only by turning a blind eye on the contradictions between their professions and their practice.[1]

Meinecke's account of political immoralism may serve as a landmark. He saw the First World War as the first mass movement inspired by a doctrine of violence and believing in its own intellectual and moral superiority over its moralizing opponents. Yet he did not see that this war was but a ripple before the approaching storm. In tracing the growth of the ideas of Realpolitik he does not even mention Marxism. So he could not suspect the total instability of moral principles in politics which was to manifest itself in the revolutions of the twentieth century.

## 11. THE MAGIC OF MARXISM

The propagandistic appeal of Marxism is the most interesting case of (what might be called) the moral force of immorality. For it is the most precisely formulated system having such a paradoxical appeal, and this self-contradiction actually seems to supply the main impulse of the Marxian movement. Isaiah Berlin, in his biography of Marx, shows him in the act of exercising his propagandistic genius by means of this

---

[1] F. Meinecke, *Machiavellism*, London and New Haven, 1957, Book 3, ch. 5.

self-contradictory principle—a prophetic idealism spurning all reference to ideals:

> The manuscripts of the numerous manifestos, professions of faith and programmes of action to which he appended his name, still bear the strokes of the pen and the fierce marginal comments, with which he sought to obliterate all references to eternal justice, the equality of man, the rights of individuals or nations, the liberty of conscience, the fight for civilization, and other such phrases which were the stock in trade . . . of the democratic movements of his time; he looked upon these as so much worthless cant, indicating confusion of thought and ineffectiveness in action.[1]

And indeed, it is not *in spite of* this contempt for justice, equality and liberty, but *because of* it that Soviet Russia is accepted by many as the true champion of these same ideals in the fight against the very nations openly professing them. As Hannah Arendt has rightly observed 'Bolshevik assurances inside and outside Russia that they do not recognize ordinary moral standards, have become a mainstay of Communist propaganda. . . .'[2]

Why should so contradictory a doctrine carry such supreme convincing power? The answer is, I believe, that it enables the modern mind, tortured by moral self-doubt, to indulge its moral passions in terms which also satisfy its passion for ruthless objectivity. Marxism, through its philosophy of 'dialectical materialism', conjures away the contradiction between the high moral dynamism of our age and our stern critical passion which demands that we see human affairs objectively, i.e. as a mechanistic process in the Laplacean manner. These antinomies, which make the liberal mind stagger and fumble, are the joy and strength of Marxism: for the more inordinate our moral aspirations and the more completely amoral our objectivist outlook, the more powerful is a combination in which these contradictory principles mutually reinforce each other.

Marxism achieves this sophisticated union by a primitive mental operation which Levy Brühl has called 'participation'.[3] For primitive thought, in a lion tearing a villager to pieces there participates the man's envious neighbour; plagues and fatalities are always endowed with the evil inten-

---

[1] I. Berlin, *Karl Marx*, Oxford, 1939, p. 10.

[2] Hannah Arendt, *The Burden of Our Time*, London, 1951, p. 301. See also G. A. Almond (*The Appeals of Communism*, Princeton, 1954, p. 22), where a quantitative analysis of the chief propagandistic writings of Lenin and Stalin shows that 94 per cent to 99 per cent of the references to the communist party and its activities describe it as seizing, manipulating and consolidating power. This is true even of Stalin's *History*, which covers a substantial part of the period during which the party has held power in the Soviet Union. The self-contradictions of Marxists mentioned in the text have been frequently pointed out, but it has not been realized that it is these very contradictions that generate the convincing power of the doctrine. For some recent references see notes to pp. 230 and 239. Throughout this chapter the term 'Marxism' is used rather for describing a current ideology, than the hypothetical beliefs of Marx himself.

[3] 'Participation' as defined by Levy-Brühl and 'immanence' identified with it in my text, are merely extensions of the semantic relation between something that means another thing and the other thing meant by it. In this case the meaningful thing is not a symbol but a striking event which 'assimilates' what it means to the extent of affirming the presence of this thing within itself.

tions of someone who sent them. The higher religions sometimes interpret misfortunes as God's retribution for past offences. More recently,
historicism has replaced God by an Historic Necessity, credited with the
easier (if even more inscrutable) role of achieving what is historically
fitting. In each case we have an active principle immanent in a manifest
event; the relation between the immanent and the manifest being the same
as between a purpose and its fulfilment, except that the connection is
here either supernatural or otherwise left undefined.

To this general type of operation—and in particular its modern variant,
historicism—Marxism adds two features which greatly enhance its scope
and convincing power. First, the active principle in this case is an aggregate
of limitless moral demands, demands which have suddenly spread all over
the globe, finding response even among millions of people who hitherto
had lived in immemorial acceptance of exploitation and squalor—while
at the same time, a strictly 'scientific' verdict is invoked to identify the
events which are to realize and fulfil these demands. Secondly, the mechanism of Marxism is amplified by working in two opposite and yet mutually
correlated directions. In a class society it is material interests which are
regarded as immanent in moral aspirations: while in a socialist state the
opposite holds: morality is immanent in the material interests of the
proletariat.

This duality may look like yet another paradoxical feature of Marxism,
but actually it can be seen to arise directly from the process by which immanent principles are injected into manifest events. To see this happening
you must imagine that you are filled from the start—as Marx was—with a
passion for Socialism and a horror of Capitalism. Looking in this light
on the ideals of liberty, justice, brotherhood, you will observe, for example,
that the Code Napoléon, based on these principles, was supremely
effective in destroying the feudal order and in opening the way for the bourgeoisie with its system of private enterprise throughout Europe. You will
also note that it has remained the guardian of the capitalist order ever since.
Bourgeois ideals will appear, therefore, as a mere superstructure of capitalism, in its opposition both to a feudalism whose rule it has subverted and
to the proletariat, whose enslavement it tries to perpetuate. Bourgeois
interests will appear to be immanent in bourgeois moral ideals. This is the
first kind of immanence, the *negative* branch of Marxism.

Think now, on the other hand, of Socialist revolutionary action. You are
filled with a passionate desire to see the workers overthrow Capitalism and
establish a realm of liberty, justice and brotherhood. But you cannot
demand this in the name of liberty, justice and brotherhood, for you
despise such emotional phrases. So you must convert Socialism from a
Utopia into a Science. You do so by affirming that the appropriation of
the means of production by 'the proletariat' will release a new flow of
wealth now entrammelled by Capitalism. This affirmation satisfies the moral
aspirations of Socialism, and is accepted therefore as a scientific truth by

229

those filled with these aspirations. Moral passions are thereby cast in the form of a scientific affirmation. This is the second kind of immanence, the *positive* branch of Marxism. By covering them with a scientific disguise it protects moral sentiments against being deprecated as mere emotionalism and gives them at the same time a sense of scientific certainty; while on the other hand it impregnates material ends with the fervour of moral passions.

One can now see that both branches of Marxism operate by denying to morality any intrinsic force of its own and that they yet both appeal in this very act to moral passions. In the first case we are presented with an analysis of bourgeois ideals in terms of immanent bourgeois interests, and because the hidden motivation of this analysis is a condemnation of capitalism, the analysis turns into an *unmasking* of bourgeois hypocrisy. Since this analysis of moral claims in terms of material interests applies quite generally, it might be thought to discredit also the moral motives of those who do the unmasking. But these motives are safe against unmasking, since they remain undeclared. Indeed, acting through the unmasking of bourgeois ideologies, they arouse powerful moral passions in others—without ever pronouncing any moral judgment. Their propagandistic effect is achieved precisely by enunciating the unmasking in purely scientific terms, which are thus immune against suspicion of a moralizing purpose.

These supposedly scientific assertions are, of course, accepted only because they satisfy certain moral passions. We have here a *self-confirmatory reverberation* between the *theory* of bourgeois ideologies and the concealed *motives* which underlie it. This is the characteristic structure of what I shall call a dynamo-objective coupling. Alleged scientific assertions, which are accepted as such because they satisfy moral passions, will excite these passions further, and thus lend increased convincing power to the scientific affirmations in question—and so on, indefinitely. Moreover, such a dynamo-objective coupling is also potent in its own defence. Any criticism of its scientific part is rebutted by the moral passions behind it, while any moral objections to it are coldly brushed aside by invoking the inexorable verdict of its scientific findings. Each of the two components, the dynamic and the objective, takes it in turn to draw attention away from the other when it is under attack.

We can see that this structure underlies also a logical fallacy exposed by the academic critics of Marxism, and explains why the fallacy survives its exposure. The critics say that no political programme can be derived from the Marxian prediction of the inevitable destruction of Capitalism at the hands of the proletariat. For it is senseless to enlist fighters for a battle which is said to be already decided; while if the battle is not yet decided, you cannot predict its issue.[1] But within a dynamo-objective

---

[1] This is about how A. J. Ayer has recently put the matter (*Encounter*, 5 (1955), p. 32). A year before that John Plamenatz had summed up his analysis epigrammatically in *German Marxism and Russian Communism* (London, 1954), p. 50, as follows: '. . . what-

coupling, the logical objection against using a historical prediction as an appeal to fight for the certain outcome of history no longer arises. For the prediction is accepted only because we believe that the Socialist cause is just; and this implies that Socialist action is right. The prediction implies therefore a call to action.

But there is more to be added here. If our sense of bad faith merely caused us to disguise our thirst for righteousness in the erudite terms of a specious sociology, the masquerade would perhaps be merely pitiful. Unfortunately, moral passions undergo a fateful change when decked out as scientific statements. I have hinted at this change already when saying that any moral objection against Marxist action can be brushed aside by pointing to its 'scientific' correctness. We can see what has happened here: when transposed into equivalent scientific affirmations, the moral motive of Socialism was torn from its original moral context. It became an isolated passion, inaccessible to moral considerations. This is fanaticism—a fanaticism fixed upon the materialistic equivalents of the original moral passion, that is upon 'the interests of the working class' or, more precisely, upon the coercive powers of those who are held to represent the interests of the working class. It is a fanatical cult of power.

This explains not only the deliberate unscrupulousness of modern totalitarianism, but also the moral appeal of its declared resolve to act unscrupulously. For this resolve is taken to certify that its power embodies righteousness, and may therefore acknowledge no higher obligation than that of defending its own supremacy, which it must do at all costs. Those who rule in its name are entitled to scorn mercy and honesty, not simply for reasons of expediency (as Machiavelli would already have allowed them to do) but on account of their moral superiority over the emotionalism, hypocrisy, and general woolliness of their moralizing opponents. Thus sceptics who deny with contempt the reality of all moral motivations will rally fanatically to the moral support of naked power.

Once accepted, Marxism eliminates the eternally menacing discrepancy between the universalist claims of morality and its actual dependence on power and profit. Marxism does so by denying to morality its claims *qua* morality, while offering it instead an immanent form of operation within a specified political force. Universality is to be achieved by this inherently righteous force through its inevitable conquest of the world.

We see then that Marxism is falsely accused of materialism: its materialism is a disguise for its moral purpose. It is true that by their materialistic

ever the relation of science and socialism as parts of the life of one man, what he can never be is a scientific socialist. Not even if his science predicts what his socialism approves. "Scientific socialism" is a logical absurdity, a myth, a revolutionary slogan, the happy inspiration of two moralists who wanted to be unlike all moralists before them.' In a recent book, *The Illusion of an Epoch* (London, 1955), Professor H. B. Acton has re-examined the whole question once more in great detail, only to conclude: 'The Marxist can derive moral precepts from his social science only to the extent that they already form, because of the vocabulary used, a concealed and unacknowledged part of it' (p. 190).

disguise these aspirations are torn out of their moral context, and are harnessed to the service of material aggrandizement and political violence. But this does not transform the underlying Socialist dynamics into a desire for comfort. The fervour of social enterprise has remained the emotional justification of Communist governments. Hence their persistent efforts to fill all economic activities with high moral significance; hence their gigantomania, their neglect of the most desperate popular needs—e.g. for better housing—in favour of ornate skyscrapers and underground marble halls; hence their whole curious economic system which revels in production and shies away from consumption. We in the West watch hopefully for every sign of true materialism in Soviet Russia. For if the regime once really consented to pursue material advantages, it would have lost its fanaticism; love of comfort may be ignoble, but one may trust it to be accommodating.

The moral appeal of immorality has been effective also in other mass movements of our time. Meinecke had detected an early form of it in Pan-Germanism, and Hitler's rise has confirmed that diagnosis in a diabolically complete way. Hitler greatly profited from the Bolshevik example, but his movement was rooted primarily in German Romantic nihilism. This doctrine taught that an outstanding individual is a law unto himself and may, as a statesman, unscrupulously impose his will on the rest of the world, and that a nation has likewise the right and the duty to fulfil its 'historic destiny' irrespective of moral obligations. Such teachings contradict the universal claim of morality, just as the Marxist-mechanistic image of man does. They identify morality with the self-fulfilment of the individual or the nation, and this emotionally charged utilitarianism can unite with a fierce patriotism all the inordinate social hopes of our age. It was able eventually, therefore, to embody both in the aim of a German world government under Hitler.

The immanence of great moral passions in Hitler's programme explains the strong moral appeal which it made by its very unscrupulousness—for example to many members of the German Youth Movement.[1] Whenever fanaticism combines with cynicism we must suspect a dynamo-objective coupling, and its presence is confirmed if we find that cynicism is making a moral appeal. Hitler's frenzy was primarily evil, but its appeal to the German youth was moral: they accepted evil actions as a moral duty. Their response was determined by the same convictions which Marx had held about the nature of moral motives in public life. They believed that such motives were mere rationalizations of power, and that power alone was real. Hence their disgust of moralizing, and their moral passion for unscrupulous violence.

In a tentative study published some years ago, I called this principle a

---

[1] Crankshaw, *op. cit.*, p. 28 quotes Himmler's highly moral exhortations to massacre all Jews. The author concludes his book (p. 247) by calling this attitude, widespread among the Gestapo, 'idealism gone rotten'.

'moral inversion'.[1] Such inversion can, of course, never be completely realized. No regime, however fanatical, can act without accepting any overt moral restraints. I have referred to this already in describing the way naked power is bound to support—and at the same time to limit—itself by the exercise of persuasion. On the other hand, an element of moral inversion may be thought to be operating in every harsh exercise of power. If 'hard cases make bad law', then it would seem that the best government must occasionally commit injustices. This is true, but occasional concessions to expediency leave unimpaired the moral principles from which they deviate—just as the principles of moral inversion are not denied by the mere fact that occasional concessions are made to overt morality.

## 12. Spurious Forms of Moral Inversion

We must guard also against assuming that a materialistic interpretation of moral motives must always result in moral inversion. Far from it. Spurious forms of moral inversion are quite common. Men may go on talking the language of positivism, pragmatism, and naturalism for many years, yet continue to respect the principles of truth and morality which their vocabulary anxiously ignores.

Take for example the text of Freud, in which he interprets culture in the light of his psychology.[2] Towards the end he writes emphatically: 'This alone I know with certainty, namely that man's value judgments are guided absolutely by their desire for happiness, and are therefore merely an attempt to bolster up their illusions by arguments.'[3] But at the opening of the same essay he had expressed his deep respect for Romain Rolland, for spurning the false standards commonly applied by men who seek power, success and wealth, and who admire these achievements in others, while they fail to appreciate the true values of life;[4] and again at another point he had declared himself for the ideal of a generous society in which 'all work together for the happiness of all'.[5]

We can see the dynamo-objective coupling operating here on the same lines as in Marxism. A utilitarian interpretation of morality accuses all moral sentiments of hypocrisy, while the moral indignation which the writer thus expresses is safely disguised as a scientific statement. And on other occasions, these concealed moral passions reassert themselves, affirming ethical ideals either backhandedly as a tight-lipped praise of social dissenters, or else disguised in utilitarian terms.

---

[1] *The Logic of Liberty*, Chicago and London, 1951, p. 106. I have shown there also that moral inversion is but the consolidation of a pseudo-substitution, i.e. the transformation of a spurious moral inversion into an actual one. The view that a free society is one accepting the service of truth and justice, and that totalitarianism is the outcome (by inversion) of a scepticism denying intrinsic force to the ideas of truth and justice, was first outlined in my *Science, Faith and Society* (1946).

[2] S. Freud, *Das Unbehagen in der Kultur.*  [3] Freud, *ibid.*, Section VIII.
[4] Freud, *ibid.*, Section I.  [5] Freud, *ibid.*, Section II.

This prevarication of the critical mind in its encounter with morality can be traced back to antiquity. Thucydides records unwittingly how the Athenians at one moment affirm that there is only one law of God and man which is 'to rule wherever one can'; how they sneer at the hypocrisy of the Spartans who likewise pursue their self-interest, but cloak it with the mantle of justice and honour—while at the next moment the same Athenians draw a sharp contrast between the path of self-interest which leads to safety, and the path of justice and honour which involves danger. Pitifully groping for tangible terms, the great Athenian's love of Athens' greatness falls back (in the Funeral Oration) on vaunting the unrivalled size of its enterprises.

Since the eighteenth century we have again seen many hardened utilitarians nobly upholding their logically unaccountable moral convictions—but only in the twentieth century has popular thought been permeated by this internal contradiction. Today our moral judgments are quite generally without theoretical protection. They may disguise themselves as a sociology of 'aggressiveness' or 'competitiveness' or of 'social stability', etc., and may advocate in these terms more kindness, generosity, tolerance, and brotherhood among men. The public, taught by the sociologist to distrust its traditional morality, is grateful to receive it back from him in a scientifically branded wrapping. Indeed, a writer who has proved his hard-headed perspicacity by denying the existence of morality will always be listened to with especial respect when he does moralize in spite of this. Thus the scientific disguise of our moral aspirations may not only protect their substance against destruction by nihilism, but even allow them to operate effectively by stealth. This is how great reformers like Bentham or Dewey have been able to use their utilitarianism for moral purposes.

To recognize the existence of moral inversion is to acknowledge moral forces as primary motives of man; it is to deny that 'sublimation' underlies (as Freud thought) the creation of culture. Of course, moral forces are elicited and shaped by education, even as man's intelligence or artistic talent is evoked by education. But this does not imply that morality is a mere rationalization of self-interest, or that science is a 'sublimation' of sexual curiosity. On the contrary, the Freudian interpretation of morality is itself but a spurious form of moral inversion. It forms part of the expurgation of modern language which substitutes objectivist—and preferably appetitive—terms for candidly moral ones.

But it is dangerous to rely on it that men will continue indefinitely to pursue their moral ideals within a system of thought which denies reality to them. Not because they might lose their ideals—which is rare, and usually without serious public consequence—but because they might slip into the logically stabler state of complete moral inversion. For the objectivist masquerade can go on only so long as the moral convictions whose internal instability it bolsters up, remain comparatively peaceable. A great upsurge of moral demands on social life, such as arose at the end

of the eighteenth century and has since flooded the whole world, must seek
a more forcible expression. When injected into a utilitarian framework it
transmutes both itself and this framework. It turns into the fanatical force
of a machinery of violence. This is how moral inversion is completed:
man masked as a beast turns into a Minotaur.

### 13. The Temptation of the Intellectuals

The moral appeal of a declared contempt for moral scruples is explained here in terms of a moral inversion. An analogous explanation
will resolve yet another paradox: the fact that Stalin's regime was acclaimed by eminent Western writers and painters whose very works were
condemned and suppressed by that regime. And indeed—as Czeslaw
Milosz has shown—its appeal was actually due in part to its proclaimed
disgust with modern art and literature, and to its determination to make
all cultural pursuits subservient to the state. Milosz records from his own
experience in Poland that these sentiments and policies formed part of
the temptation offered by Marxism to the Polish intellectuals.[1]

To understand this we must consider, first, that unmasking and impreg-
nating—the negative and positive operations of Marxism—can be applied
to every form of thought in the transition from Capitalism to Socialism.
Just as the bourgeois ideals of freedom and democracy are unmasked,
while a party-dictatorship is endowed instead with the quality of being
intrinsically free and democratic, so also bourgeois art and literature are
unmasked, and the glorification of Socialism is endowed instead with the
values of art and literature. All cultural life is subjected to a similar transformation, which surrenders it totally to the interests of the Socialist state
at the discretion of its absolute rulers. This process accords with the logic
of inversion. But this fact does not explain altogether why such an inversion has appealed to a notable number of intellectuals in free countries,
pursuing vocations which totalitarianism discredits and suppresses.

A first clue to this enigma is suggested by the word 'unmasking'.
Socialism was not alone in rebelling against bourgeois domination in the
nineteenth century, nor was scientism the only weapon for attacking
bourgeois ideals. Allied to these was a general alienation of the intellectuals. The joint effects of the romantic and scientific movements engendered a modern cultural nihilism which repudiated the existing society
as comprehensively as Marxism did. This happened when the excessive
moral aspirations of modern man were disappointed by the normal complacency, selfishness and hypocrisy of man, and these shortcomings were
accounted for by interpreting morality as something which people obey
only if they cannot evade it. Once more—as in Marxism—moral nihilism

---

[1] Czeslaw Milosz, *La Grande Tentation*, published by the Congress for Cultural
Freedom, Paris, 1952. The argument is enlarged in Czeslaw Milosz, *The Captive Mind*,
New York, 1953.

is the mark here of exceptionally strong moral passions. Turgenev portrayed this in the student Bazarov, the literary archetype of philosophic nihilism.

Though philosophic nihilists were radical individualists, they naturally tended to sympathize with revolutionary movements aiming at the total destruction of society. Even so, the fact that many of them would go so far as to give their fervent support to totalitarian governments hostile to their own vocation as intellectuals, still remains to be explained. It can be understood only in its historical setting.

We must acknowledge that personal nihilism has served for a century as an inspiration to literature and philosophy, both by itself and by provoking a reaction to itself. A loathing of bourgeois society, a rebellious immoralism and despair, have been prevailing themes of great fiction, poetry and philosophy on the continent of Europe since the middle of the nineteenth century. Anti-philistinism, which bred the modern bohemian, has also stimulated in him a fierce originality which has renewed the fine arts by a profusion of masterpieces unsurpassed in any previous period of history.

But these triumphs left their authors mortified by self-doubt. Their hatred of the established culture had spread (as in Marxism) into an attack on the very status of man and of human thought. Peer Gynt, at the end of his pilgrimage of pretences, recognizes himself in the image of an onion: leaf after leaf of self-dramatization is peeled off, leaving nothing at the core. The bourgeois encyclopaedists Bouvard and Pécuchet lose themselves in a labyrinth of inanities. Musil's 'Man Without Properties' has ceased to be, for he thinks about life instead of living it. The futile regress of 'thoughts, thoughts, about thoughts, thoughts about thoughts about thoughts' exhausts Sartre's Mathieu in *L'Age de Raison*; yet a totally unreflecting man like the 'Stranger' of Camus is equally cut off from reality, imprisoned in his private world. The destruction of all meaning in *La Nausée* is the ultimate point of this progression.

We can then no longer say anything in good faith, and all rational action becomes a lifeless banality; violence alone is still honest, but only gratuitous violence is authentic action. Having arrived at this stage, the modern intellectual will include himself in his nauseated contempt for the moral and cultural futility of his time. Having rendered the universe utterly meaningless, he himself dissolves in a universal wasteland.

If the intellectual is now attacked from the flank by Marxist unmaskers, who will lump him together with the bourgeoisie, his position is very precarious. His own growing consciousness of living in a spiritual desert tends to re-echo the Marxian analysis of his own art and science as mere super-structures of a contemptible Capitalism. Moreover, any resistance to this attack would tend to prove its justice by forcing him into partnership with the bourgeoisie, and it would also threaten to deprive him of that anti-bourgeois status on which his self-respect is founded. This dilemma suffices by itself to account for the surrender of men like Sartre, Picasso and Bernal, to a philosophy which denied the very existence of their intellectual pursuits; the more so, since—under the protection of their own bourgeois governments—they could happily continue to cultivate these pursuits for the time being.

And here we reach the turning point. The philosophic nihilist's hidden moral passions are always available for political action if this can be based on nihilistic assumptions. He can safely indulge his moral passions by accepting the intrinsic righteousness of an unscrupulous revolutionary power. Injected into the engines of violence, his humane aspirations can at last expand without danger of self-doubt and his whole person responds joyfully to a civic home of such acid-proof quality. At last, he is engaged, he is safe.

Admittedly, the artist or scientist will still find it difficult to accept the dreary cultural aims of a Communist dictatorship as the true fulfilment of his vocation. Yet he may try to overcome his revulsion for reasons that are not altogether base. For he is relieved thereby from belonging to 'the dying culture of a rotten society' or from not belonging to any society. He may also feel that a subservient role in a Communist society can be only temporary. For ultimately, the triumph of Historic Necessity must fulfil the needs of the mind as those of the body; and even in the meantime, often no more is demanded of him than an occasional lip-service to the official cultural policy.

Besides, the temptation is great to replace the standards which the artist has set himself, by an objective rightness immanent in historic necessity. Such rightness will appear self-evident. For within a dynamo-objective coupling, a power may prove its historic necessity by the mere fact of its victory, and a cultural standard ordained by such a power must appear inherently right. Its teachings could be doubted only by breaking up the fundamental dynamo-objective coupling on which the whole Communist universe rests. These teachings, therefore, offer a firm framework to the intellectual's yearning for objective standards safe against self-doubt.[1]

## 14. MARXIST-LENINIST EPISTEMOLOGY

Since the rise of Greek philosophy in the fifth century B.C., men have been considering the possibility of systematically doubting what they believed in. Marxism is a relatively stable structure in which moral aspirations can be saved from self-doubt at the price of fixing them to the pursuit of a certain set of material ends. But a similar fixation seems to operate less successfully for artistic passions. While the people of the Soviet Union under Stalin did not lack a sense of moral purpose, they

---

[1] 'The successes of Communism among the intellectuals were due mainly to their desire to have value guaranteed, if not by God, at least by history.' Czeslaw Milosz in *Confluence* (Harvard), **5** (1956), p. 14.

were bored with the official artistic products. And the attempt to identify the search for truth with the advancement of Soviet Communism has met with even greater difficulties. This had many reasons.

In spite of Hume's scepticism and its antecedents, going back to ancient Pyrrhonism, there was no self-doubt among scientists in the modern free societies of the twentieth century. On the contrary, belief in science stood supreme as the only belief that remained practically unchallenged. Indeed, according to the positivist view widely disseminated since Comte, all human thought was seen engaged in a humble pilgrimage towards scientific perfection, and to Marx and Engels natural science was the archetype of objective truth: to them science was definitely *not* an ideology to be un-masked now, and to be later identified with the victory of Socialism. But once the fundamental dynamo-objective coupling is firmly established for moral passions, it inevitably tends to become extended to science on the same lines as to artistic pursuits. The neo-Marxian theory of science first rose to importance around 1930 and became within the subsequent decade the official doctrine of the U.S.S.R. under Stalin. At first it was limited to re-interpreting the history of science, showing that each step in its progress occurred in response to practical needs. To claim indepen-dent status for pure science was ridiculed as mere snobbery.[1] Then, from the unmasking of science as being really technology, there followed the glorification of technology as being really science. And since technology achieves material welfare, it was accepted as part of progress and of Socialism itself; so that the pursuit of science became at last embodied in the advancement of Socialism.

So far this was harmless nonsense. But soon unmasking became more virulent. It began with scattered sniping at the more modern developments of 'bourgeois science', in relativity, quantum mechanics, astronomy, psychology, and it culminated in the campaign against Mendelism. The new position was finally established when in August, 1948 Lysenko tri-umphantly announced to the Academy of Science that his biological views had been approved by the Central Committee of the Communist Party and members rose as one man to acclaim this decision.

The universality of science was now definitely repudiated. The claims of bourgeois science to universal validity were unmasked as deceptive ideology, while Soviet Science was directed to rely frankly on its partisan or class character. Owing to the dual mechanism of Marxism, the doctrine that all science is class science served simultaneously both to discredit bourgeois science and to accredit socialist science. Moreover, in serving the Party, science recovers—in a new sense—a claim to univer-sality: the universality of truth is replaced by the inherently righteous and therefore historically inevitable victory of a future Communist world-government.

[1] Bukharin explained to the author, on the occasion of a visit to Moscow in March, 1935, that pure science, as distinct from technology, can exist only in a class society.

The dual meanings of 'objectivity' and 'partisanship' in this method of accrediting Soviet science are self-consistent. The claims of bourgeois science to objectivity and universal validity are unmasked as false pretences on the grounds that no affirmation of science, history, or philosophy can be objective and that in reality they are always partisan weapons. At the same time, Marxism claims to have made politics into a science that bases every political action on a strictly objective assessment of the social conditions in which it has to operate, and the unmasking of bourgeois objectivity as partisan is itself an example of this Marxist objectivity. But such objectivity does not claim universality, for it would contradict itself if it claimed—for example—that the bourgeoisie could be persuaded to accept it as objective. Marxism claims for itself therefore to be objective only in the sense of being a weapon of proletarian partisanship. Neither 'objectivity' nor 'partisanship' is either right or wrong, it is only Socialism that is right (i.e. rising) and Capitalism that is wrong (i.e. decaying). The demand made by Stalin's regime on Soviet scholars to eschew objectivity (in the sense of universal validity) and to be guided instead by Socialist partisanship, is therefore quite consistent with the Marxist's own claims to objectivity.[1]

The strict application of this theory of knowledge would suppress natural science, with the exception of the narrow area in which pure science overlaps with technology. I have spoken of this outcome of the Laplacean programme in more general terms before. We can see now that the radical utilitarianism resulting from an objectivist view of man does not produce this result by itself. For its logic often remains mercifully suspended. Only when great moral aspirations, aiming at the radical transformation of society, are injected into the mechanistic idea of man, are engines of power engendered which press for the fulfilment of this logic. Even so, the attempt may remain abortive: the intellectual passions of scientists may rebel successfully, and reduce the effects of totalitarianism over scientific thought to a verbal disguise of its proper standards. In fact, —even in biology—a brief incantation of Marxist phrases has usually sufficed to secure immunity from Marxist doctrine for the substance of a scientific paper in Soviet Russia.

## 15. Matters of Fact

We see now that throughout my previous text, in which I spoke confidently of such things as science and art as forming part of our culture, and of law and morality preserving justice and decency, I have been

---

[1] Bochenski, *Der Sowietrussische Dialektische Materialismus*, Bern, 1950, quotes (p. 142) the Soviet writer M. D. Kammari (1947 and 1948) for the view that Marxism is objectively true because the true interests of science coincide with the interests of the proletariat and with the objective movement of history. But Bochenski himself condemns Marxism as manifestly self-contradictory (pp. 156-7). Sidney Hook, in *Marx and the Marxists* (New York, 1955), points out the same self-contradiction (pp. 45-6).

begging some decisive questions. I was referring to 'bourgeois' science, 'bourgeois' art, and generally 'bourgeois' culture, law, morality, justice, etc., which are not acknowledged as genuine science, art, culture, law, morality, justice, etc. by their Marxist-Stalinist critics, but are condemned by them as corrupt, objectivist, idealist, cosmopolitan, formalist or undemocratic. They deny the whole set of standards which I took for granted when speaking of science, art, culture, law and morality, and reduce the intellectual and moral passions upholding these standards, which I have agreed to share, to the status of an illusory subjectivity. The instability of these standards in the light of critical reflection is to them no source of anxiety but of triumphant satisfaction. The consummation of this instability, which looms to me as the final self-destruction of the human mind, would be to them but the final unmasking of my idealistic deceptions. Within a society based on Dialectical Materialism, the forces of coercion, anchored to the centre of a supreme power, become in fact the agents of valid appreciation. If, then, standards are seen to be upheld by force, this no longer makes them appear questionable but marks them instead with the stamp of authenticity.

Nor does this process of mental inversion stop here altogether. It inevitably undermines the very conception of facts—of ordinary matters of fact. Remember that the overwhelming part of our factual beliefs are held at second hand through trusting others; and that in the great majority of cases our trust is placed in the authority of certain persons, either by virtue of their public office or as our chosen intellectual leaders. The establishment of public facts outside science is entrusted, in free societies, to newspapers, parliaments, and law courts. Their fact-finding is continuous with that of sociologists, historians, and scientists, and is granted strong presumptive credit also by the whole of society, even though there are always dubious cases in which rival affirmation will compete for public acceptance. As in science, this system of shared beliefs relies on a chain of overlapping areas, within each of which a few authoritative persons can keep watch over each other's integrity and their sense of what is important. A society affiliated to such a network of mutual confidence may be said to maintain a certain standard of 'factuality'—provided that one accepts its methods of fact-finding.[1]

We know, of course, that even people whose conceptions of the nature of things otherwise coincide, may be fundamentally divided in respect to the reality of certain facts. Antagonists on either side of a great scientific controversy do not accept the same facts as real and significant. A society believing in magic, witchcraft and oracles, will agree on a whole system of facts which modern men regard as fictitious. Similar logical gaps could be found between standards of factuality prevailing in different periods of European history. But I will keep here to the effects of contemporary political dynamics on the accrediting of matters of fact.

[1] 'Factuality' is Hannah Arendt's expression.

240

The widely extended network of mutual trust, on which the factual consensus of a free society depends, is fragile. Any conflict which sharply divides people will tend to destroy their mutual trust and make universal agreement on facts bearing on the conflict difficult to achieve. In France the Third Republic was shaken to its foundations by a question of fact: the question whether Captain Dreyfus had written the 'bordereau'. In Britain the dispute over the genuineness of the 'Zinoviev Letter', as in the U.S. the trial of Alger Hiss, aroused popular conflicts which made it impossible to agree universally on the facts of these matters.

Such temporary and partial failures of factuality may of course be excused as passing excesses of political passion. But under totalitarianism we can see factuality reduced to the extent of allowing the State to fashion public facts almost at will, as it suits its own interests. These powers to spread falsehoods are due to some extent simply to the government's monopoly of public utterances, backed by terror; but such coercive powers do not account for the currency gained by these falsehoods abroad. Any willing acceptance of these facts is evidence of a persuasiveness of their own, which must be assumed to be effective also in gaining currency for them within the territories under coercion. This indicates a corruption of the very principles of factual evidence, involving a wholesale shift of the usual presumptions which underlie the process of fact-finding. It is only when our sense of reality has already been gravely impaired by such a shift, that we become receptive to downright clumsy falsifications.

A modern revolutionary government aiming at the total renewal of society inevitably sets off this change by severing all ties with its opponents. Whoever is not its unconditional supporter is held to be its mortal enemy. The dictatorship thus creates a situation in which any dissenter must in fact become its mortal enemy, and this justifies unlimited suspicion. When all open dissent is eliminated, disaffection can manifest itself only in trifles, and hence the secret police must be allowed to construe trifles as potential conspiratorial acts. The presuppositions of such investigations become analogous to those governing the Freudian analysis of a neurotic. On the assumption of an Oedipus complex, the patient's every word and action, whether uttered or unspoken, done or undone (and even events in which he became involved by accident), can be interpreted as expressing his hidden hostility to his father. Similarly, once you assume that any trifles may be interpreted as a sign of disaffection which, in its turn, may be construed into an act of high treason, the methods of fact-finding practised in Stalin's prisons will appear to have been altogether appropriate to the purpose. Even the exercise of physical pressure bordering on torture will become inevitable—for the same reasons which made torture indispensable to the Inquisition. Accusations concerning a man's hidden intentions cannot be regarded as firmly established unless the accused eventually admits them, and for this he must be morally, intellectually and physically broken. The extorted confessions of others confront those still resisting pressure with an increased persuasive force and thus extend further the fictitious universe established by violence or sophistry.

This process of fashioning public facts in the interests of the state will naturally receive support from a scholarship conceived as a political weapon. Historians will supplement accusations of recent subversive activities by suitably re-interpreting the part played by the accused at earlier times of history. Stories which otherwise might seem fantastic will be rendered plausible in view of the ruling party's conspiratorial experience. There is nothing absurd in accusing a veteran communist of having always been in the pay of the police, since a Malinovsky, for many years Lenin's most trusted fellow conspirator and leader of the Bolshevik fraction in the Duma, could turn out later to have been a police spy during all that time.[1]

In every modern country, national prejudice tends to obfuscate the establishment of public facts of political interest. In a free society this tendency is counteracted by the rivalry of opinions, which will maintain a universe of true facts so long as people can mutually trust each other to observe a proper level of factuality in drawing their conclusions from contradictory arguments. The élite of a modern revolutionary party is trained, on the contrary, to exercise its political bias to the utmost. 'Its members' whole education' (writes Hannah Arendt) 'is aimed at abolishing their capacity for distinguishing between truth and fiction. Their superiority consists in the ability immediately to dissolve every statement of fact into a declaration of purpose.'[2] Such dynamism, backed up by terror, would suffice by itself to loosen the roots in reality of all officially alleged facts, and to separate revolutionary opinion by a logical gap from that of their opponents. Yet this propaganda would remain comparatively ineffectual, but for the parallel effects of terror and secrecy in creating situations which lend colour to every conceivable suspicion. At this point facts relevant to politics cease altogether to exist, in the sense that one can only choose between either accepting no facts, or accepting some at one's pleasure on manifestly insufficient evidence.

My own insistence on the reality of facts in public life implies therefore that I am speaking from inside a free society to which I give my

[1] See Bertram D. Wolfe, *Three Who Made a Revolution*, New York, 1948, pp. 534-57. 'There was something', he writes, 'in the Russian temperament and scene which engendered these men of ambivalent spirit and double role, these Gapons, Azevs, Kaplinskys, Bagrovs and Malinovskys—figures without parallel in the police and revolutionary movements of other lands.' But these figures actually tend to reappear wherever two secret organizations oppose each other. Since only few initiates know the identity of the members it is comparatively easy to plant spies among them, and these spies will tend to play a double role. They will earn their pay by occasionally denouncing some terrorists and accredit themselves with the revolutionary side by taking part in acts of violence against government officials. When this double-crossing has been practised for many years, as in the case of Malinovsky—who went on with it from 1902 until his execution in 1918—it is no longer possible to say, even with a complete knowledge of the facts, which side the man betrayed and which he served.

[2] Hannah Arendt, *op. cit.*, p. 372.

allegiance,[1] just as my insistence on the independent status of science, art and morality implies such participation and allegiance.

## 16. POST-MARXIAN LIBERALISM

No regime has carried out the implications of modern revolutionary dynamism to its logical limits. In fact, it may seem quite impracticable even to approach the complete subordination of all thought to the service of one specific centre of power. The artificial universe encased in the official Soviet jargon had always to be supplemented by natural human sentiments expressed in normal language, and occasionally whole masses of them have been re-introduced. This happened in the 1930's when the Kremlin decided to restore national sentiments and their traditional heroes into Russian historic consciousness, abandoning thereby the hitherto uncontested doctrine of M. N. Pokrovsky (1868-1932), who had turned historiography into an abstract sociological analysis on Marxist lines. On another occasion, when—in 1950—Stalin repudiated the absurd doctrine of N. Y. Marr (1864-1934), according to which all language was class-language, the dictator gave a vivid picture of how this orthodoxy had hitherto trampled on Soviet linguistics—freely drawing for this purpose on the vocabulary of liberalism, and aptly using its principles for the condemnation of this case of totalitarian thought-control which he had himself imposed until that moment.[2] Despite all Lenin's admonitions that party-spirit is the only true objectivity, the concepts of truth, and of the freedom of thought essential to the establishment of truth, were by no means extinguished even in the period of the strictest ideological dictatorship.

The gradual humanization of the Soviet regime that has occurred since Stalin's death may be due to the leaking out of the generous passions buried alive under the Marxist armoury of violence. Indeed, by a kind of inverted Freudian ab-reaction this captive zeal for righteousness may yet be gradually released from its pathological repressions and enter once more into the context of consciously declared moral aspirations.

The first step in that direction was taken immediately after Stalin's death, when his successors released the thirteen Kremlin doctors who had confessed to the assassination of Zsdanov. On the day in March, 1953, when this took place, the systematic transmutation of Communist dignitaries

[1] George Orwell, in *Nineteen-eighty-four*, London, 1949, p. 250, has already said that belief in reality is a subversive principle under totalitarianism.
[2] 'The slightest criticism of the state of affairs in Soviet linguistics, even the most timid attempt to criticize the so-called "new doctrine" in linguistics, was persecuted and suppressed by the leading linguistic circles. Valuable workers and researchers in linguistics were dismissed from their posts or demoted for being critical of N. Y. Marr's heritage, or expressing the slightest disapproval of his teachings. Linguistic scholars were appointed to leading posts not on their merits, but because of their unqualified acceptance of N. Y. Marr's theories' (I. V. Stalin, *Concerning Marxism in Linguistics*, Soviet News Booklet, London, 1950, p. 22).

into self-confessed spies—publicly begging to be hanged for their infamous crimes—was brought to an end. The new masters did not trust altogether the universe of deception and self-deception erected by Stalin, and tried to consolidate their rule by abandoning its worst distortions of the truth; they hoped to gain in persuasive power what they renounced in coercive force.

The liberation of thought that has since been going on and has culminated so far in the Hungarian and Polish revolutions of October, 1956, has been called the Revolution of Truth. The designation is apt, if the meaning of truth is taken to comprise the fruits of all independent thought. For the rights of art, morality, religion and patriotism were restored to some extent along with the right to the knowledge of facts.

The Hungarian insurgents revived the slogans of 1848, and various writers have declared that the movement re-asserted the belief in absolute values as held in the eighteenth century. Others declared that the liberal revolutions had to be fought all over again. But this description is misleading. Compare Wazyk's *Poem for Adults*, published in August, 1955, with the *Marseillaise*, written by Rouget de Lisle in April, 1792; compare Petöfi's flaming sentiments with the cold incisiveness of a Jozsef Attila. The background of 1848 was the French Revolution, which had challenged an immemorial static order by proclaiming the right of society to perfect itself according to reason, and the liberalism of the nineteenth century fought for this aim and against that order. But when the appetitive conception of man denied the reality of moral motives in public life, the ideals of liberalism were inverted into the doctrines of modern totalitarianism. Liberalism had then to fight its way back to a position which had proved disastrously unstable in the light of modern philosophies. This is why Wazyk speaks of 'vomiting' the lies swallowed under Stalin's rule; and why every insurgent Communist speaks of a period during which he connived with growing reluctance at a soul-destroying tyranny which he accepted as the only authentic instrument of human progress.

Can a revulsion against the consequences of modern totalitarianism restore a set of beliefs, on the logical weakness of which the doctrines of totalitarianism itself were founded? Can the beliefs of liberalism, no longer believed to be self-evident, be upheld henceforth in the form of an orthodoxy? Can we face the fact that, no matter how liberal a free society may be, it is also profoundly conservative?

For this is the fact. The recognition granted in a free society to the independent growth of science, art and morality, involves a dedication of society to the fostering of a specific tradition of thought, transmitted and cultivated by a particular group of authoritative specialists, perpetuating themselves by co-option. To uphold the independence of thought implemented by such a society is to subscribe to a kind of orthodoxy which, though it specifies no fixed articles of faith, is virtually unassailable within the limits imposed on the process of innovation by the cultural leadership of a free society. If this is what Lenin meant by saying that 'The absence of party spirit (partinost) in philosophy is nothing but despicable and disguised servility towards idealism and fideism',[1] we cannot deny the charge. And we must face also the fact that this orthodoxy, and the cultural authorities which we respect, are backed by the coercive power of the state and financed by the beneficiaries of office and property. The institutions by which their authority is exercised, the schools, universities, churches, academies, law courts, newspapers and political parties, are under the protection of the same policemen and soldiers who guard the wealth of the landowners and capitalists.

Must this institutional framework be accepted as the civic home of a free society? Is it true that the absolute right of moral self-determination, on which political liberty was founded, can be upheld only by refraining from any radical action towards the establishment of justice and brotherhood? That indeed, unless we agree that within our lifetime we must no more than loosen the ties of a free society, however iniquitous they may be, we shall inevitably precipitate men into abject servitude?

For my part, I would say: Yes. I believe that, on the whole, these limitations are imperative. Unjust privileges prevailing in a free society can be reduced only by carefully graded stages; those who would demolish them overnight would erect greater injustices in their place. An absolute moral renewal of society can be attempted only by an absolute power which must inevitably destroy the moral life of man.

This truth is unpalatable to our conscience. Does it follow that we must suppress our conscience, or else accept the totalitarian teaching that violence alone is honest? I said in the introduction to this chapter that I would renew within a social setting the question, how we can keep holding beliefs that can conceivably be doubted. The attempt made in this book to stabilize knowledge against scepticism, by including its hazardous character in the conditions of knowledge, may find its equivalent, then, in an allegiance to a manifestly imperfect society, based on the acknowledgment that our duty lies in the service of ideals which we cannot possibly achieve.

[1] *Extracts from the Soviet Philosophical Dictionary*, p. 18 (Published by the Congress for Cultural Freedom, Paris, 1953).

PART THREE

# THE JUSTIFICATION OF
# PERSONAL KNOWLEDGE

# 8

# THE LOGIC OF AFFIRMATION

## 1. INTRODUCTION

**B**Y now I have surveyed a series of facts which seriously suggest a reappraisal of our capacity to acquire knowledge. This reappraisal demands that we credit ourselves with much wider cognitive powers than an objectivist conception of knowledge would allow, but at the same time it reduces the independence of human judgment far below that claimed traditionally for the free exercise of reason. It is useless to accumulate more evidence unless we can first master what has been given so far. I shall now try, therefore, to give a firmer outline to the conception of personal knowledge. The argument will be focussed for this purpose once more on the narrow range of knowledge, forming the hard core of greatest certainty. Only if we can find simple formulations which define the indeterminacy and existential dependence of such knowledge, can we hope to devise a stable framework within which any kind of knowledge can be justified.

## 2. THE CONFIDENT USE OF LANGUAGE

An object alleged to be a tool is not a tool if our conception of its alleged use is altogether mistaken (in the way conceptions of a *perpetuum mobile* are mistaken) or if it otherwise fails to serve its alleged purpose; it is an error to rely on a tool in such a case. Similarly, it is an error to rely on a descriptive word if either the conception which it conveys is false, or the word does not properly cover the subject matter in question.

One can use a tool tentatively, or merely show that it is useless. Similarly, we can use a descriptive word sceptically by putting it in quotation marks. Suppose a paper is published under the title: *An Explanation of 'Extra-sensory Perception'*, and another in reply to it, entitled: *An 'Explanation' of Extra-sensory Perception*. Guided by the quotation marks we

249

recognize immediately that the first paper regards extra-sensory perception as spurious, while the second accredits it as genuine and discredits, on the contrary, the explanation suggested for it in the first paper.

Descriptive words written down as part of a sentence without quotation marks around them are confidently relied upon: they accredit the substantial character of the conception which they convey and its appositeness to the matter in hand. I shall call this the *confident* or *direct* use of a word. By contrast, a descriptive word used in quotation marks (as part of a sentence not concerning that word)[1] is used in a *sceptical* or *oblique* fashion. Such use calls in question either the reality of the conception evoked by the word, or its applicability to the case in point. Since a word remains the same whether used directly or obliquely, the difference between uttering it confidently or sceptically must lie wholly in the tacit coefficient of its utterance. This difference identifies formally the unspecifiable personal coefficient attached to the confident use of a descriptive term.

### 3. THE QUESTIONING OF DESCRIPTIVE TERMS

We may try to eliminate the indeterminate residue of a meaning by explaining it in words aided by demonstrations. Such verbal definitions operate in the same way as the analysis of a skill or the axiomatization of a scientific method of enquiry; they disclose certain rules of art which we have hitherto practised tacitly, and help to consolidate and improve their use. Accordingly, in formulating a definition we must rely on watching the way the art of using a word is authentically practised; or more precisely, watch ourselves applying the term to be defined in ways that *we regard* as authentic. 'Ostensive definitions' are merely a suitable extension of this watching. They call the listener's attention to examples believed to be particularly clear, supplementing as it were, the explanation of a clever feat by showing how it is done. The formalization of meaning relies therefore *from the start* on the practice of unformalized meaning. It necessarily does so also *in the end*, when we are using the undefined words of the definitions. Finally, the practical interpretation of a definition must rely *all the time* on its undefined understanding by the person relying on it. Definitions only shift the tacit coefficient of meaning; they reduce it but cannot eliminate it.

The tacit coefficient is an act of confidence, and all confidence can be conceivably misplaced. I have spoken of this risk before, when showing that all articulation is rooted in the kind of comprehension by which animals make sense of their situation. We have seen how passionate is this confidence, how inventive and persuasive; how it is shared, fostered and disciplined, by a society dedicated to its cultivation; and how, to this extent, our confidence in the meaning of words is an act of social allegi-

---

[1] This should exclude the use of a word in quotes as a name of the same word, e.g. when we say that the word 'cat' designates a cat.

ance. All these tacit commitments appeared self-satisfying, irreversible and hence unspecifiable. They seemed to face us with an immensely ramified system of wholly indeterminate uncertainties which we have to accept blindly, if we are ever to speak at all.

By contrasting the oblique use of words with their direct use, we can now show formally that these risks of confident utterance are unavoidable. We may place a word in quotation marks, while using language confidently through the rest of a sentence. But the questioning of *each* word *in turn* would never question *all at the same time*. Accordingly, it would never reveal a comprehensive error which underlies our entire descriptive idiom. We can of course write down a text and withdraw our confidence from all its words simultaneously, by putting each descriptive word between quotation marks. But then none of the words would mean anything and the whole text would be meaningless. The hazards of confidence inherent in the act of attaching a meaning at least to some set of descriptive terms are ineradicable.

## 4. Precision

I have also said before that we must accept the risks of semantic indeterminacy, since only words of indeterminate meaning can have a bearing on reality and that for meeting this hazard we must credit ourselves with the ability to perceive such bearing.[1] This decision would eliminate precision of meaning as an ideal, and raise the question in what sense (if any) we may apply the term 'precise' or 'imprecise' to the meaning of a descriptive term.

I suggest that the term 'precise' is applicable to a descriptive word in the same way as to a measured quantity, a map, or any other description, in so far as the word appears to match experience. Precision or imprecision is a property that can be predicated of *a designation* when it is tested by matching it against something which is *not a designation*. but is the situation on which the designation bears.

This test itself cannot be tested in the same sense. It is a tacit performance, and as such lacks the duality which makes the confrontation and the matching of two things—the designation and the designate— logically possible. Therefore, when we say of a descriptive term that it is precise, we declare the result of a test which itself cannot be said to be precise in the same sense. Of course, *the application of the term 'precise'* might once more be said to be precise, or imprecise, when we confront it with the test from which it was derived; but this second confrontation would have to rely once more on a personal appraisal which cannot be said to be precise in the sense in which a description can be. The precision of a word will ultimately always rely, therefore, on a test which is not precise in the same sense as the word is said to be.

[1] P. 95.

The indefinite and futile regress on which we enter when asking whether the application of the term 'precise' is itself precise, suggests that such a question should be avoided by denying to the word 'precise' the character of a descriptive term. When we say that a word is precise (or apt, or fitting, or clear, or expressive), we approve of an act of our own which we have found satisfying while carrying it out. We are satisfied by something we *do* in the same way as when we make sense of blurred sights or faint noises; or when we find our way or recover our balance. We properly declare the outcome of this personal comprehension of our own, by saying that a word which we are using is precise. The indefinite regress arises only when we disguise this announcement of our self-satisfaction as a descriptive term designating a property of another descriptive term.

We shall avoid this fallacy by fully acknowledging that only a speaker or listener can mean something *by* a word, and a word *in itself* can mean nothing. When the act of meaning is thus brought home to a person exercising his understanding of things by the use of words which describe them, the possibility of performing the act of meaning according to strict criteria appears logically meaningless. For any strictly formal operation would be impersonal and could not therefore convey the speaker's personal commitment. The analysis of the term 'precise' applied to descriptive terms which can mean something real only if they are not strictly precise, reveals therefore the self-reliant act of the speaker uttering the term and assessing its precision.

## 5. The Personal Mode of Meaning

If, then, it is not words that have meaning, but the speaker or listener who means something by them, let me declare accordingly my true position as the author of what I have written so far, as well as of what is still to follow. I must admit now that I did not start the present reconsideration of my beliefs with a clean slate of unbelief. Far from it. I started as a person intellectually fashioned by a particular idiom, acquired through my affiliation to a civilization that prevailed in the places where I had grown up, at this particular period of history. This has been the matrix of all my intellectual efforts. Within it I was to find my problem and seek the terms for its solution. All my amendments to these original terms will remain embedded in the system of my previous beliefs. Worse still, I cannot precisely say what these beliefs are. I can say nothing precisely. The words I have spoken and am yet to speak mean nothing: it is only *I* who mean something *by them*. And, as a rule, I do not focally know what I mean, and though I could explore my meaning up to a point, I believe that my words (descriptive words) must mean more than I shall ever know, if they are to mean anything at all.

This prospect may sound deplorable, but a programme that accepts it may at least claim to be self-consistent, while any philosophy that sets up strictness of meaning as its ideal is self-contradictory. For if the active participation of the philosopher in meaning what he says is regarded by it as a defect which precludes the achievement of objective validity, it must reject itself by these standards. Nor is the consistency of an objectivist philosophy restored by confessing that words have an open texture. For such words, as we have seen (p. 113), have no meaning except through our accrediting of the speaker's sense of fitness. Therefore, without the explicit acknowledgment and endorsement of the philosopher's personal judgment as an integral part of his philosophy, a philosophy expressed in terms of 'open texture' is also meaningless.

While impersonal meaning is self-contradictory, the justification of personal meaning is self-justifying, if only it admits its own personal character. It licenses certain conditions of articulation which are bound to become apparent when we reflect on this process of licensing, but which cannot be held to invalidate it, since they should be acceptable in the very light of this licensing. If I agree that every word I confidently utter as meaningful is so uttered as a personal commitment of my own, then I may agree also that the words used for making this statement itself are similarly employed to mean what I myself mean by them. Thus if I cannot speak except from inside a language, I may at least speak of my language in a manner consistent with this situation.

But consistency is not enough. There must also be some significance left in my programme. Can I ever justify saying anything at all, if the moment I start speaking I am accrediting the indefinitely ramified implications of a particular vocabulary, and any subsequent justification of these implications would still be necessarily couched in the very idioms which I seek to justify? It might seem that we have saved the concept of meaning from destruction by depersonalization, only to expose it to being reduced to the status of dogmatic subjectivity.

Here I must leave my enquiry temporarily suspended. For the justification of the personal mode of meaning as described in this section can be attempted only later, in conjunction with the kindred problems arising from the fiduciary mode of assertion.

## 6. Assertions of Fact

Denis de Rougemont has remarked that man alone among animals can lie. It may be more accurate to say that man can deceive others most effectively, because he alone can tell them a falsehood. Every conceivable assertion of fact can be made in good faith or as a lie. The statement remains the same in both cases, but its tacit components are different. A truthful statement commits the speaker to a belief in what he has asserted: he embarks in it on an open sea of limitless implications. An untruthful statement withholds this belief, launching a leaking vessel for others to board and sink in it.

Unless an assertion of fact is accompanied by some heuristic or persuasive feeling, it is a mere form of words saying nothing. Any attempt to eliminate this personal coefficient, by laying down precise rules for making or testing assertions of fact, is condemned to futility from the start. For we can derive rules of observation and verification only from examples of factual statements that we have accepted as true *before* we knew these rules; and *in the end* the application of our rules will necessarily fall back once more on factual observations, the acceptance of which is an act of personal judgment, unguided by any explicit rules. And besides, the application of such rules must rely *all the time* on the guidance of our own personal judgment. This argument formally confirms the participation of the speaker in any sincere statement of fact.

How can we take this coefficient into account in our conception of the truth? What can we mean by saying that a factual statement is true?

An articulate assertion is composed of two parts: a sentence conveying the content of what is asserted and a tacit act by which this sentence is asserted.[1] The articulate assertion can be tested by separating its two parts and tentatively cancelling the act of assertion, while the unasserted sentence is being confronted with experience. If as a result of this test we decide to renew the act of assertion, the two parts are reunited and the sentence is reasserted. This reassertion may be made explicit by saying that the originally asserted sentence is true.

The act of assertion itself does not, of course, consist of two parts— one tacit, the other articulate—of which the first can be cancelled while the second, now unasserted, can be tested by confrontation with the facts. It is an act of tacit comprehension, which relies altogether on the self-satisfaction of the person who performs it. It can be repeated, improved or cancelled, but not tested or said to be true, in the sense in which a factual statement can be tested and said to be true.

Therefore, if '*p* is true' expresses my assertion or reassertion of the sentence *p*, then '*p* is true' cannot be said to be true or false in the sense in which a factual sentence can. '*p* is true' declares that I identify myself with the content of the factual sentence *p*, and this identification is something I am doing, and not a fact that I am observing. The expression '*p* is true' is therefore not itself a sentence but merely the assertion of (an otherwise unasserted) sentence, the sentence *p*. To say that '*p* is true' is to underwrite a commitment or to sign an acceptance, in a sense akin to the commercial meaning of such acts. Hence we cannot assert the expression '*p* is true', any more than we can endorse our own signature; only a sentence can be asserted, not an action.

The misleading form of the expression '*p* is true' which disguises an act of commitment in the form of a sentence stating a fact leads to logical paradoxes. If the assertion of the sentence *p* has to be followed up by

---

[1] R. M. Hare, *The Language of Morals* (Oxford, 1952), p. 18, calls the unasserted statement the 'phrastic' and its assertion the 'neustic' part of the asserted statement.

saying '*p* is true' and '*p* is true' is itself a sentence, then this sentence brings in its sequel ' "*p* is true" is true' and so on, indefinitely. This insatiable regress does not arise, if we realize that '*p* is true' is not a sentence.

The Paradox of the Liar is eliminated on similar grounds. We may write this paradox in the form: 'the sentence on top of page 10 of this book is false', in which the word 'sentence' designates (as we discover by looking up the top of page 10) 'the sentence on top of page 10 of this book is false'. Denote the sentence just quoted by *p*; then *p* is true if and only if the sentence on top of page 10 of this book is false, i.e. *p* is true if and only if *p* is false. But if '*p* is false' merely declares that the speaker denies acceptance to *p*, then '*p* is false' is not a sentence and the paradox does not arise, for there is then no sentence to be found on top of page 10 of the book in question.

The fact that we can eliminate an infinite regress and a notorious self-contradiction, by reinterpreting the expressions '*p* is true' and '*p* is false' as expressing an act of assertion or doubt, substantially strengthens this interpretation. By generalizing our distinction between the confident use of language for primary purposes, and the class of expressions which merely endorse our confidence in what we have said, a whole range of persistent philosophic problems can be eliminated.[1]

## 7. TOWARDS AN EPISTEMOLOGY OF PERSONAL KNOWLEDGE

We have re-defined the word 'true' as expressing the asseveration of the sentence to which it refers. This is closely akin to Tarski's definition of 'true' which implies, for example: ' "snow is white" is true if and only if snow is white'. But Tarski's definition now appears to equate a sentence with an action. This anomaly may be eliminated by revising the definition as follows: 'I shall *say* that "snow is white" is true if and only if I *believe* that snow is white'. Or perhaps more reasonably: 'If I believe snow is white I shall say that "snow is white" is true'. This expression, admittedly, suggests a difference in emphasis between asserting a sentence and saying that it is true: the first stressing the personal character of our knowledge, the second its universal intent. But they both remain personal endorsements of the statement.

Earlier on—in the chapter on Probability[2]—I have denied the possibility of expressing the act of placing my confidence in a statement of fact by a statement of the probability of this fact. I suggested that the act of endorsing a sentence should be indicated in writing by the prefix '⊢'

---

[1] My re-definition of 'truth' is reminiscent of Max Black's 'No truth theory' of truth (*Language and Philosophy*, Ithaca, N.Y., 1949, pp. 104–5) and is in accord also with P. F. Strawson's critique of the semantic theory ('The Semantic Theory of Truth', *Analysis*, **9** (1949), No. 6). But the purpose of both these authors is to eliminate the problem arising from the definition of truth, and not to accredit the use of 'truth' as part of an a-critical act of affirmation.

[2] P. 29.

used by Frege as an assertion sign, which should be read as 'I believe', or as some equivalent expression of endorsement. Such a prefix should not function as a verb, but as a symbol determining the modality of the sentence. The transposition of an assertion of fact into the 'fiduciary mode' would correctly reflect the fact that such an assertion is necessarily attributable to a definite person at a particular place and time: for example, to the writer of the assertion at the moment of putting it to paper, or to the reader when he reads and accepts what is written.

This transposition considerably modifies the situation within which we have to account for making assertions of fact. So long as we ascribed to declaratory sentences the properties of being true or false, we had to account for these properties in the same sense as we would explain what makes green leaves green. Such self-speaking sentences appeared to possess the quality of being true or false impersonally, and this would have to be accounted for again in terms of impersonal criteria: which is of course impossible. We might have a better chance of achieving the purpose of epistemological reflection if we asked ourselves instead why we do believe certain statements of fact, or why we believe certain classes of statements, such as those of science. For having recognized that an 'impersonal allegation' is a contradiction in terms—just as an 'anonymous cheque' would be —we shall no longer try to arrive at any justification of our allegations which would not in its turn be composed of personal allegations of our own. It should not be too difficult to justify my scientific beliefs, in particular, in terms of some logically antecedent beliefs of my own, this justification itself being acknowledged once more to involve a fiduciary act of my own. The trouble is in fact that this may appear so easy as to be quite pointless. For it will be objected that 'You can believe what you like'; which brings us back once more to the paradox of self-set standards; if the criteria of reasonableness, to which I subject my own beliefs, are ultimately upheld by my confidence in them, the whole process of justifying such beliefs may appear but a futile authorization of my own authority.

Yet so be it. Only this manner of adopting the fiduciary mode is consonant with itself: the decision to do so must be admitted to be itself in the nature of a fiduciary act. Indeed, the same must apply to the whole of this enquiry and to all conceivable conclusions to be derived from it. While I shall continue to argue a series of points and adduce evidence for my proposed conclusions, I shall always wish it to be understood that in the last resort my statements affirm my personal beliefs, arrived at by the considerations given in the text in conjunction with other not specifiable motives of my own. Nothing that I shall say should claim the kind of objectivity to which in my belief no reasoning should ever aspire; namely that it proceeds by a strict process, the acceptance of which by the expositor, and his recommendation of which for acceptance by others, include no passionate impulse of his own.

I hope to consolidate this decision later. Meanwhile, I have yet to face some further dilemmas arising from the objectivist urge to depersonalize our intelligent mental processes.

## 8. INFERENCE

Our intellectual superiority over the animals is almost entirely due to our powers of symbolic operations; it is only by relying on these that we are able to carry out any process of consecutive reasoning. No wonder then that the movement bent on the ideal of impersonal thought has consistently aimed at reducing this central agency of human intelligence to operations governed by strict rules. This hope was recently heightened by the construction of highly effective automatic devices for various complex purposes. Anti-aircraft guns were equipped with predictors automatically governed by the gunner's initial readings. Once the sights were set on a plane, the machines computed the course of the swiftly moving target, as well as that of the projectile ready to be sent out, and aimed the gun so as to assure a hit. There followed the construction of automatic pilots and of guided missiles, and the comprehensive automation of work in office and factory. Here were instruments carrying out complex feats of intelligence without any intervention of man. This clearly offered new prospects for attaining the ideal of completely detached thought.

Since I have spoken already (and shall yet say more later) of the impossibility of formalizing the process of empirical inference, I shall deal here only with the attempt to depersonalize the process of deductive inference.

We have seen before that deductive reasoning may be altogether ineffable and that even the most completely formalized logical operations must include an unformalized tacit coefficient. We have seen how the passionate force of this coefficient actuates discovery, inflames controversy, and sustains the student's efforts to understand what he is being taught; we have seen how these passions are shared between mathematicians working in different fields, so that they will always be guided by common standards which they enforce on each other by their professional consensus. I shall refer in the following to the widely ramified operations of these tacit coefficients in the deductive sciences only in brief formal terms. By compendiously designating this whole aggregate of personal commitments, I shall lend to my argument sufficient rigidity for carrying the burden I shall presently place upon it.

The operations of digital computers *as machines of logical inference* coincide with the operations of symbolic logic. We may therefore identify the formalization involved in the construction and the use of machines, operating in this particular way, with the procedure governing the construction of a deductive system. This procedure is threefold. (1) It designates undefined terms; (2) it specifies unproven asserted formulae (axioms); and (3) it prescribes the handling of such formulae for the purpose of writing down new asserted formulae (proofs). This result is achieved by a sustained effort to eliminate what are called 'psychological' elements—the factors which I call 'tacit'. The undefined terms are intended to stand without signifying anything, complete in themselves as marks on paper; unproven asserted formulae are to replace statements believed to be self-evident; operations constituting 'formal proof' are similarly intended to replace 'merely psychological' proof.

However, this attempt to eliminate the personal participation of the logician must leave at each of these points an irreducible residue of mental operations, on which the operations of the formalized system itself will continue to rely. (1) The acceptance of a mark on paper as a symbol implies that (a) we believe that we can identify the mark in various instances of it and (b) that we know its proper symbolic use. In both these beliefs we may be mistaken, and they constitute therefore commitments of our own. (2) In agreeing to regard an aggregate of symbols as a formula, we accept it as something that can be asserted. This implies that we believe that such an aggregate says something about something. We expect to recognize things which satisfy a formula, as distinct from other things which fail to do so. Since the process by which our axioms will be satisfied is necessarily left unformalized, our countenancing of this process constitutes an act of commitment on our part. (3) The handling of symbols according to mechanical rules cannot be said to be a proof, unless it carries the conviction that whatever satisfies the axioms from which the operation starts will also satisfy the theorems arrived at. No handling of symbols to which we refuse to award the success of having convinced us that an implication has been demonstrated can be said to be a proof. And again, this award is an unformalized process which constitutes a commitment.

Thus, at a number of points, a formal system of symbols and operations can be said to function as a deductive system only by virtue of unformalized supplements, to which the operator of the system accedes: symbols must be identifiable and their meaning known, axioms must be understood to assert something, proofs must be acknowledged to demonstrate something, and this identifying, knowing, understanding, acknowledging, are unformalized operations on which the working of a formal system depends. We may call them the *semantic functions* of the formal system. These are performed by a person with the aid of the formal system, when the person relies on its use.[1]

[1] Formalization can be carried beyond this point, but only for an 'object theory' described within a metatheory which is itself informal. The following passage from S. C. Kleene, *Introduction to Metamathematics* (Amsterdam, 1952) p. 62, vividly describes this position: 'The metatheory belongs to intuitive and informal mathematics. ... [It] will be expressed in ordinary language, with mathematical symbols ... introduced according to need. The assertions of the metatheory must be understood. The deductions must carry conviction. They must proceed by intuitive inferences, and not, as the deductions in the formal theory, by applications of stated rules. Rules have been stated to formalize the object theory, but now we must understand without rules how

258

It is, indeed, logically absurd to say of a logical inference machine that it draws inferences of its own. *By itself* an inference machine is merely an 'inference machine' and can only draw 'inferences'. The omission of the quotation marks expresses our accrediting of the machine, and hence our acceptance of the conclusions arrived at by its operations, as our own inferences. The legitimate purpose of formalization lies in the reduction of the tacit coefficient to more limited and obvious informal operations; but it is nonsensical to aim at the total elimination of our personal participation.

This conclusion will be found applicable in this general form to all kinds of automatic mechanisms. It can be precisely elaborated for the time being only for the process of logical inferences and for machines carrying out logical inferences; but this will prove instructive also for the logical analysis of any kind of automatic machinery used for intelligent purposes.

The most important theorems limiting the formalization of logical thought are due to Gödel. They are based on the fact that within any deductive system which includes arithmetic (such as for example the system of *Principia Mathematica*) it is possible to construct formulae—i.e. sentences—which are demonstrably undecidable within that system, and that such a sentence—the famous Gödelian sentence—may say of itself that it is undecidable within the system. We can then go further by informally matching the sentence with the situation on which it bears, that is, with the demonstration of its own undecidability. We shall now find that what the sentence says is true and decide accordingly to assert it in that sense. Thus asserted, the sentence represents an additional axiom, which is independent of the axioms from which the unasserted sentence was derived.[1]

This process reveals both that any formal system (of sufficient richness) is necessarily incomplete and that our personal judgment can reliably add new axioms to it. It offers a model of conceptual innovation in the deductive sciences, which illustrates in principle the inexhaustibility of mathematical heuristics and also the personal and irreversible character of the acts which continue to draw upon these possibilities.

Gödel has also shown that the sentence which is demonstrably undecidable may say that the axioms of the system cannot be proved to be consistent. This shows (as I have mentioned before) that we never know altogether what our axioms mean, since if we knew, we could avoid the possibility of asserting in one axiom what another axiom denies. This uncertainty can be eliminated for any particular deductive system by shifting it unto a wider system of axioms, within which we may be able to prove the consistency of the original system. But any such proof will still remain uncertain, in the sense that the consistency of the wider system will always remain undecidable.

those rules work. An intuitive mathematics is necessary even to define the formal mathematics.'        [1] K. Gödel, *Monatsh. Math. Phys.*, **38** (1931), pp. 173–98.

In a logical demonstration closely akin to the proof of Gödel's theorems, Tarski has shown that any formal system in which we could assert a sentence and also reflect on the truth of its assertion must be self-contradictory. Thus, in particular, the assertion that any theorem of a given formal language is true, can be made only by a sentence that is meaningless within that language. Such an assertion forms part of a richer language than that which comprises the sentences whose truth it asserts.[1]

The construction of the Gödelian sentence shows that a process of deductive inference can produce a situation which irresistibly suggests an assertion not formally implied in its premisses. Tarski's theorem that the assertion of truth belongs to a logically richer (formal) language than the (formal) language of the sentences asserted to be true, shows that the question whether a previously asserted sentence is true evokes a similar expansion. It arises in both cases from a reflection on what has been said. In the Gödelian process we add to a formally undecided statement of ours a tacit interpretation of our own. The act of innovation consists here in realizing that what we had just said was true in this new sense. The Tarski process is based on the 'duality' of asserted sentences; the formal innovation being due here to our capacity for calling in question a hitherto tacit assent of ours and renewing our assent in explicit terms. In both processes we establish something new by an inescapable act of our own, induced—but not performed—by formal operations.

I have described before (Part Two, ch. 5, p. 131) how the mathematician works his way towards discovery, by shifting his confidence from intuition to computation, and back again from computation to intuition, while never releasing his hold on either of the two. These shifts are usually gradual. The matching of the Gödelian sentence with the facts to which it refers and the subsequent reassertion of the Gödelian sentence, determine jointly a precise point at which tacit thought takes over control for the crossing of a logical gap.[2]

We find a similar alternation involved in the method of 'mathematical induction' which Poincaré regarded as the prototype of all mathematical

---

[1] A. Tarski, 'The Semantic Conception of Truth and the Foundations of Semantics', *Philosophy and Phenomenological Research*, **4** (1944), pp. 341–76. Tarski shows that by keeping the two languages apart, the Paradox of the Liar is avoided. We arrived at the same result by showing that if a factual assertion is made by a sentence $p$, '$p$ is true' is not a sentence. For the purpose of the present argument this result may be taken to be expressed by Tarski's theorem that '$p$ is true' belongs to a different language than $p$—a language in which to every asserted sentence of the original language there corresponds the name of that sentence, i.e. the same sentence in quotation marks.

[2] The tacit component of any formal process of inference performs a similar function in detaching the consequent. (See H. Jeffreys, *Brit. Journ. Phil. Sci.*, **5** (1955), p. 283, who supports the argument by Lewis Carroll in 'What the Tortoise said to Achilles', *Mind*, N.S., **4** (1895), p. 278.) The same tacit operation was also implied in Tarski's definition of truth in the transition from the sentence, ' "Snow is white" is true' to the act of asserting that snow is white. The rejection of a null hypothesis rendered improbable on statistical evidence is another case where a tacit decision is induced by virtually compelling circumstances.

innovations.[1] It starts by proving a series of theorems which apply to successive whole numbers, each consecutive theorem being derived from the previous one, and proceeds to conclude hence that the theorem is true generally for all numbers. To draw such inferences the mind must look back upon a series of demonstrations and generalize the principle of its own past operations. In Part Two, ch. 6 (p. 185) I have quoted an account by Daval and Guilbaud, showing how the conception of continuity was discovered by such a process of reflection.

The analogy between the Gödelian process of innovation and the grammar of discovery outlined by Poincaré lends support to the continuity between the informal act of assertion and the equally informal act of discovery. The difference between the two lies in the width of the logical gap that is being crossed. The gap to be crossed for the reassertion of the Gödelian sentence is extremely narrow—almost imperceptible—while in true acts of discovery it may be as large as any human mind can hope to overcome. The act of assent proves once more to be logically akin to the act of discovery: they are both essentially unformalizable, intuitive mental decisions.

## 9. AUTOMATION IN GENERAL

The proliferation of axioms discovered by Gödel offers manifest proof that a person operating a logical inference machine can achieve informally a range of knowledge which no operations of such a machine can demonstrate, even though its operations suggest an easy access to it. It proves that the powers of the mind exceed those of a logical inference machine. But we have yet to face the wider problem raised by gunsight predictors, automatic pilots, etc., that is, by machines whose performances range far beyond logical inferences. A. M. Turing has shown[2] that it is possible to devise a machine which will both construct and assert as new axioms an indefinite sequence of Gödelian sentences. Any heuristic process of a routine character—for which in the deductive sciences the Gödelian process is an example—could likewise be carried out automatically. A routine game of chess can be played automatically by a machine, and indeed, all arts can be performed automatically to the extent to which the rules of the art can be specified. While such a specification may include random elements, like choices made by spinning a coin, no unspecifiable skill or connoisseurship can be fed into a machine.

We shall not be able to circumscribe the scope of automatic operations in general by such formal criteria as apply to logical inference machines. Yet the necessary relatedness of machines to persons does essentially

[1] L. E. J. Brouwer agrees with Poincaré on this. See H. Weyl, *Philosophy of Mathematics and the Natural Sciences*, Princeton, 1949, p. 51.

[2] In a communication to a Symposium held on 'Mind and Machine' at Manchester University in October, 1949. This is foreshadowed in 'Systems of Logic Based on Ordinals', *Proc. London Maths. Soc.*, Series 2, **45** (1938-9), pp. 161-228.

restrict the independence of a machine and reduce the status of automata in general below that of thinking persons. For a machine is a machine only for someone who relies on it (actually or hypothetically) for some purpose, that he believes to be attainable by what he considers to be the proper functioning of the machine: it is the instrument of a person who relies on it. This is the difference between machine and mind. A man's mind can carry out feats of intelligence *by aid of* a machine and also *without* such aid, while a machine can function only as the extension of a person's body under the control of his mind. Accordingly, the machine can exist as a machine only within a tripartite system

| I | II | III |
|---|---|---|
| mind | machine | functions, purposes etc. entertained by the mind. |

Since the control exercised over the machine by the user's mind is—like all interpretations of a system of strict rules—necessarily unspecifiable, the machine can be said to function intelligently only by aid of unspecifiable personal coefficients supplied by the user's mind.

### 10. NEUROLOGY AND PSYCHOLOGY

Neurology is based on the assumption that the nervous system—functioning automatically according to the known laws of physics and chemistry—determines all the workings which we normally attribute to the mind of an individual. The study of psychology shows a parallel tendency towards reducing its subject matter to explicit relationships between measurable variables; relationships which could always be represented by the performances of a mechanical artefact.

This raises the question whether in view of the logical analysis of 'a machine in use' we can accept a neurological model (or an analogous psychological model) as the representation of an individual's mind. In answering this question we must take into account an obvious difference between an automatic neurological model and a machine operated for intelligent purposes; namely, that the neurological model is not supposed to operate for purposes of the neurologist, but for purposes attributed to its operations by the neurologist on behalf of the subject whose mind it represents. The tripartite system accordingly becomes:

| I | II | III |
|---|---|---|
| Mind (of neurologist) | Neurological model of subject | Intellectual purposes attributed to the subject by the neurologist. |

But the informal mental functions briefly indicated under III are those of the neurologist's mind, since the informal, and hence personal, functions of the *subject's* mind are in fact not represented at all in the tripartite system. For the neurological model is—like a machine—strictly impersonal and can account for none of the unspecifiable propensities of the subject.

These personal powers include the capacity for understanding a meaning, for believing a factual statement, for interpreting a mechanism in relation to its purpose, and on a higher level, for reflecting on problems and exercising originality in solving them. They include, indeed, every manner of reaching convictions by an act of personal judgment. The neurologist exercises these powers to the highest degree in constructing the neurological model of a man—to whom he denies in this very act any similar powers. The same is true of a psychologist who reduces the mental manifestations of man to specifiable relations of measured quantities, for as such these can always be represented by the performances of a robot.

This disparity between the powers which the interpreting mind is confidently exercising in the very act of denying them to the subject interpreted by it, is justified, so long as the observer is concerned only with the automatic responses of his subject. When a physiologist records the reflexes of a person, he is rightly claiming for himself powers of judgment which are absent in the faculties he is examining in another person. To the extent to which mental illness deprives those suffering from it of control over their thoughts, a psychiatrist will also observe the pathological mechanism in question from the superior position assumed by him towards his subject.

By contrast, to acknowledge someone as a sane person is to establish a reciprocal relation to him. By virtue of our own art of comprehension we experience another person's similar faculties as the presence of that person's mind. Our capacity for knowing things either focally or subsidiarily is decisive here. Mind is not the aggregate of its focally known manifestations, but is that on which we focus our attention while being subsidiarily aware of its manifestations. This is the way (to be analysed further in Part Four) by which we acknowledge a person's judgment and share also other forms of his consciousness. This manner of knowing a person qualifies him fully for the functions of a mind in position I of a tripartite system controlled by a mind; while the aggregate of his focally known manifestations do not qualify him for these functions.

According to these definitions of 'mind' and 'person', neither a machine, nor a neurological model, nor an equivalent robot, can be said to think, feel, imagine, desire, mean, believe or judge something. They may conceivably simulate these propensities to such an extent as to deceive us altogether. But a deception, however compelling, does not qualify thereby as truth: no amount of subsequent experience can justify us in accepting as identical two things known from the start to be different in their nature.[1]

---

[1] I dissent therefore from the speculations of A. M. Turing (*Mind*, N.S., **59** (1950), p. 433) who equates the problem: 'Can machines think?' with the experimental question, whether a computing machine could be constructed to deceive us as to its own nature as successfully as a human being could deceive us in the same respect.

Our theory of knowledge is now seen to imply an ontology of the mind. Objectivism requires a specifiably functioning mindless knower. To accept the indeterminacy of knowledge requires, on the contrary, that we accredit a person entitled to shape his knowing according to his own judgment, unspecifiably. This notion—applied to man—implies in its turn a sociology in which the growth of thought is acknowledged as an independent force. And such a sociology is a declaration of loyalty to a society in which truth is respected and human thought is cultivated for its own sake.[1] This ontology—which flows from my theory of knowledge—will be outlined further in Part Four.

## 11. ON BEING CRITICAL

All kinds of articulate affirmations can be made more or less critically—and indeed quite uncritically. Where there is criticism, what is being criticized is, every time, *the assertion of an articulate form*. It is our personal acceptance of an articulate form that is judged to have been critical or uncritical, and this judgment expresses our appraisal of the tests to which we have subjected the articulate form or articulate operation before accepting it. It is the mind granting this acceptance which is said to have been acting critically or uncritically. The process of logical inference is the strictest form of human thought, and it can be subjected to severe criticism by going over it stepwise any number of times. Factual assertions and denotations can also be examined critically, although their testing cannot be formalized to the same extent.

In the sense just specified, tacit knowing cannot be critical. Animals, like men, may be alert against delusions. A young dog is more rash than an old fox. The hesitations of a chimpanzee in solving a problem may impose a strain upon him. But systematic forms of criticism can be applied only to articulate forms, which you can try out afresh again and again. We should not apply, therefore, the terms 'critical' or 'uncritical' to any process of ʾacit thought *by itself*; any more than we would speak of the critical or uncritical performance of a high-jump or a dance. Tacit acts are judged by other standards and are to be regarded accordingly as *a-critical*. The significance of this distinction should become clearer in the next two chapters.[2]

## 12. THE FIDUCIARY PROGRAMME

Our tacit powers decide our adherence to a particular culture and sustain our intellectual, artistic, civic and religious deployment within its framework. The articulate life of man's mind is his specific contribu-

---

[1] See p. 142 and p. 219.
[2] In my account of tradition on p. 53, 'uncritical' should now be replaced by 'a-critical'.

tion to the universe; by the invention of symbolic forms man has given birth and lasting existence to thought. But though our thinking has contrived these artifices, yet they have power to control our own thought. They speak to us and convince us, and it is precisely in their power over our own minds that we recognize their justification and their claim to universal acceptance.

Yet who convinces whom here? If man died, his undeciphered script would convey nothing. Seen in the round, man stands at the beginning and at the end, as begetter and child of his own thought. Is he speaking to himself in a language he alone can understand?

In the beginning many words were held to be sacred. The law was respected as divine, and religious texts were revered as revealed by God. Christians worshipped the word made flesh. What the Church taught required no verification by man. When accepting its doctrine man was not speaking to himself, and in his prayers he could address the very source of the doctrine.

Later, when the supernatural authority of laws, churches and sacred texts had waned or collapsed, man tried to avoid the emptiness of mere self-assertion by establishing over himself the authority of experience and reason. But it has now turned out that modern scientism fetters thought as cruelly as ever the churches had done. It offers no scope for our most vital beliefs and it forces us to disguise them in farcically inadequate terms. Ideologies framed in these terms have enlisted man's highest aspirations in the service of soul-destroying tyrannies.

What then can we do? I believe that to make this challenge is to answer it. For it voices our self-reliance in rejecting the credentials both of medieval dogmatism and modern positivism, and it asks our own intellectual powers, lacking any fixed external criteria, to say on what grounds truth can be asserted in the absence of such criteria. To the question, 'Who convinces whom here?' it answers simply, 'I am trying to convince myself.'

I have insisted on this before on diverse occasions: pointing out repeatedly that we must accredit our own judgment as the paramount arbiter of all our intellectual performances, and claiming that we are competent to pursue intellectual excellence as a token of a hidden reality. I shall yet try to elaborate the structure of this ultimate self-reliance, to which this entire book shall bear witness. Let me observe now only that this self-accrediting is itself a fiduciary act of my own, which legitimizes in its turn the transposition of all my ultimate assumptions into declarations of my own beliefs.

When I gave this book the sub-title 'Towards a Post-Critical Philosophy' I had this turning point in mind. The critical movement, which seems to be nearing the end of its course today, was perhaps the most fruitful effort ever sustained by the human mind. The past four or five centuries, which have gradually destroyed or overshadowed the whole medieval cosmos, have enriched us mentally and morally to an extent unrivalled by any period of similar duration. But its incandescence had fed on the combustion of the Christian heritage in the oxygen of Greek rationalism, and when this fuel was exhausted the critical framework itself burnt away.

Modern man is unprecedented; yet we must now go back to St. Augustine to restore the balance of our cognitive powers. In the fourth century A.D., St. Augustine brought the history of Greek philosophy to a close by inaugurating for the first time a post-critical philosophy. He taught that all knowledge was a gift of grace, for which we must strive under the guidance of antecedent belief: *nisi credideritis, non intelligitis*.[1] His doctrine ruled the minds of Christian scholars for a thousand years. Then faith declined and demonstrable knowledge gained superiority over it. By the end of the seventeenth century Locke distinguished as follows between knowledge and faith:

> How well-grounded and great soever the assurance of faith may be wherewith it is received; but faith it is still and not knowledge; persuasion and not certainty. This is the highest the nature of things will permit us to go in matters of revealed religion, which are therefore called matters of faith; a persuasion of our own minds, short of knowledge, is the result that determines us in such truths.[2]

Belief is here no longer a higher power that reveals to us knowledge lying beyond the range of observation and reason, but a mere personal acceptance which falls short of empirical and rational demonstrability. The mutual position of the two Augustinian levels is inverted. If divine revelation continues to be venerated, its functions—like those of the Kings and Lords in England—are gradually reduced to that of being honoured on ceremonial occasions. All real power goes to the nominally Lower House of objectively demonstrable assertions.

Here lies the break by which the critical mind repudiated one of its two cognitive faculties and tried completely to rely on the remainder. Belief was so thoroughly discredited that, apart from specially privileged opportunities, such as may be still granted to the holding and profession of religious beliefs, modern man lost his capacity to accept any explicit statement as his own belief. All belief was reduced to the status of subjectivity: to that of an imperfection by which knowledge fell short of universality.

We must now recognize belief once more as the source of all knowledge. Tacit assent and intellectual passions, the sharing of an idiom and of a cultural heritage, affiliation to a like-minded community: such are the impulses which shape our vision of the nature of things on which we rely for our mastery of things. No intelligence, however critical or original, can operate outside such a fiduciary framework.

---

[1] St. Augustine, *De libero arbitrio*, Book I, par. 4: 'The steps are laid down by the prophet who says, "Unless ye believe, ye shall not understand".'
[2] Locke, *A Third Letter on Toleration*.

While our acceptance of this framework is the condition for having any knowledge, this matrix can claim no self-evidence. Although our fundamental propensities are innate, they are vastly modified and enlarged by our upbringing; moreover, our innate interpretations of experience may be misleading, while some of our truest acquired beliefs, though clearly demonstrable, may be most difficult to hold. Our mind lives in action, and any attempt to specify its presuppositions produces a set of axioms which cannot tell us why we should accept them. Science exists only to the extent to which there lives a passion for its beauty, a beauty believed to be universal and eternal. Yet we know also that our own sense of this beauty is uncertain, its full appreciation being limited to a handful of adepts, and its transmission to posterity insecure. Beliefs held by so few and so precariously are not indubitable in any empirical sense. Our basic beliefs are indubitable only in the sense that we believe them to be so. Otherwise they are not even beliefs, but merely somebody's states of mind.

This then is our liberation from objectivism: to realize that we can voice our ultimate convictions only from within our convictions—from within the whole system of acceptances that are logically prior to any particular assertion of our own, prior to the holding of any particular piece of knowledge. If an ultimate logical level is to be attained and made explicit, this must be a declaration of my personal beliefs. I believe that the function of philosophic reflection consists in bringing to light, and affirming as my own, the beliefs implied in such of my thoughts and practices as I believe to be valid; that I must aim at discovering what I truly believe in and at formulating the convictions which I find myself holding; that I must conquer my self-doubt, so as to retain a firm hold on this programme of self-identification.

An example of a logically consistent exposition of fundamental beliefs is St. Augustine's *Confessions*. Its first ten books contain an account of the period before his conversion and of his struggle for the faith he was yet lacking. Yet the whole of this process is interpreted by him from the point of view which he reached after his conversion. He seems to acknowledge that you cannot expose an error by interpreting it from the premises which lead to it, but only from premises which are believed to be true. His maxim *nisi credideritis non intelligitis* expresses this logical requirement. It says, as I understand it, that the process of examining any topic is both an exploration of the topic, and an exegesis of our fundamental beliefs in the light of which we approach it; a dialectical combination of exploration and exegesis. Our fundamental beliefs are continuously reconsidered in the course of such a process, but only within the scope of their own basic premises.

Similarly, the decision which I have now stated, to give deliberate expression to the beliefs I find myself truly holding, was duly anticipated all during the previous parts of this book. As I surveyed the operations of the tacit coefficient in the art of knowing, I pointed out how everywhere the mind follows its own self-set standards, and I gave my tacit or explicit endorsement to this manner of establishing the truth. Such an endorsement is an action of the same kind as that which it accredits and is to be classed therefore as *a consciously a-critical statement*.

This invitation to dogmatism may appear shocking; yet it is but the corollary to the greatly increased critical powers of man. These have endowed our mind with a capacity for self-transcendence of which we can never again divest ourselves. We have plucked from the Tree a second apple which has for ever imperilled our knowledge of Good and Evil, and we must learn to know these qualities henceforth in the blinding light of our new analytical powers. Humanity has been deprived a second time of its innocence, and driven out of another garden which was, at any rate, a Fool's Paradise. Innocently, we had trusted that we could be relieved of all personal responsibility for our beliefs by objective criteria of validity — and our own critical powers have shattered this hope. Struck by our sudden nakedness, we may try to brazen it out by flaunting it in a profession of nihilism. But modern man's immorality is unstable. Presently his moral passions reassert themselves in objectivist disguise and the scientistic Minotaur is born.

The alternative to this, which I am seeking to establish here, is to restore to us once more the power for the deliberate holding of unproven beliefs. We should be able to profess now knowingly and openly those beliefs which could be tacitly taken for granted in the days before modern philosophic criticism reached its present incisiveness. Such powers may appear dangerous. But a dogmatic orthodoxy can be kept in check both internally and externally, while a creed inverted into a science is both blind and deceptive.

# 9

# THE CRITIQUE OF DOUBT

## 1. The Doctrine of Doubt

**M**Y resolve to make philosophy the declaration of my ultimate beliefs will have yet to be stated systemically. But we must first get rid of a prejudice which otherwise will undermine the morale of our whole enterprise.

It has been taken for granted throughout the critical period of philosophy that the acceptance of unproven beliefs was the broad road to darkness, while truth was approached by the straight and narrow path of doubt. We were warned that a host of unproven beliefs were instilled in us from earliest childhood. That religious dogma, the authority of the ancients, the teaching of the schools, the maxims of the nursery, all were united to a body of tradition which we tended to accept merely because these beliefs had been previously held by others, who wanted us to embrace them in our turn. We were urged to resist the pressure of this traditional indoctrination by pitting against it the principle of philosophic doubt. Descartes had declared that universal doubt should purge his mind of all opinions held merely on trust and open it to knowledge firmly grounded in reason. In its stricter formulations the principle of doubt forbids us altogether to indulge in any desire to believe and demands that we should keep our minds empty, rather than allow any but irrefutable beliefs to take possession of them. Kant said that in mathematics there was no room for mere opinion, but only for real knowledge, and that short of possessing knowledge we must refrain here from all judgment.[1]

The method of doubt is a logical corollary of objectivism. It trusts that the uprooting of all voluntary components of belief will leave behind unassailed a residue of knowledge that is completely determined by the objective evidence. Critical thought trusted this method unconditionally for avoiding error and establishing truth.

I do not say that during the period of critical thought this method has

[1] Kant, *Critique of Pure Reason*, B 851.

been always, or indeed ever, rigorously practised—which I believe to be impossible—but merely that its practice has been avowed and emphatic, while its relaxation was marginal and acknowledged only in passing. Admittedly, Hume was fairly frank in this respect; he openly chose to brush aside the conclusions of his own scepticism at those points where he did not think he could honestly follow them. Even so he failed to acknowledge that by so doing he was expressing his own personal beliefs; nor did he claim his right and accept his duty to declare such beliefs, when this amounted to the silencing of doubt and the abandonment of strict objectivity. His dissent from scepticism was strictly unofficial, forming no explicit part of his philosophy. Kant, however, took this contradiction seriously. He rallied to a superhuman effort to meet the situation exposed by Hume's critique of knowledge, without admitting any relaxation of doubt. 'The root of these disturbances', he wrote in respect of such difficulties,

> which lies deep in the nature of human reason, must be removed. But how can we do so, unless we give it freedom, nay, nourishment, to send out shoots so that it may discover itself to our eyes, and that it may then be entirely destroyed? We must, therefore, bethink ourselves of objections which have never yet occurred to any opponent, and indeed lend him our weapons, and grant him the most favourable position which he could possibly desire. We have nothing to fear, but much to hope for; namely, that we may gain for ourselves a possession which can never again be contested.[1]

Kant's hopes of an incontestable estate of reason has long since proved too high; but the fervour of doubting was transmitted up to our day. Popular thought in the nineteenth century was dominated by writers who, with an eye on the natural sciences, declared with complete assurance that they accepted no belief whatever that had not passed the test of unrestricted doubt. As a distinguished example for a thousand lesser ones, take this eloquent declaration of the principle of doubt by J. S. Mill:

> The beliefs which we have most warrant for have no safeguard to rest on, but a standing invitation to the whole world to prove them unfounded. If the challenge is not accepted, or is accepted and the attempt fails, we are far enough from certainty still; but we have done the best that the existing state of human reason admits of; we have neglected nothing that could give the truth a chance of reaching us; if the lists are open, we may hope that if there be a better truth, it will be found when the human mind is capable of receiving it; and in the meantime we may rely on having attained such approach to truth as is possible in our own day. This is the amount of certainty attainable by a fallible being, and this the sole way of attaining it.[2]

No proclamation of intellectual integrity could be more sincere; yet its words are devoid of any definite meaning, and their ambiguity conceals precisely the kind of personal convictions which they so loudly repudiate. For we know that J. S. Mill and other writers standing in the Liberal tradition of philosophic doubt held—and hold today—a wide range of

---

[1] Kant, *Critique of Pure Reason*, B 805 6.
[2] J. S. Mill, *On Liberty*, ch. 2 (Everyman edn., p. 83).

beliefs in science, ethics, politics, etc., which are by no means unquestioned. If they regard these as not having been 'proved unfounded', this merely reflects their decision to reject the arguments which are or were advanced against them. At no time could the beliefs of Liberalism be regarded as irrefutable in any other sense. But in this sense all fundamental beliefs are irrefutable as well as unprovable. The test of proof or disproof is in fact irrelevant for the acceptance or rejection of fundamental beliefs, and to claim that you strictly refrain from believing anything that could be disproved is merely to cloak your own will to believe your beliefs behind a false pretence of self-critical severity.

This complacency is not lessened, but further enhanced by humbly acknowledging the uncertainty of our own conclusions. For when we admit that the proofs on which our beliefs are supposed to be founded may conceivably be incomplete, we effectively cover up the brute fact that we can have no proof at all to warrant them. Indeed, the emphatic admission of our fallibility only serves to reaffirm our claim to a fictitious standard of intellectual integrity and to bring out the shining qualities of our open mind, in contrast to the hidebound attitude of those who openly profess their beliefs as their final personal commitment.

Doubt has been acclaimed not only as the touchstone of truth, but also as the safeguard of tolerance. The belief that philosophic doubt would appease religious fanaticism and bring about universal tolerance goes back to Locke, and this belief is still vigorously alive in our own day. Its most influential representative, Lord Russell, expressed it eloquently many times, as for example in this passage:

> Arians and Catholics, Crusaders and Muslims, Protestants and adherents of the Pope, Communists and Fascists, have filled large parts of the last 1600 years with futile strife, when a little philosophy would have shown both sides in all these disputes that neither had any good reason to believe itself in the right. Dogmatism . . . in the present age as in former times, is the greatest of the mental obstacles to human happiness.[1]

It remains deeply ingrained in the modern mind—as I find even in my own mind—that though doubt may become nihilistic and imperil thereby all freedom of thought, to refrain from belief is always an act of intellectual probity as compared with the resolve to hold a belief which we could abandon if we decided to do so. To accept a belief by yielding to a voluntary impulse, be it my own or that of others placed in a position of authority, is felt to be a surrender of reason. You cannot teach the necessity for doing this without incurring—even in your own heart—the suspicion of obscurantism. At every step in quest of a post-critical philosophy the warning of the critical age will echo in our minds. In the words of Kant:

> Reason must in all its undertakings subject itself to criticism; should it limit freedom of criticism by any prohibitions, it must harm itself, drawing

---

[1] Bertrand Russell, *Universities Quarterly*, 1 (1946), p. 38.

upon itself a damaging suspicion. Nothing is so important through its usefulness, nothing so sacred, that it may be exempted from this searching
examination, which knows no respect for persons. Reason depends on this
freedom for its very existence.[1]

I shall not feel reassured in advocating an attitude of a-critical belief, unless
I have first fully met this warning by a critical examination of the principle
of doubt.

## 2. EQUIVALENCE OF BELIEF AND DOUBT

We may speak of doubt in a very wide sense. A moment of hesitancy such
as may be observed in the behaviour of any animal possessing a glimmer of
intelligence, could be described as doubt. A marksman taking aim may be
in doubt until he pulls the trigger. The renewed attempts of a poet to get a
line right are filled with such hesitations.[2] A measure of such tacit doubt is
present in all articulate forms of intelligence within the act of assertion,
throughout its many variants. It is the *only* kind of doubt that applies to
the acceptance of an articulate framework as a dwelling place, and it
controls therefore at their source the range and manner of our mental
existence. But before examining this deeper doubt I shall deal briefly with
*explicit* forms of doubt: that is, the questioning of explicit statements of
fact either asserted by others or previously asserted by ourselves.

The first point in my critique of doubt will be to show that the doubting
of any explicit statement merely implies an attempt to deny the belief
expressed by the statement, in favour of other beliefs which are not
doubted for the time being.

Suppose somebody says 'I believe $p$' where $p$ stands for 'planets move
along elliptic orbits', or else for 'all men are mortal'. And I reply 'I doubt
$p$'. This may be taken to mean that I contradict $p$, which could be expressed
by 'I believe not-$p$'. Alternatively, I may be merely objecting to the
assertion of $p$ as true, by denying that there are sufficient grounds to
choose between $p$ or not-$p$. This may be expressed by saying 'I believe $p$
is not proven'. We may call the first type of doubt 'contradictory' and the
second 'agnostic'.

It is immediately apparent that an expression of contradictory doubt 'I
believe not-$p$' is of the same character as the affirmation 'I believe $p$' which
it calls in question. For between $p$ and not-$p$ there is no other difference
than that they refer to different matters of fact. 'I believe not-$p$' could
stand for the allegation that planets move along orbits which are not
elliptical.

The history of science offers many illustrations for the logical equivalence
of affirmation and contradiction. In mathematics a problem may often be

---

[1] Kant, *Critique of Pure Reason*, B 766.
[2] Well illustrated by Stephen Spender, *The Making of a Poem*, London, 1955, pp.
51–2.

set for a time in the positive form and then turned round into its opposite, namely to prove the impossibility of finding a solution for it. The squaring of the circle and the trisection of an angle by aid of ruler and compass were both inverted after a time in that sense; these constructions have been proved to be impossible. In mechanics, centuries of misplaced ingenuity having been spent on solving the problem of perpetual motion, eventually the impossibility of constructing such a machine was established as a fundamental law of nature. The Second and Third Laws of Thermodynamics, the theory of Chemical Elements, the Principles of Relativity and of Indeterminacy, as well as the Pauli Principle, were all formulated in terms of negations. Eddington based his whole system of nature on the assumption of a series of impossibilities. In all these cases the difference between a positive statement and the denial of a positive statement is merely a matter of wording, and the acceptance and rejection of either form of allegation are both decided by similar tests.

Agnostic doubt is somewhat more complex, as it is composed of two halves of which the second is not always clearly implied. The first half of an agnostic doubt is a contradictory doubt, which can be either temporary or final. A temporary agnostic doubt ('I believe $p$ is not proven') leaves open the possibility that $p$ may yet be demonstrated in future; while in its final form ('I believe $p$ cannot be proven') agnostic doubt denies that $p$ can ever be demonstrated. But neither of these denials alleges, strictly speaking, anything concerning the credibility of $p$, and hence they represent only a first and so far inconclusive part of agnostic doubt.

There are, in fact, various instances in which the first half of agnostic doubt is raised without prejudice to the credibility of the affirmation that is called in doubt. Suppose we want to consider the possibility of forming a deductive system with $p$ as one of its axioms. For this it would be necessary that $p$ should be consistent with the other axioms and independent of them; which means that neither $p$ nor not-$p$ should be provable within the proposed system of axioms when set out short of $p$ itself. If this has been successfully demonstrated, we are at liberty to include $p$ as one of our axioms, or else reject it, depending on reasons which in general will be quite independent of the demonstration in question. Only the Gödelian sentence which affirms its own undecidability within a given formal system emerges as true, once its undecidability has been demonstrated. Not so otherwise. Take, for example, the proof given by Gauss that the Fifth Postulate of Euclid cannot be derived from his first four Postulates: it served as a justification for considering the Fifth Postulate as optional and replacing it by newly invented non-Euclidean alternatives.

Yet even though in such cases the agnostic suspension of belief in respect to a particular statement says nothing about its credibility, it still has a fiduciary content. It implies the acceptance of certain beliefs concerning the possibilities of proof. Kant's demand that, in pure mathematics, unless we *know*, we must abstain from all acts of judgment, would therefore make agnostic doubt itself untenable. For this demand is based on affirming 'I believe $p$ is not proven' or 'not provable', which implies the acceptance of some not strictly indubitable framework within which $p$ can be said to be proven or not-proven, provable or not-provable. Kant would of course not have recognized this contradiction, since he held that the foundations of mathematics, including the axioms of Euclid, were indubitable *a priori*; but this view has proved to be mistaken.

We shall presently explore the scope of agnostic doubt further in the natural sciences, the law courts and in religious matters.

### 3. REASONABLE AND UNREASONABLE DOUBT

The fiduciary character of doubt is revealed by the limitation to 'reasonable doubt' characteristic of law and also of sceptical philosophy. To urge that doubt must be reasonable, is to rely on something that cannot reasonably be doubted—that is, in legal phrase, a 'moral certainty'.[1] I shall illustrate this by the example of scientific doubt.

Natural scientists can be said to be more critical than astrologers only in so far as we regard their conception of stars and men as truer than that of the astrologers. More precisely speaking: when we disregard the evidence for the veridicity of horoscopes, we express the belief that this evidence can be explained within the scientific view of stars and men, as being merely accidental or otherwise invalid. During the seventeenth and eighteenth centuries scientific beliefs have thus opposed and discredited a whole system of supernatural beliefs and the authorities which taught these beliefs. We may regard this sceptical movement as altogether reasonable and be unaware of its fiduciary character until we are confronted with its blunders, for example in the scepticism of scientists concerning meteorites, of which I spoke before.[2] Ordinary people were convinced of the fall of a meteorite, when an incandescent mass struck the earth with a crash of thunder a few yards away, and they tended to attach supernatural significance to it. The scientific committees of the French Academy disliked this interpretation so much that they managed, during the whole of the eighteenth century, to explain the facts away to their own satisfaction. It was again scientific scepticism which brushed aside all the instances of hypnotic phenomena occurring in the form of miraculous cures and spellbinding, and which—even in the face of the systematic demonstrations of hypnosis by Mesmer and his successors—denied for another century after Mesmer's first appearance the reality of hypnotic phenomena. When the medical profession ignored such palpable facts as the painless amputation of human limbs, performed before their own eyes in hundreds of successive cases, they acted in a spirit of scepticism, convinced that they were defending science

---

[1] C. S. Kenny, *Outlines of Criminal Law*, 12th edn., Cambridge, 1926, pp. 389–90.
[2] See p. 138 above.

against imposture.[1] We regard these acts of scepticism as unreasonable and indeed preposterous today, for we no longer consider the falling of meteorites or the practice of mesmerism to be incompatible with the scientific world view. But other doubts, which we now sustain as reasonable on the grounds of our own scientific world view, have once more only our beliefs in this view to warrant them. Some of these doubts may turn out one day to have been as wanton, as bigoted and dogmatic as those of which we have now been cured. My critique of objectivism has already pointed to some of these perverse doubts which scepticism expresses today.

## 4. SCEPTICISM WITHIN THE NATURAL SCIENCES

In the natural sciences the proof of an allegation cannot be as rigorous as it normally is in mathematics. We often refuse to accept an alleged scientific proof largely because on general grounds we are reluctant to believe what it tries to prove. It was the presumption of Wöhler and Liebig against the idea that fermentation was due to living cells which made them disregard the evidence in its favour. The kind of evidence produced by van't Hoff for the asymmetrical carbon atom was condemned by Kolbe as worthless by the very nature of its argumentation. Pasteur's evidence for the absence of spontaneous generation was rejected by his opponents by interpreting it in their own way, and even Pasteur admitted that this possibility could not be excluded.[2]

Inexplicable things continue to happen in a laboratory. For example, traces of helium or traces of gold may unaccountably turn up in sealed vessels and the effect may be reproducible. At a time when the artificial transmutation of elements first appeared vaguely possible, a number of scientists accepted such observations as evidence that transmutation had taken place. But once the true conditions for transmutation had

[1] Mesmer (1734–1815) was denounced as an impostor; Esdaile (1808–59) carried out about 300 major operations painlessly under mesmeric trance in India, but neither in India nor in Great Britain could he get medical journals to print accounts of his work. His results were explained by the assumption that natives liked to be operated upon and tried to please Esdaile. In England in 1842 W. S. Ward amputated a leg painlessly under mesmeric trance and reported the case to the Royal Medical and Chirurgical Society. 'The Society however refused to believe. Marshall Hall, the pioneer in the study of reflex action, urged that the patient must have been an impostor, and the note of the paper's having been read was stricken from the minutes of the Society . . . Eight years later, Marshall Hall informed the Society that the patient had confessed to an imposition, but that the source of his information was indirect and confidential. The patient, however, then signed a declaration that the operation had been painless.' Elliotson (1791–1868) was Professor of Medicine in the University College, London, founder of University College Hospital, practised mesmerism there, mainly for therapeutical purposes, until in 1837 the Council of University College forbade this practice, whereupon he resigned his chair. (This account is based on and the quotations are from E. Boring, *History of Experimental Psychology*, New York, 2nd edn., 1950. It is supported in greater detail in respect to Elliotson's career in Harley Williams, *Doctors Differ*, London, 1946. See Part One, ch. 4, p. 52 above.)

[2] See above Part Two, ch. 6, p. 157.

been elucidated, observations of this kind were no longer heeded by scientists.[1]

In an earlier book I have mentioned a paper published by Lord Rayleigh in June, 1947, in the *Proceedings of the Royal Society*, describing a simple experiment which demonstrated that a hydrogen atom impinging on a metal wire released energies ranging up to a hundred electron volts.[2] This conclusion, if correct, would have been of immense importance. Physicists whom I consulted could find no fault with the experiment, yet they ignored its results, and did not even think it worth while to repeat it. A possible explanation of it is suggested by a recent experiment of R. H. Burgess and J. C. Robb.[3] They have shown that in the presence of traces of oxygen (0·22–0·94 mm.), hydrogen atoms will cause a rise in temperature on a metal wire many times exceeding the heat of recombination of H atoms on the wire. If this is the explanation, physicists were well advised to ignore this work.

A scientist must commit himself in respect to any important claim put forward within his field of knowledge. If he ignores the claim he does in fact imply that he believes it to be unfounded. If he takes notice of it, the time and attention which he diverts to its examination and the extent to which he takes account of it in guiding his own investigations are a measure of the likelihood he ascribes to its validity. Only if a claim lies totally outside his range of responsible interests can the scientist assume an attitude of completely impartial doubt towards it. He can be strictly agnostic only on subjects of which he knows little and cares nothing.

## 5. Is Doubt a Heuristic Principle?

We have seen that the practice of scientific scepticism in respect to allegations rejected by science consists in upholding the current scientific view of their subject matter, and we have seen this kind of scepticism also directed against fellow scientists in a fundamental controversy within science. But is there not a kind of rebellious scientific achievement which requires the power to doubt hitherto accepted beliefs of science? To be sure, every scientific discovery is conservative in the sense that it maintains and expands science as a whole, and to this extent confirms the scientific view of the world and strengthens its hold on our minds; but no major discovery can fail also to modify the outlook of science, and some have changed it profoundly. A number of revolutionary discoveries, like those of the heliocentric system, of genes, of quanta, of radioactivity or of relativity, come readily to mind. Might it not be that the process of assimilating fresh topics to the existing system *merely* conserves science, while true innovations include a revolutionary change by which the whole framework of science is reformed?

[1] See my *Science, Faith and Society*, Oxford, 1946, pp. 75–6.
[2] *The Logic of Liberty*, London and Chicago, 1951, p. 12.
[3] R. H. Burgess and J. C. Robb, *Trans. Far. Soc.*, **53** (1957).

This sounds plausible, but it is not true. The power to expand hitherto accepted beliefs far beyond the scope of hitherto explored implications is itself a pre-eminent force of change in science. It is this kind of force which sent Columbus in search of the Indies across the Atlantic. His genius lay in taking it literally and as a guide to practical action that the earth was round, which his contemporaries held vaguely and as a mere matter for speculation. The ideas which Newton elaborated in his *Principia* were also widely current in his time; his work did not shock any strong beliefs held by scientists, at any rate in his own country. But again, his genius was manifested in his power of casting these vaguely held beliefs into a concrete and binding form. One of the greatest and most surprising discoveries of our own age, that of the diffraction of X-rays by crystals (in 1912) was made by a mathematician, Max von Laue, by the sheer power of believing more concretely than anyone else in the accepted theory of crystals and X-rays. These advances were no less bold and hazardous than were the innovations of Copernicus, Planck or Einstein.

There exists, accordingly, no valid heuristic maxim in natural science which would recommend either belief or doubt as a path to discovery. Some discoveries are prompted by the conviction that something is fundamentally lacking in the existing framework of science, others by the opposite feeling that there is far more implied in it than has yet been realized. The first conviction may be regarded as more sceptical than the second, but it is precisely the first which is more likely to be hampered by doubt—owing to excessive adherence to the existing orthodoxy of science.

Besides, as there is no rule to tell us at the moment of deciding on the next step in research what is truly bold and what merely reckless, there is none either for distinguishing between doubt which will curb recklessness and thus qualify as true caution, and doubt which cripples boldness and will stand condemned as unimaginative dogmatism. Vesalius is praised as a hero of scientific scepticism for boldly rejecting the traditional doctrine that the dividing wall of the heart was pierced by invisible passages; but Harvey is acclaimed for the very opposite reason, namely for boldly assuming the presence of invisible passages connecting the arteries with the veins.

## 6. Agnostic Doubt in Courts of Law

The procedure of the law courts prescribes the observance of strictly impartial agnostic doubt in respect to a specified range of topics. There are a number of matters which would normally be considered relevant to a criminal charge, into which the court may not enquire. If a man, having just witnessed a murder, described it to a party of people and later collapsed and died, the murderer might go scot free without any member of the party being allowed to report in court what the man who witnessed the deed had told him. Much other information, for example evidence concerning the character of the accused, that would normally be relevant, may not be raised by the prosecution. If any information excluded from judicial notice is inadvertently brought up, the jury are directed to forget it. By the enforcement of such rules which restrict the usual range of interest to which the members of the court would respond in connection with the case before them, the law succeeds in keeping out of their minds a certain number of allegations *p* and their contradictories not-*p* which they would otherwise entertain. By suppressing the voicing of either of these alternatives the law hopes to achieve a strictly agnostic attitude in respect to them. This is equivalent to the establishment of the first half of agnostic doubt in respect to the *p*'s in question, without any subsequent decision as to the credibility of these *p*'s. In such a case the range of beliefs entertained is effectively reduced, but only to the extent to which we are prevented from knowing of the matters to which they refer.

On the other hand, the questions relevant to the issue which are admitted in court must be decided one way or another. If, after the evidence is exhausted it is found that both *p* and not-*p* are consistent with it, the presumptions laid down by law decide in favour of one of the two alternatives. The most widely known legal presumptions are perhaps those which grant the benefit of the doubt to the accused in criminal proceedings. If the allegations *p* and not-*p* are both consistent with the evidence, the court will as a rule presume—i.e. believe—the alternative which does not prejudice the innocence of the accused. But there is no sweeping presumption in this respect. In the absence of proof to the contrary an accused is presumed sane although this tells against him. There are numerous legal presumptions of a particular kind which prevail both in civil and criminal suits and bear no relation to the distinction between the two contesting sides. Such presumptions serve largely to avoid deadlock and to decide as reasonably as possible important issues for which there is no evidence that would be normally regarded as adequate. The judge will find for example that, of a married couple who were drowned together, the older one will have died first, even though he knows nothing whatever about it.

To take into consideration any matter which the court must not notice, or to form beliefs that are contrary to the proper legal presumptions, or quite generally, to form any legally unreasonable beliefs, is condemned as bias or caprice. In so far as these rules exclude the forming of certain beliefs to which we would normally be prone, they enforce a doubt or a state of agnosticism in respect to these beliefs. But once more, as in the scientific interpretation of experience, the system of beliefs which displaces here the beliefs of the man-in-the-street is no less definite and comprehensive than that which would be held otherwise. The law which orders that a man be presumed innocent until he is found guilty, does not impose an open mind on the court, but tells it on the contrary what to believe at the start: namely that the man is innocent. Even the legal exclusion of normally relevant matter may be interpreted as the prescription of specific beliefs, namely that they are in fact irrelevant to the issue. In all these respects the supposedly open mind of an unbiassed court can be sustained only by a much *stronger* will to believe than the usual beliefs of a person discharging no judicial responsibility. The former beliefs are much less plausible than the latter, and to this extent they may be said to be dogmatically imposed for the occasion. This seems to have been one of the reasons why Western observers were at first inclined to take lightly the patent omission of legal safeguards in the Moscow trials. Proper legal procedure does not appeal to common sense.

The dogmatic and often arbitrary character of legally imposed beliefs is justified by the peculiar context in which they are established and affirmed. The court does not try to find out the truth about certain interesting events, but only to find—by a legally prescribed procedure—the facts relevant to a certain legal issue. The will to believe these affirmations, even when they are not justifiable in themselves, originates in the will to do justice by making these affirmations and acting upon them. There is therefore, strictly speaking, no possible contradiction between the factual findings of a court of law and those of scientific and ordinary experience. They by-pass each other. The relation between observed facts and legal facts is similar in principle to that between factual experience and an art based on such experience, or between empirical facts and mathematical conceptions. In all these cases experience serves *as a theme* for an intellectual activity which develops one aspect of it into a system that is established and accepted on the grounds of its internal evidence. The system of legal facts is accepted as part of a social life shaped by the corresponding legal framework.

## 7. RELIGIOUS DOUBT

The belief in the efficacy of doubt as a solvent of error was sustained primarily—from Hume to Russell—by scepticism about religious dogma and the dislike of religious bigotry. This has been the dominant passion of critical thought for centuries, in the course of which it has completely transformed man's outlook on the universe. It must, accordingly, form the main subject of my critique of doubt. I shall limit the argument to religious doubts in respect of the Christian faith, picking up the thread from the point reached in my chapter on Intellectual Passions.

Religion, considered as an act of worship, is an indwelling rather than an affirmation. God cannot be observed, any more than truth or beauty can be observed. He exists in the sense that He is to be worshipped and obeyed, but not otherwise; not as a fact—any more than truth, beauty or justice exist as facts. All these, like God, are things which can be apprehended only in serving them. The words 'God exists' are not, therefore, a statement of fact, such as 'snow is white', but an accreditive statement, such as ' "snow is white" is true', and this determines the kind of doubt to which the statement 'God exists' can be subjected.[1] For since ' "snow is white" is true' stands for an a-critical act of assertion made by the speaker, it is not a descriptive sentence and cannot be the subject of explicit doubt. It can merely be uttered with varying degrees of confidence, and what its assertion may lack in perfect assurance might then be regarded as the doubt attached by the speaker to his own assertion. This would be a tacit doubt, an inarticulate hesitancy, like that of a marksman dubiously pulling the trigger, and the words 'God exists' can also be doubted only in the sense of a tacit hesitancy.

But this formulation somewhat exaggerates the sharpness of the distinction between acts of faith which imply (in a sense yet to be explored) the existence of God, and the meaning of the words 'God exists'. It is true that these words form no part of worship and can mean nothing beyond the endorsement of an act of faith by which the speaker has surrendered to God, yet it is not possible to separate this act of acceptance as sharply from that which it accepts, as we can the acceptance of factual statements from the accepted statements.

We shall turn for guidance instead, therefore, to the more general relationship which obtains between our tacit powers of comprehension and the spoken words and empirical particulars controlled by our comprehension. (See Part Two, ch. 5, p. 92.) This will lead us back to the conception of religious worship as a heuristic vision and align religion in turn also with the great intellectual systems, such as mathematics, fiction and the fine arts, which are validated by becoming happy dwelling places of the human mind. We shall see then that in spite of its a-critical character, the force of religious conviction does depend on factual evidence and can be affected by doubt concerning certain facts. Let me develop this programme.

In the chapter on Intellectual Passions I have described the Christian faith as a passionate heuristic impulse which has no prospect of consummation. A heuristic impulse is never without a sense of its possible inadequacy, and what it lacks in absolute assurance may be described as its inherent doubt. But the sense of inadequacy inherent in the Christian faith goes beyond this, for it is part of the Christian faith that its striving can never reach an endpoint at which, having gained its desired result, its continuation would become unnecessary. A Christian who reached his spiritual endpoint in this life would have ceased to be a Christian. A sense of its own imperfection is essential to his faith. 'Faith embraces itself and the doubt about itself,' writes Tillich.[2]

---

[1] On the difficulty of affirming 'God exists', see Paul Tillich, *Systematic Theology*, 1, London, 1953, pp. 227-33, 262-3.
[2] Paul Tillich, *Biblical Religion and the Search for Ultimate Reality*, London, 1955, p. 61.

Yet according to the Christian faith this inherent dubiety of the true faith is sinful and this sin is an ineradicable source of anguish. Take away doubt, sin and anguish, and Christian faith turns into a caricature of itself. It becomes a set of inaccurate, often false and largely meaningless statements, accompanied by conventional gestures and complacent moralizing. This is the forbidden endpoint of all Christian endeavour: its relapse into emptiness.

A heuristic impulse can live only in the pursuit of its proper enquiry. The Christian enquiry is worship. The words of prayer and confession, the actions of the ritual, the lesson, the sermon, the church itself, are the clues of the worshipper's striving towards God. They guide his feelings of contrition and gratitude and his craving for the divine presence, while keeping him safe from distracting thoughts.

As a framework expressing its acceptance of itself as a dwelling place of the passionate search for God, religious worship can say nothing that is true or false. Words of prayer are addressed to God, and while other parts of the service *speak of* God, they are mostly declarations of interpersonal relations—such as the praise of God. Some parts of worship, like the credo, admittedly make theological assertions, and the lessons from the Bible are couched in plainly narrative language. But the accent of the credo lies on the words: 'I believe' which emotionally endorse worship, while the extracts from the Bible are not quoted in the course of a Christian religious service in order to convey information, but as starting points for teachings that sustain the faith. All such statements function as subsidiaries to worship.

But the doctrines of theology and the records of the Bible are also taught in themselves. Can their statements *then* be said to be true or false and be subjected to explicit doubt? The answer is neither yes nor no, and it can be given here only in outline.

Only a Christian who stands in the service of his faith can understand Christian theology and only he can enter into the religious meaning of the Bible. Theology and the Bible together form the context of worship and must be understood in their bearing on it; but we shall see that this bearing is different in the two cases.

A theological statement, like 'God exists', may be little more than the endorsement of an act of worship in descriptive terms; something like saying ' "Snow is white" is true' after having confidently said 'Snow is white'. To this extent the expression 'God exists' is a-critical and not explicitly dubitable. But theology as a whole is an intricate study of momentous problems. It is a theory of religious knowledge and a corresponding ontology of the things thus known. As such, theology reveals, or tries to reveal, the implications of religious worship, and it can be said to be true or false, but only as regards its adequacy in formulating and purifying a pre-existing religious faith. While theological attempts to prove the existence of God are as absurd as philosophical attempts to prove the

281

premisses of mathematics or the principles of empirical inference, theology pursued as an axiomatization of the Christian faith has an important analytic task. Though its results can be understood only by practising Christians, it can greatly help them to understand what they are practising.

Theological accounts of God must, of course, appear meaningless and often blatantly self-contradictory if taken to claim validity within the universe of observable experience. Such a result is inevitable, whenever a language that is apposite to one subject matter is used with reference to another altogether different matter. The comparatively modest attempt to describe atomic processes in terms of classical electro-magnetics and mechanics has led to self-contradictions which appeared no less intolerable until we eventually got accustomed to them. Today physicists enjoy these apparent absurdities which they alone can comprehend, even as Tertullian seems to have enjoyed the startling paradoxes of his faith. Far from raising doubts in my mind concerning the rationality of Christian beliefs, the paradoxes of Christianity will serve me as examples for an analogous framing and stabilizing of other beliefs by which man strives to satisfy his own self-set standards.

Theology comprises biblical exegesis and the principles of biblical exegesis, and in this context it deals also with the question which I have set myself here, namely how religious faith depends on observable facts, or—more precisely—on the truth or falsity of statements concerning observable facts. I shall therefore have to trespass now for a short stretch on the domain of theology.

I have described Christian religious service as a framework of clues which are apt to induce a passionate search for God. I have spoken of the tacit act of comprehension which originates faith from such clues. The capacity for such skilful religious knowing seems universal, at least in children. Once acquired, the skill is hardly ever lost, but it is rarely mastered at an advanced age without some previous training in childhood. Divine service can mean nothing to a person completely lacking the skill of religious knowing.

The power of a framework composed of words and gestures to elicit its own religious comprehension in a receptive person will depend partly on the non-religious significance of its elements. The framework must impress a child or an unbeliever in the first place by the appeal made by its dogma, its narratives, its morality and its ritual exercise, before these have been religiously comprehended by him. Historical evidence confirming some decisive event recorded by the Gospels will, therefore, augment the strength of Christian teaching. And conversely, Biblical criticism and the progress of science which weakened or destroyed the extra-religious plausibility of many Biblical narratives and discredited the supposed magical powers of some Christian ritual, were bound to shake a faith implemented by the assertion of such teachings and the performance of such rituals. Modern theology has accepted these attacks as its guide for re-interpreting and consolidating the Christian faith in a truer form. In the following I shall try to state this result in my own terms.[1]

Let us take in for this purpose the whole of experience, including—but extending of course far beyond—the reading of the Bible, and let us observe its religious effects on a person's mind in the process of his conversion. All this experience, which so far is still non-religious, may supply the mind with clues to the Christian faith, even as all kinds of knowledge, whether culled from scientific books or obtained by direct observation, may serve as clues for a scientific outlook. Both kinds of comprehension establish their own heuristic vision which asserts no specific fact. They are forms of highly personal knowledge which subsidiarily comprise a set of relatively impersonal experiences. This relation of factual clues to a heuristic vision is similar to the relation of factual experience to mathematics and to works of art. The analogy brings religious faith into line with these great articulate systems which are also based on experience, but which the mind can yet inhabit without asserting any definite empirical facts. External experience is indispensable both to mathematics and art, *as their theme*, but to a person prepared to inhabit their framework, mathematics or art convey their own internal thought, and it is for the sake of this internal experience that his mind accepts their framework as its dwelling place.

Religion stands in a similar relation to non-religious experience. Secular experiences are its raw material: religion uses such experience as its theme for building up its own universe. The universe of every great articulate system is constructed by elaborating and transmuting one particular aspect of anterior experience: the Christian faith elaborates and renders effective the supernatural aspect of anterior experience in terms of its own internal experience. The convert enters into the articulate framework of worship and doctrine by surrendering to the religious ecstasy which their system

[1] Although I should not venture to declare that my argument in the present section agrees entirely with the views of any one theological writer, I find my own conception of the scope and method of a progressive Protestant theology confirmed by many passages in the writings of Paul Tillich. See, for example, his *Systematic Theology*, **1**, London, 1953, p. 130: 'Science, psychology, and history are allies of theology in the fight against the supranaturalistic distortions of genuine revelation. Scientific and historical criticism protect revelation; they cannot dissolve it, for revelation belongs to a dimension of reality for which scientific and historical analysis are inadequate. Revelation is the manifestation of the depth of reason and the ground of being. It points to the mystery of existence and to our ultimate concern. It is independent of what science and history say about the conditions in which it appears; and it cannot make science and history dependent on itself. No conflict between different dimensions of reality is possible. Reason receives revelation in ecstasy and miracles; but reason is not destroyed by revelation, just as revelation is not emptied by reason.' Or on 'the dynamics of revelation' (*ibid.*, p. 140): 'It is true that "Jesus Christ . . . the same yesterday, today and forever" is the immovable point of reference in all periods of church history. But the act of referring is never the same, since new generations with new potentialities of reception enter the correlation and transform it.' On the other hand (p. 144): 'Knowledge of revelation, although it is mediated primarily through historical events, does not imply factual assertion, and it is therefore not exposed to critical analysis by historical research. Its truth is to be judged by criteria which lie within the dimension of revelatory knowledge.'

evokes and accredits thereby its validity. This is again analogous to the process of validation by which men learn to enjoy and pursue mathematics or to contemplate with pleasure—and sometimes even produce—works of art.[1]

I have shown how natural science, mathematics and technology mutually interpenetrate each other. All the arts are similarly interwoven; while the arts and the methods of science penetrate each other in the domain of the humanities. Religion has even more comprehensive affinities: it can transpose all intellectual experiences into its own universe, and has also served, in reverse, most other intellectual systems as their theme. The relation of Christianity to natural experience, in which we are interested here, is but one thread in this network of mutual penetrations.

The two kinds of findings, the religious and the natural, by-pass each other in the same way as the findings of law courts by-pass ordinary experience. The acceptance of the Christian faith does not express the assertion of observable facts and consequently you cannot prove or disprove Christianity by experiments or factual records. Let me apply this to the belief in miracles. Ever since the attacks of philosophers like Bayle and Hume on the credibility of miracles, rationalists have urged that the acknowledgment of miracles must rest on the strength of factual evidence. But actually, the contrary is true: if the conversion of water into wine or the resuscitation of the dead could be experimentally verified, this would strictly disprove their miraculous nature. Indeed, to the extent to which any event can be established in the terms of natural science, it belongs to the natural order of things. However monstrous and surprising it may be, once it has been fully established as an observable fact, the event ceases to be regarded as supernatural. Recent biological suggestions, for example, that virgin birth *might* take place in exceptional circumstances would, if accepted as the explanation of the birth of Christ, not confirm, but totally destroy the doctrine of the Virgin Birth. It is illogical to attempt the proof of the supernatural by natural tests, for these can only establish the natural aspects of an event and can never represent it as supernatural. Observation may supply us with rich clues for our belief in God; but any scientifically convincing observation of God would turn religious worship into an idolatrous adoration of a mere object, or natural person.

Of course, an event which has in fact never taken place can have no supernatural significance; and whether it has taken place or not must be established by factual evidence. Hence the religious force of biblical criticism, shaking or, alternatively, corroborating certain facts which form the main themes of Christianity. But evidence that a fact has not occurred may sometimes leave largely unimpaired the religious truth conveyed by a narrative describing its occurrence. The book of Genesis and its great pictorial illustrations, like the frescoes of Michelangelo, remain a far more intelligent account of the nature and origin of the universe than the representation of the world as a chance collocation of atoms. For the

[1] Cf. pp. 192–5.

biblical cosmology continues to express—however inadequately—the significance of the fact that the world exists and that man has emerged from it, while the scientific picture denies any meaning to the world, and indeed ignores all our most vital experience of this world. The assumption that the world has some meaning which is linked to our own calling as the only morally responsible beings in the world, is an important example of the supernatural aspect of experience which Christian interpretations of the universe explore and develop. In chapter 13, I shall show how we can arrive by continuous stages from the scientific study of evolution to its interpretation as a clue to God.

Christianity is a progressive enterprise. Our vastly enlarged perspectives of knowledge should open up fresh vistas of religious faith. The Bible, and the Pauline doctrine in particular, may be still pregnant with unsuspected lessons; and the greater precision and more conscious flexibility of modern thought, shown by the new physics and the logico-philosophic movements of our age, may presently engender conceptual reforms which will renew and clarify, on the grounds of modern extra-religious experience, man's relation to God. An era of great religious discoveries may lie before us.

Let me sum up my conclusions about religious doubt before going further. The Christian faith can be attacked by doubt in two ways. Its internal evidence can be doubted in the sense in which conceptual innovations in mathematics or novel works of art can be held to be unsound. We may refuse, or at least hesitate, to enter on the mental life which they offer, either for want of appreciation for it or—more forcibly—for fear of losing our hold on reality. A similar kind of doubt applies to every heuristic vision: we are always conscious of some hazard attached to it. This kind of doubt is the hesitancy of an acceptance. Our reluctance to accept the habitation offered to our minds may be craven or wise, and so we may prove eventually to have been dull or rash. Yet we can apply to our action no test of the kind to which we can appeal for proving or disproving an explicit declaratory statement. There is therefore no possibility either for doubting what we do (or declare we do) in the sense in which an explicit statement can be doubted. Our doubt must remain intrinsic to a mental act of our own.

It is also part of the Christian faith that its striving is unfulfillable. It must always remain painfully conscious of its inherent dubiety. But since this is part of the faith, it does not derogate from it. Yet this indispensable internal dubiety of the Christian faith can be increased, even to the point of destroying our faith altogether, by explicit critical tests of the articulate framework on which we rely for deploying our faith. For the power of this framework to induce its own comprehension in terms of a surrender to God depends to an important extent on the convincing power of the statements which are its elements, in the same way as the power of a set of clues for inducing a heuristic vision based on these clues will depend on the reliability of the facts used as clues. Doubts directed against the clues as facts may thus shake the internal evidence of the system relying on them. Explicit doubts may intensify the intrinsic doubts of our acceptance to the point of converting it into a complete rejection.

The weakening of religious beliefs under the impact of advancing historical and scientific knowledge during the past 300 years represents, therefore, a case in which the effect of doubt was substantial. It destroyed the religious meaning of things without fully compensating for this loss by a different meaning, and the total volume of belief, from which all meaning flows, was effectively reduced. If the universe were in fact meaningless, the destruction of religious beliefs would have been fully justified. Since I do not believe that the universe is meaningless, I can admit only that the rejection of religion was reasonable in view of the grounds on which religious doctrines were asserted at the time. Today we should be grateful for the prolonged attacks made by rationalists on religion for forcing us to renew the grounds of the Christian faith. But this does not remotely justify the acknowledgment of doubt as the universal solvent of error which will leave truth untouched behind. For all truth is but the external pole of belief, and to destroy all belief would be to deny all truth. Though religious beliefs are often formulated more dogmatically than other beliefs, this is not essential. The extensive dogmatic framework of Christianity arose from ingenious efforts, sustained through many centuries, to axiomatize the faith already practised by Christians. In view of the high imaginative and emotional powers by which Christian beliefs control the whole person and relate him to the universe, the specification of these beliefs is much more colourful than are the axioms of arithmetic or the premises of natural science. But they belong to the same class of statements, performing kindred fiduciary functions.

We owe our mental existence predominantly to works of art, morality, religious worship, scientific theory and other articulate systems which we accept as our dwelling place and as the soil of our mental development. Objectivism has totally falsified our conception of truth, by exalting what we can know and prove, while covering up with ambiguous utterances all that we know and *cannot* prove, even though the latter knowledge underlies, and must ultimately set its seal to, all that we *can* prove. In trying to restrict our minds to the few things that are demonstrable, and therefore explicitly dubitable, it has overlooked the a-critical choices which determine the whole being of our minds and has rendered us incapable of acknowledging these vital choices.

## 8. Implicit Beliefs

The limitations of doubting as a principle can be elaborated further by extending our enquiry to the beliefs held in the form of our conceptual framework, as expressed in our language. Our most deeply ingrained convictions are determined by the idiom in which we interpret our experience and in terms of which we erect our articulate systems.[1] Our formally declared beliefs can be held to be true in the last resort only because of our logically anterior acceptance of a particular set of terms, from which all our references to reality are constructed.

The fact that primitive people hold distinctive systems of beliefs inherent in their conceptual framework and reflected in their language was first stated with emphasis by Lévy-Brühl earlier in this century. The more recent work of Evans-Pritchard on the beliefs of Azande[2] has borne out and has given further precision to this view. The author is struck by the intellectual force shown by the primitive African in upholding his beliefs against evidence which to the European seems flagrantly to refute them. An instance in point is the Zande belief in the powers of the poison-oracle. The oracle answers questions through the effects on a fowl of a poisonous substance called *benge*. The oracle-poison is extracted from a creeper gathered in a traditional manner, which is supposed to become effective only after it has been addressed in the words of an appropriate ritual. Azande—we are told—have no formal and coercive doctrine to enforce belief in witch-doctors and their practice of the poison-oracle, but their belief in these is the more firmly held for being embedded in an idiom which interprets all relevant facts in terms of witchcraft and oracular powers. Evans-Pritchard gives various examples of this peculiar tenacity of their implicit belief.

Suppose that the oracle in answer to a particular question says 'Yes', and immediately afterwards says 'No' to the same question. In our eyes this would tend to discredit the oracle altogether, but Zande culture provides a number of ready explanations for such self-contradictions. Evans-Pritchard lists no less than eight secondary elaborations of their beliefs by which Azande will account for the oracle's failure. They may assume that the wrong variety of poison had been gathered, or a breach of taboo committed, or that the owners of the forest where the poisonous creeper grows had been angered and avenged themselves by spoiling the poison; and so on.

Our author describes further the manner in which Azande resist any suggestion that *benge* may be a natural poison. He often asked Azande, he tells us, what would happen if they were to administer oracle-poison to a fowl without delivering an address, or if they were to administer an extra portion of poison to a fowl which has recovered from the usual doses. 'The Zande'—he continues—'does not know what would happen and is not interested in what would happen; no one has been fool enough to waste good oracle-poison in making such pointless experiments which only a

---

[1] Cf. p. 80.
[2] E. E. Evans-Pritchard, *Witchcraft, Oracles and Magic Among the Azande*, Oxford, 1937.

European could imagine. . . . Were a European to make a test which in his view proved Zande opinion wrong they would stand amazed at the credulity of the European. If the fowl died they would simply say that it was not good *benge*. The very fact of the fowl dying proves to them its badness.' [1]

This blindness of Azande to the facts which to us seem decisive is sustained by remarkable ingenuity. 'They reason excellently in the idiom of their beliefs,' (says Evans-Pritchard), 'but they cannot reason outside, or against, their beliefs because they have no other idiom in which to express their thoughts.' [2]

Our objectivism, which tolerates no open declaration of faith, has forced modern beliefs to take on implicit forms, like those of Azande. And no one will deny that those who have mastered the idioms in which these beliefs are entailed do also reason most ingeniously within these idioms, even while—again like Azande—they unhesitatingly ignore all that the idiom does not cover. I shall quote two passages to illustrate the high stability of two modern interpretative frameworks, based on these principles:

> My party education had equipped my mind with such elaborate shock-absorbing buffers and elastic defences that everything seen and heard became automatically transformed to fit a preconceived pattern. (A. Koestler, in *The God that Failed*, London, 1950, p. 68.)

> The system of theories which Freud has gradually developed is so consistent that when one is once entrenched in them it is difficult to make observations unbiased by his way of thinking. (Karen Horney, *New Ways of Psychoanalysis*, London, 1939, p. 7.)

The first of these statements is by a former Marxist, the second by a former Freudian writer. At the time when they still accepted as valid the conceptual framework of Marx or of Freud—as the case may be—these writers would have regarded the all-embracing interpretative powers of this framework as evidence of its truth; only when losing faith in it did they feel that its powers were excessive and specious. We shall see the same difference reappear in our appraisals of the interpretative power of different conceptual systems, as part of our acceptance or rejection of these systems.

## 9. Three Aspects of Stability

The resistance of an idiom of belief against the impact of adverse evidence may be regarded under three headings, each of which is illustrated by the manner in which Azande retain their beliefs in the face of situations which in our view should invalidate them. Analogous cases can be adduced from other systems of beliefs.

The stability of Zande beliefs is due, in the first place, to the fact that

[1] E. E. Evans-Pritchard, *Witchcraft, Oracles and Magic Among the Azande*, Oxford, 1937, pp. 314–15.      [2] *ibid.*, p. 338.

objections to them can be met one by one. This power of a system of implicit beliefs to defeat valid objections one by one is due to the circularity of such systems. By this I mean that the convincing power possessed by the interpretation of any particular new topic in terms of such a conceptual framework is based on past applications of the same framework to a great number of other topics not now under consideration, while if any of these other topics were questioned now, their interpretation in its turn would similarly rely for support on the interpretation of all the others. Evans-Pritchard observes this for Zande beliefs in mystical notions. 'The contradiction between experience and one mystical notion is explained by reference to other mystical notions.'[1]

So long as each doubt is defeated in its turn, its effect is to strengthen the fundamental convictions against which it was raised. 'Let the reader consider (writes Evans-Pritchard) any argument that would utterly demolish all Zande claims for the power of the oracle. If it were translated into Zande modes of thought it would serve to support their entire structure of belief.'[2] Thus the circularity of a conceptual system tends to reinforce itself by every contact with a fresh topic.

The circularity of the theory of the universe embodied in any particular language is manifested in an elementary fashion by the existence of a dictionary of the language. If you doubt, for example, that a particular English noun, verb, adjective or adverb has any meaning in English, an English dictionary dispels this doubt by a definition using other nouns, verbs, adjectives and adverbs, the meaningfulness of which is not doubted for the moment. Enquiries of this kind will increasingly confirm us in the use of a language.

Remember also what we have found about the axiomatization of mathematics; namely that it merely declares the beliefs implied in the practice of mathematical reasoning. The axiomatized system is therefore circular: our anterior acceptance of mathematics lends authority to its axioms, from which we then deduce in turn all mathematical demonstrations. The division of mathematical formulae, or of the asserted sentences of any deductive system, into axioms and theorems is indeed largely conventional, for we can usually replace some or all of the axioms by theorems and derive from these the previous axioms as theorems. Every assertion of a deductive system can be demonstrated by, or else shown to be implied as axioms of, the others. Therefore, if we doubt each assertion in its turn each is found confirmed by circularity, and the refutation of each consecutive doubt results in strengthening our belief in the system as a whole.

Circularity operates by divided roles when a number of persons holding the same set of pre-suppositions mutually confirm each other's interpretation of experience. Take the following story of a South African explorer, L. Magyar, collected by Lévy-Bruhl who regards it as typical.[3]

---

[1] *ibid.*, p. 339.    [2] *ibid.*, p. 319.
[3] Cf. Lévy-Bruhl, *The 'Soul' of the Primitive*, London, 1928, pp. 44–8.

Two African natives, S. and K., went to the wood to gather honey. S. found four big trees full of honey, whilst K. could find only one. K. went home bewailing his ill luck, while S. had been so fortunate. Meanwhile S., having returned to the wood to bring away the honey, was attacked by a lion and torn to pieces.

The relatives of the lion's victim at once went to the soothsayer to discover who was responsible for his death. The soothsayer consults the oracle several times and declares that K., jealous of S.'s rich harvest of honey, assumed the form of a lion in order to avenge himself. The accused denied his guilt strenuously and the chieftain ordered the matter to be settled by the ordeal of poison. 'Matters then followed their usual course'— says the explorer's account—'the ordeal was unfavourable to the accused, he confessed and succumbed to torture. . . . The accusation appears quite natural to the soothsayer who formulates it, the prince who orders the trial by ordeal, the crowd of bystanders and to K. himself who had been transformed into a lion, in fact to everybody except the European who happens to be present.' [1]

It is clear to us that K. had not actually experienced turning into a lion and tearing S. to pieces, and so at first he denied having done so. But he is confronted with an overwhelming case against himself. The interpretative framework which he shares with his accusers does not include the conception of accidental death; if a man is devoured by a lion there must be some effective reason behind it, such as the envy of a rival. This makes him an obvious suspect and when the oracle, which he has always trusted, confirms the suspicion, he can no longer resist the evidence of his guilt and he confesses having turned into a lion and having devoured S. This closes the circle of the argument and confirms the magical framework in which it was conducted, and it thus enhances the powers of this framework for assimilating the next case which will come under its purview.

Communists who have experienced the procedure which leads to confessions in Russian sabotage trials have described a similar circularity. The prisoner will usually resist the accusation to start with, but when it is persistently borne in upon him from all sides by the examining magistrate and by the evidence extorted from his former associates, he begins to give way to the convincing power of the case against himself. [2] On the grounds on which he had habitually condemned others he tends now to condemn

---

[1] *loc. cit.*
[2] Cf. A. Weissberg, *Conspiracy of Silence*, London, 1952, pp. 128, 202, 318, 352; F. Beck and W. Godin, *Russian Purge*, London, 1950, p. 179. The effect of self-condemnation in eliciting Communist confessions was first described by Arthur Koestler in *Darkness at Noon*. Since the accounts of some former prisoners have failed to confirm Koestler's theory, I have discussed the matter at some length with Mr. and Mrs. Paul Ignotus, who broadly endorsed the theory from their own extensive experience. The resistance of Communist prisoners against the accusations made against them is much reduced by their continued acceptance of Marxism-Leninism, and some prisoners go so far as utterly to doubt their own reason rather than to call in question the judgment of the Party.

himself—and thus close the circle which once more confirms these grounds and makes them stronger than ever for the next occasion.

A second aspect of stability arises from an automatic expansion of the circle in which an interpretative system operates. It readily supplies elaborations of the system which will cover almost any conceivable eventuality, however embarrassing this may appear at first sight. Scientific theories which possess this self-expanding capacity are sometimes described as epicyclical, in allusion to the epicycles that were used in the Ptolemaic and Copernican theory to represent planetary motions in terms of uniform circular motions. All major interpretative frameworks have an epicyclical structure which supplies a reserve of subsidiary explanations for difficult situations. The epicyclical character of Zande beliefs was shown above by the ready availability of eight different subsidiary assumptions for explaining a point-blank self-contradiction in two consecutive answers of an oracle.

The stability of Zande beliefs is manifested, thirdly, in the way it denies to any rival conception the ground in which it might take root. Experiences which support it could be adduced only one by one. But a new conception, e.g. that of natural causation, which would take the place of Zande superstition, could be established only by a whole series of relevant instances, and such evidence cannot accumulate in the minds of people if each of them is disregarded in its turn for lack of the concept which would lend significance to it. The behaviour of Azande whom Evans-Pritchard tried to convince that *benge* was a natural poison which owed none of its effectiveness to the incantations customarily accompanying its administration, illustrates the kind of contemptuous indifference with which we normally regard things of which we have no conception. 'We feel neither curiosity nor wonder', writes William James, 'concerning things so far beyond us that we have no concepts to refer them to or standards by which to measure them.' The Feugians in Darwin's voyage, he recalls, wondered at the small boats, but paid no attention to the big ship lying at anchor in front of them.[1] A more recent instance of this occurred when Igor Gouzenko, cypher clerk of the Soviet Embassy in Canada, tried in vain for two days in succession (September 5th and 6th, 1945) to attract attention to the documents concerning Soviet atomic espionage which he was showing round in Ottawa at the risk of his life.

This third defence mechanism of implicit beliefs may be called the principle of suppressed nucleation. It is complementary to the operations of circularity and self-expansion. While these protect an existing system of beliefs against doubts arising from any adverse piece of evidence, suppressed nucleation prevents the germination of any alternative concepts on the basis of any such evidence.

Circularity, combined with a readily available reserve of epicyclical elaborations and the consequent suppression in the germ of any rival

---

[1] William James, *Principles of Psychology*, **2**, New York, 1890, p. 110.

conceptual development, lends a degree of stability to a conceptual framework which we may describe as the measure of its completeness. We may acknowledge the completeness or comprehensiveness of a language and the system of conceptions conveyed by it—as we do in respect to Azande beliefs in witchcraft—without in any way implying that the system is correct.

## 10. The Stability of Scientific Beliefs

We do not share the beliefs of Azande in the power of poison-oracles, and we reject a great many of their other beliefs, discarding mystical conceptions and replacing them by naturalistic explanations. But we may yet deny that our rejection of Zande superstitions is the outcome of any general principle of doubt.

For the stability of the naturalistic system which we currently accept instead rests on the same logical structure. Any contradiction between a particular scientific notion and the facts of experience will be explained by other scientific notions; there is a ready reserve of possible scientific hypotheses available to explain any conceivable event. Secured by its circularity and defended further by its epicyclical reserves, science may deny, or at least cast aside as of no scientific interest, whole ranges of experience which to the unscientific mind appear both massive and vital.[1]

The restrictions of the scientific outlook which I summed up as objectivism have been recurrent themes throughout this book. My attempt to break out of this highly stabilized framework and to enter avenues of legitimate access to reality from which objectivism debars us will be presently pursued further. At the moment I only wish to give some illustrations to show how, *within science itself*, the stability of theories against experience is maintained by epicyclical reserves which suppress alternative conceptions in the germ; a procedure which in retrospect will appear right in some instances and wrong in others.

The theory of electrolytic dissociation proposed in 1887 by Arrhenius assumed a chemical equilibrium between the dissociated and the undissociated forms of an electrolyte in solution. From the very start, the measurements showed that this was true only for weak electrolytes like acetic acid, but not for the very prominent group of strong electrolytes, like common salt or sulphuric acid. For more than thirty years the discrepancies were carefully measured and tabulated in textbooks, yet no one thought of calling in question the theory which they so flagrantly contradicted. Scientists were satisfied with speaking of the 'anomalies of strong electrolytes', without doubting for a moment that their behaviour was in fact governed by the law that they failed to obey. I can still remember

[1] I have described similar stabilities before, when showing that two alternative systems of scientific explanation are separated by a logical gap and thus give rise to passionate controversy in science. See pp. 150–9, also pp. 112–13.

my own amazement when, about 1919, I first heard the idea mooted that the anomalies were to be regarded as a refutation of the equilibrium postulated by Arrhenius and to be explained by a different theory. Not until this alternative conception (based on the mutual electrostatic interaction of the ions) was successfully elaborated in detail, was the previous theory generally abandoned.

Contradictions to current scientific conceptions are often disposed of by calling them 'anomalies'; this is the handiest assumption in the epicyclical reserve of any theory. We have seen how Azande make use of similar excuses to meet the inconsistencies of poison-oracles. In science this process has often proved brilliantly justified, when subsequent revisions of the adverse evidence or a deepening of the original theory explained the anomalies. The modification of Arrhenius's theory for strong electrolytes is a case in point.

Another example may illustrate how a series of observations which at one time were held to be important scientific facts, were a few years later completely discredited and committed to oblivion, without ever having been disproved or indeed newly tested, simply because the conceptual framework of science had meanwhile so altered that the facts no longer appeared credible. Towards the end of the last century numerous observations were reported by H. B. Baker[1] on the power of intensive drying to stop some normally extremely rapid chemical reactions and to reduce the rate of evaporation of a number of commonly used chemicals. Baker went on publishing further instances of this drying effect for more than thirty years.[2] A large number of allegedly allied phenomena were reported from Holland by Smits[3] and some very striking demonstrations of it came from Germany.[4] H. B. Baker could render his samples unreactive sometimes only by drying them for periods up to three years; so when some authors failed to reproduce his results it was reasonable to assume that they had not achieved the same degree of desiccation. Consequently, there was little doubt at the time that the observed effects of intensive drying were true and that they reflected a fundamental feature of all chemical change.

Today these experiments, which aroused so much interest from 1900 to 1930, are almost forgotten. Textbooks of chemistry which thoughtlessly go on compiling published data still record Baker's observations in detail, merely adding that their validity 'is not yet certainly established',[5] or that

---

[1] *Journal of the Chemical Society of London*, 1894, 65, 611.

[2] Cf. *ibid.*, 1922, 121, 568; 1928, Part One, 1051.

[3] Smits, *The Theory of Allotropy* (1922). Baker's experiments are referred to (p. vii) as 'the most beautiful means of establishing the complexity of unary phases' postulated by the author.

[4] Coehn and Tramm, *Ber. deutsch. Chem. Ges.*, 56 (1923), 456; *Zeitschr. f. Phys. Chem.*, 105 (1923), 356, 110 (1924), 110; and Coehn and Jung, *Ber. deutsch. Chem. Ges.*, 56 (1923), 695. These authors reported the stopping of the photochemical combination of hydrogen and chlorine by intense drying.

[5] F. A. Philbrick, *Textbook of Theoretical and Inorganic Chemistry*, revised edition, London, 1949, p. 215.

'some (of his) findings are disputed by later workers, but the technique is difficult'.[1] But active scientists no longer take any interest in these phenomena, for in view of their present understanding of chemical processes they are convinced that most of them must have been spurious, and that, if some were real, they were likely to have been due to trivial causes.[2] This being so, our attitude towards these experiments is now similar to that of Azande towards Evans-Pritchard's suggestion of trying out the effects of oracle-poison without an accompanying incantation. We shrug our shoulders and refuse to waste our time on such obviously fruitless enquiries. The process of selecting facts for our attention is indeed the same in science as among Azande; but I believe that science is often right in its application of it, while Azande are quite wrong when using it for protecting their superstitions.[3]

I conclude that what earlier philosophers have alluded to by speaking of coherence as the criterion of truth is only a criterion of *stability*. It may equally stabilize an erroneous or a true view of the universe. The attribution of truth to any particular stable alternative is a fiduciary act which cannot be analysed in non-committal terms. I shall return to this point in my next chapter. At the moment it only serves to make it clear that there exists no principle of doubt the operation of which will discover for us which of two systems of implicit beliefs is true—except in the sense that we will admit decisive evidence against the one we do not believe to be true, and not against the other. Once more, the admission of doubt proves here to be as clearly an act of belief as does the non-admission of doubt.

## 11. UNIVERSAL DOUBT

What meaning can we attach in this light to a principle of universal doubt? So long as the reconsideration of any single belief is undertaken against an overwhelming background of unquestioned beliefs, the beliefs forming this background cannot simultaneously be alleged to be doubtful. Though every element of our belief can conceivably be confronted in its turn with all the rest, it is inconceivable that all should be subject simultaneously to this operation. But this is not to say that a system of beliefs can never be doubted as a whole. Euclidean geometry was called in question

[1] J. R. Partington, *General and Inorganic Chemistry*, 1946, p. 483. Thorpe's *Dictionary of Applied Chemistry*, Article 'Benzene and its Homologues' (1947), reports Baker's 'interesting discovery' without any qualification.

[2] Other examples of this procedure in which its result subsequently proved erroneous were given before in sec. 4. of this chapter to illustrate the equivalence of belief and doubt.

[3] The wise neglect of awkward facts may be of value even for the development of the deductive sciences. Greek mathematicians allowed themselves to be discouraged from developing algebra by the impossibility of representing the ratio of two incommensurable line segments in terms of whole numbers. B. L. Van der Waerden (*Science Awakening*, Groningen, 1954, p. 266) says that 'it does honour to Greek mathematics that it adhered inexorably to such logical consistency'. But had their successors been as exacting in their logical scruples, mathematics would have died of its own rigor.

as a whole and reduced to an optional status by the establishment of non-Euclidean geometry. We might conceivably feel inclined to reconsider one day our acceptance of mathematics as a whole. I have admitted already that the decline of religious faith entailed a genuine reduction in the volume of our beliefs.

Such speculations may serve to indicate a meaning of universal doubt which is free from self-contradiction. We may imagine an indefinite extension of the process of abandoning hitherto accepted systems of articulation, together with the theories formulated in these terms or implied in our use of them. This kind of doubt might eventually lead to the relinquishing, without compensation, of all existing means of articulation. It would make us forget all hitherto used idioms and dissolve all concepts which these idioms conveyed. Our articulate intellectual life, which operates by the handling of denotable concepts, would thus be reduced to abeyance for the time being.

Such an interpretation of universal doubt would certainly be repudiated by the adherents of the principle of doubt, but I can see no grounds on which they can dissent from it. This is the only manner of doubting which could truly liberate our minds from uncritically acquired preconceived beliefs. If we cannot accept the justification of holding beliefs uncritically, then our only logical alternative is to wipe out all such preconceived beliefs. And if this proves difficult in practice, we must at least recognize it as our ideal of perfection. We must accept the virgin mind, bearing the imprint of no authority, as the model of intellectual integrity.

At the risk of labouring the obvious, it should be made quite clear what exactly is implied in this assumption of a mind which could shape its judgment on all questions without any preconceived opinions. It cannot mean the mind of a newborn child, since this yet lacks sufficient intelligence to grasp any problems and discover any solutions for them. A virgin mind must be allowed to mature until the age at which it reaches its full natural powers of intelligence, but would have to be kept unshaped until then by any kind of education. It must be taught no language, for speech can be acquired only a-critically, and the practice of speech in one particular language carries with it the acceptance of the particular theory of the universe postulated by that language.

An entirely untutored maturing of the mind would, however, result in a state of imbecility. The emotional and appetitive impulses that are inherent in animal life will of course pour into such channels as are available to them. In the absence of a rational conceptual framework to guide them, their manifestation will not be sceptically restrained but frantic and inchoate. We have observed this already in animals well below the human level. I have mentioned how chickens brought up in isolation were perplexed and behaved confusedly, reflecting desperate consternation, when confronted for the first time with other chicks.[1]

[1] See Part Two, ch. 7, above, p. 210.

Yet even such dumb creatures would not be prevented from forming conceptions which seriously prejudiced a critical detachment. We have seen already how the mind actively participates in our sensory awareness of things (p. 97). Sometimes this way of seeing things is mistaken, and such instinctive error may gravely hamper the progress of philosophy and science. The contrast between a body 'at rest' and 'in motion' is compelling in all visual perception. We see the earth at absolute rest, with the sun, the moon and the stars swinging round it as their centre. The geocentric world view has a firm support in our most primitive perceptive prejudices. Indeed, even in Newtonian mechanics, the solar system was regarded in its turn as fixed, with the rest of the universe moving around it; and this prejudice was finally discarded only in Einstein's general theory of relativity. Today the Newtonian framework is condemned as the product of uncritical thought; yet its error can be traced back to the lowest level of visual perception and would therefore be committed even by children raised among wolves or nursed to maturity in the solitude of an incubator.

If, therefore, the ideal of a virgin mind is to be pursued to its logical limit, we have to face the fact that every perception of things, particularly by our eyes, involves implications about the nature of things which could be false. Whether we see an object as black or white is not determined by the amount of light it sends into our eye. Snow seen at dusk appears white, a dinner jacket seen in sunshine appears black, though the jacket in this case sends more light into the eyes than the snow. It is said that black is black and white is white—yet whether we see an object as black or white is decisively affected by the whole context in which the light from the object reaches our eyes. The way we embody this context in our perception of the colours, sizes, distances and shapes of the perceived object is determined by our innate physiological inclinations and their subsequent development under the influence of our experience. My perceptions today as an adult are different from those which I had as a new-born baby, and much of this difference is due to the functioning of convergence, adaptation and other more complex sensory processes, which are performed according to principles which may be wrong. But if all these functions could be eliminated by training myself to look at things again with unperceiving eyes, letting their images sweep across my retina, like a motion picture which is continuously slipping through the gate of the projecting lantern, I would not feel assured of gaining access thereby to a core of indubitable virgin data. I should merely be blotting out my eyesight, just as fakirs do when they go into a trance with open eyes. Nor could I recover my powers of perception by some critically controlled process, but only by an effort to see again by using my eyes with all their complex equipment, helped by the postural adjustment of my head and combined with the awareness of sound, touch and the exploratory motions of my body—following a process which embodies a whole system of implications to which I must a-critically commit myself for the time being. While we can reduce the sum of our conscious acceptances to varying degrees, and even to nil, by reducing ourselves to a state of stupor, any given range of awareness seems to involve a correspondingly extensive set of a-critically accepted beliefs.

Thus the programme of comprehensive doubt collapses and reveals by its failure the fiduciary rootedness of all rationality.

I do not suggest, of course, that those who advocate philosophic doubt as a general solvent of error and a cure for all fanaticism would desire to bring up children without any rational guidance or contemplate any other scheme of universal hebetation. I am only saying that this would be what their principles demand. What they actually want is not expressed but concealed by their declared principles. They want their own beliefs to be taught to children and accepted by everybody, for they are convinced that this would save the world from error and strife. In his Conway Lecture of 1922, republished in 1941, Bertrand Russell revealed this in a single sentence. After condemning both Bolshevism and clericalism as two opposite dogmatic teachings, which should both be combated by philosophic doubt, he sums up by saying: 'Thus rational doubt alone, if it could be generated, would suffice to introduce the Millennium.' [1] The author's intention is clear: he intends to spread certain doubts which he believes to be justified. He does not want us to believe the doctrines of the Catholic Church, which he denies and dislikes, and he also wants us to resist Lenin's teaching of unbridled revolutionary violence. These disbeliefs are recommended as 'rational doubts'. Philosophic doubt is thus kept on the leash and prevented from calling in question anything that the septic believes in, or from approving of any doubt that he does not share. The Inquisition's charge against Galileo was based on doubt: they accused him of 'rashness'. The Pope's Encyclical 'Humani Generis', issued in 1950, continues its opposition to science on the same lines, by warning Catholics that evolution is still an unproven hypothesis. Yet no philosophic sceptic would side with the Inquisition against the Copernican system or with Pope Pius XII against Darwinism, Lenin and his successors have elaborated a form of Marxism which doubts the reality of almost everything that Bertrand Russell and other rationalists teach us to respect, but these doubts, like those of the Inquisition, are not endorsed by Western rationalists, presumably because they are not 'rational doubts'. Since the sceptic does not consider it rational to doubt what he himself believes, the advocacy of 'rational doubt' is merely the sceptic's way of advocating his own beliefs. Russell's previously quoted sentence should therefore read: 'The acceptance of rational beliefs such as my own would suffice to introduce the Millennium.' Rationalism expressed in this form would renounce its illusory principle of doubt and face up to its own fiduciary foundations.

[1] Bertrand Russell, *Let the People Think*, London, 1941, p. 27.

In the times of Montaigne and Voltaire, rationalism identified itself with doubt of the supernatural, and rationalists called this 'doubt' as opposed to 'belief'. This practice was excusable at the time, since the beliefs held by rationalists—for example, in the supremacy of reason, and in science as an application of reason to nature—had not yet been effectively challenged by scepticism. In propagating their own beliefs the early rationalists were opposing traditional authority on so wide a front that they could well regard themselves as radical sceptics. But the beliefs of rationalism have since been effectively called in question by the revolutionary doctrines of Marxism and Nazism. It is absurd to oppose such doctrines now on the ground of scepticism. For they gained their present ascendancy only recently by a sweeping rejection of Western tradition, and it is rationalism which today relies on tradition—the tradition of the eighteenth and nineteenth centuries—against them. It should also have become clear by this time that the beliefs transmitted by this now imperilled tradition are by no means self-evident. Modern fanaticism is rooted in an extreme scepticism which can only be strengthened, not shaken, by further doses of universal doubt.

# IO

# COMMITMENT

'*I believe that in spite of the hazards involved, I am called upon to search for the truth and state my findings.*' This sentence, summarizing my fiduciary programme, conveys an ultimate belief which I find myself holding. Its assertion must therefore prove consistent with its content by practising what it authorizes. This is indeed true. For in uttering this sentence I both say that I must commit myself by thought and speech, and do so at the same time. Any enquiry into our ultimate beliefs can be consistent only if it presupposes its own conclusions. It must be intentionally circular.

The last statement is itself an instance of the kind of act which it licenses. For it stakes out the grounds of my discourse by relying essentially on the very grounds thus staked out; my confident admission of circularity being justified only by my conviction, that in so far as I express my utmost understanding of my intellectual responsibilities as my own personal belief, I may rest assured of having fulfilled the ultimate requirements of self-criticism; that indeed I am obliged to form such personal beliefs and can hold them in a responsible manner, even though I recognize that such a claim can have no other justification than such as it derives from being declared in the very terms which it endorses. Logically, the whole of my argument is but an elaboration of this circle; it is a systematic course in teaching myself to hold my own beliefs.

The moment such a programme is formulated it appears to menace itself with destruction. It threatens to sink into subjectivism: for by limiting himself to the expression of his own beliefs, the philosopher may be taken to talk only about himself. I believe that this self-destruction can be avoided by modifying our conception of belief. My previous suggestion, that for the sake of precision declaratory sentences should be formulated in the fiduciary mode, with the words 'I believe' prefixed to them, was a step in this direction, as it eliminated any formal distinction between statements of belief and statements of fact. But this reform, which would link every asserted sentence to its asserter, has yet to be supplemented in order to keep the sentence linked also to its other pole, that is, to the things to which it refers. For this purpose the fiduciary mode will have to be merged in the wider framework of commitment.

The word 'commitment' will be used here in a particular sense which will be established by its usage, the practice of which should also serve to accredit my belief in the existence and justification of commitment. Thus equipped, I should be able to show that a philosophy which recognizes commitment in the sense which I have in mind can regard itself as the philosopher's commitment and nothing but his commitment, avoiding thereby both the false claim of strict impersonality and the reduction of itself, on its own showing, to an utterance having no impersonal standing.

## 2. THE SUBJECTIVE, THE PERSONAL AND THE UNIVERSAL

The personal participation of the knower in the knowledge he believes himself to possess takes place within a flow of passion. We recognize intellectual beauty as a guide to discovery and as a mark of truth.

Love of truth operates on all levels of mental achievement. Köhler has observed how chimpanzees will repeat an ingenious trick which they had invented in the first place for the purpose of getting hold of food, by way of a game in which they use it to collect pebbles instead. The anguish suffered by animals puzzled by problems (of which I shall say more later) demonstrates the correlated capacity for enjoying intellectual success. These emotions express a belief: to be tormented by a problem is to believe that it has a solution and to rejoice at discovery is to accept it as true.

The passionate aspects of intellectual commitment become more precisely circumscribed by contrasting them with other passions or pervasive conditions which are not commitments. Intense bodily pain pervades our whole person, yet the feeling of such a pain is not an action or a commitment. When someone feels hot or tired or bored, this pervasively affects his state of mind, but does not imply any affirmation beyond that of his own suffering. There exist also purely sensual pleasures which are almost as passive as these pains; but the more intense gratifications of our senses come from the satisfaction of our appetites and to this extent entail a manner of commitment.

On such grounds as these, I think we may distinguish between the personal in us, which actively enters into our commitments, and our subjective states, in which we merely endure our feelings. This distinction establishes the conception of the *personal*, which is neither subjective nor objective. In so far as the personal submits to requirements acknowledged by itself as independent of itself, it is not subjective; but in so far as it is an action guided by individual passions, it is not objective either. It transcends the disjunction between subjective and objective.

300

The structure of commitment, which serves as a logical matrix to the personal, is most clearly exemplified by the act of consciously solving a problem. Such acts emerge only at a somewhat elevated intellectual level and they tend to disappear once more at even higher degrees of sophistication. Problem-solving combines elements from the two adjoining domains below it and above it, and can be best introduced here by attending first to these two.

At the *lower end* of the intellectual scale lies the satisfaction of appetites. Processes of this kind, as for example the selection of food, may show delicate discrimination, but the capacity for this is largely non-deliberative rather than guided by conscious personal judgment. Similarly, the act of perception by which we notice and identify objects, though sometimes requiring a marked effort of intelligence, does not as a rule involve any deliberation but is brought to completion automatically. Though appetites and sensory impulses are clearly personal actions, they are those of a person within ourselves with which we may not always identify ourselves. We have often to restrain our primary desires and correct the judgment of our senses, which shows that such sub-intellectual performances do not wholly commit ourselves. At the *upper end* of the scale we find forms of intelligence in which our personal participation tends to be reduced for quite different reasons. Mathematical science is widely accepted as the most perfect of sciences, and science as the most perfect of all feats of intelligence. While these claims may be excessive or even altogether mistaken, they express the inescapable ideal of a completely formalized intelligence, which would eliminate from its manifestations every trace of personal commitment.

A conscious and persistent striving for the solution of an articulate problem lies midway between these two extremes. It canalizes the native urge for achieving coherence, which we share with the higher animals, into the heuristic manipulation of articulate thought. Science can serve here as a leading example. The distinctive ability of a scientific discoverer lies in the capacity to embark successfully on lines of enquiry which other minds, faced with the same opportunities, would not have recognized or not have thought profitable. This is his originality. Originality entails a distinctively personal initiative and is invariably impassioned, sometimes to the point of obsessiveness. From the first intimation of a hidden problem and throughout its pursuit to the point of its solution, the process of discovery is guided by a personal vision and sustained by a personal conviction.

While originality conflicts sharply with the ideal of a completely formalized intelligence, it also differs altogether from drive-satisfaction. For our appetites are *ours*, and it is ourselves they seek to satisfy, while the discoverer seeks a solution to a problem that is satisfying and compelling both for himself and everybody else.[1] Discovery is an act in which satisfaction, submission and universal legislation are indissolubly combined.

[1] Cf. p. 171.

Some discoveries obviously reveal something that already existed, as when Columbus discovered America. This does not impair the measure of the discoverer's originality; for though America was there for Columbus to discover, its discovery was still made by him. But the universal intention of a radical innovation can also be represented as a sense of its pre-existence. When a mathematician putting forward a daring new conception, like non-Euclidean geometry or the theory of sets, demands acceptance for them from his reluctant contemporaries, he shows that in his enquiries he had aimed at the satisfaction of pre-existing standards of intellectual merit and that he regards the product of his thought as the disclosure of a pre-existing possibility for the satisfaction of these standards. Even in the natural sciences, radical innovations may have to rely for acceptance on yet undeveloped sensibilities. The purely mathematical framework of modern physics was not satisfying from the point of view of previous generations, who sought for explanations in terms of mechanical models. In order to prevail, modern physicists had to educate their public to use new standards of intellectual appreciation. Yet from the start, the pioneers of modern physics assumed that the new sensibility was latent in their fellow scientists and would be developed in them in response to the possibilities of the more profound and truer outlook which appealed to this new sensibility. They undertook to revise the current standards of scientific merit in the light of more fundamental intellectual standards, which they assumed to be pre-existing and universally compelling. All this applies of course emphatically to artistic innovations.

Our appreciation of originality should make clearer the distinction between the *personal* and the *subjective*. A person may have most peculiar predilections or phobias, yet not be credited with originality. His distinctive sensibilities will be regarded as mere idiosyncrasies; even if he is altogether wrapped up in his private world his condition will not be recognized as a commitment. Instead, he will be said to be subject to obsessions and illusions, and may even be certified as insane. Originality may, of course, be mistaken for sheer madness, which has happened to modern painters and writers; and the reverse is also fairly common, namely for people labouring under delusions to believe themselves to be great inventors, discoverers, prophets, etc. But two totally different things may often be mistaken for each other. It is enough to establish here once more the principle which distinguishes them: namely, that commitment is a personal choice, seeking, and eventually accepting, something believed (both by the person incurring the commitment and the writer describing it) to be impersonally given, while the subjective is altogether in the nature of a condition to which the person in question is subject.

We observe here a mutual correlation between the personal and the universal within the commitment situation. The scientist pursuing an enquiry ascribes impersonal status to his standards and his claims, because he regards them as impersonally established by science. But his submission to scientific standards for the appraisal and guidance of his efforts is the *only sense* in which these standards can be said to pre-exist, or even to exist at all, for him. No one can know universal intellectual standards except by acknowledging their jurisdiction over himself as part of the terms on which he holds himself responsible for the pursuit of his mental efforts. I can speak of facts, knowledge, proof, reality, etc., within my commitment situation, for it is constituted by my search for facts, knowledge, proof, reality, etc., as binding on me. These are proper designations for commitment targets which apply so long as I am committed to them; but they cannot be referred to non-committally. You cannot speak without self-contradiction of knowledge you do not believe, or of a reality which does not exist. I may deny validity to some particular knowledge, or some particular facts, but then to me these are only allegations of knowledge or of facts, and should be denoted as 'knowledge' and as 'facts', to which I am not committed. Commitment is in this sense the only path for approaching the universally valid.

## 3. THE COHERENCE OF COMMITMENT

Epistemology has traditionally aimed at defining truth and falsity in impersonal terms, for these alone are accepted as truly universal. The framework of commitment leaves no scope for such an endeavour; for its acceptance necessarily invalidates any impersonal justification of knowledge. This can be illustrated by writing down a symbolic representation of the elements joined together *within* a commitment and contrasting these with the same elements, when looked upon non-committally from *outside* the commitment situation. We may, for example, represent a factual statement

$$\text{from } within \text{ as:} \left\{ \begin{matrix} \text{personal} \\ \text{passion} \end{matrix} \right. \rightarrow \begin{matrix} \text{confident} \\ \text{utterance} \end{matrix} \rightarrow \left. \begin{matrix} \text{accredited} \\ \text{facts} \end{matrix} \right\}$$

and from *outside* as:  subjective belief;  declaratory sentence;  alleged facts.

The arrows in the first row indicate the force of commitment and the brackets the coherence of the elements involved in the commitment; accordingly, in the second row both these sets of symbols are omitted.

*The fiduciary passions which induce a confident utterance about the facts are* **personal**, *because they submit to the facts as universally valid, but when we reflect on this act non-committally its passion is reduced to* **subjectivity**. At the same time the confident utterance is reduced to a sentence of unspecified modality, and the facts become merely *alleged* facts. These elements, set out in the second row, are mere *fragments* of the commitment that we had previously accredited by the symbols in the first row.

Any particular commitment may be reconsidered, and this movement of doubt would be expressed by passing from the first row to the second;

after which, having satisfied his doubts, the reflecting person would re-commit himself and move back into the situation represented by the first row. But he would find this return blocked if, having realized that this movement involves an act of his own judgment, he denied justification to it by reason of its personal character.

In such a case, the reflecting person remains faced with the fragments of his previous commitment, which as such no longer require each other: for a subjective belief cannot be accounted for by unaccredited facts, and a declaration expressing such a belief can no longer be said to correspond to the facts. If he still continues to feel that there is some consistent relation between his beliefs and the factual evidence presented to him, he will regard this (with Hume) as a mere habit, without acknowledging any justification of the convictions expressed by this habit.

The reflecting person is then caught in an insoluble conflict between a demand for an impersonality which would discredit all commitment and an urge to make up his mind which drives him to recommit himself. Hume has described most candidly the ensuing oscillation between a scepticism which admittedly lacks conviction, and a conviction which dare not consciously acknowledge its own acts and can be upheld only by neglecting the result of philosophic reflection. I shall call this the objectivist dilemma.

This dilemma has long haunted philosophy in the guise of the 'correspondence theory of truth'. Bertrand Russell, for example, defines truth as a coincidence between one's subjective belief and the actual facts;[1] yet it is impossible, in terms which Russell would allow, to say how the two could ever coincide.

The answer is this. The 'actual facts' are accredited facts, as seen within the commitment situation, while subjective beliefs are the convictions accrediting these facts as seen non-committally, by someone not sharing them. But if we regard the beliefs in question non-committally, as a mere state of mind, we cannot speak confidently, without self-contradiction, of the facts to which these beliefs refer. *For it is self-contradictory to secede from the commitment situation as regards the beliefs held within it, but to remain committed to the same beliefs in acknowledging their factual content as true.* It is nonsense to imply that we simultaneously both hold and do not hold the same belief, and to define truth as the coincidence between our actual belief (as implied in our confident reference to the facts) and our denial of the same belief (as implied in our reference to it as a mere state of our mind concerning these facts).

[1] Bertrand Russell, *The Problems of Philosophy*, 4th edn., London, 1919, p. 202: '. . . a belief is true when there is a corresponding fact, and is false when there is no corresponding fact'. Cf. *Human Knowledge: Its Scope and Limits*, London, 1948, pp. 164–70. P. 170: 'Every belief which is not merely an impulse to action is in the nature of a picture, combined with a yes-feeling or a no-feeling; in the case of a yes-feeling it is "true" if there is a fact having to the picture the kind of similarity that a prototype has to an image; in the case of a no-feeling it is "true" if there is no such fact. A belief which is not true is called "false". This is a definition of "truth" and "false-hood".'

I have mentioned before (Part Three, ch. 8) the futile regress and the logical self-contradiction in which we become involved when casting the reaffirmation of a factual statement into the form of another factual statement, and have argued that we should avoid these anomalies by denying that the utterance '*p* is true' is a sentence. We now see how the theory of knowledge is also thrown into confusion by the same objectivistic language habit. This habit *transforms an assertion coupled to an asserted sentence into two asserted sentences*: one about primary objects, the other about the truth of a sentence mentioning these objects. And this in its turn lands us with the problem of how we can be said to know this truth as if it existed by itself (like snow), outside us, though it is not something (like snow) that we can observe impersonally, but an expression recording our own judgment. The muddle can be avoided, once more, only by denying that '*p* is true' is a sentence, and acknowledging accordingly that it stands for an a-critical act of acceptance which is not something one can either assert or know. The word 'true' does not designate, then, a quality possessed by the sentence *p*, but merely serves to make the phrase '*p* is true' convey that the person uttering it still believes *p*.

Admittedly, to say '*p* is true', instead of 'I believe *p*', is to shift the emphasis within one's commitment from the personal to the external pole. The utterance 'I believe *p*' expresses more aptly a heuristic conviction or a religious belief, while '*p* is true' will be preferred for affirming a statement taken from a textbook of science. But a greater fiduciary contribution does not necessarily correspond here to a greater uncertainty of that which is affirmed. The emphasis on the personal coefficient depends on the heuristic or persuasive passion that it conveys. These passions may vary over all possible intensities, whether the statement affirmed by them is unambiguous or statistical, and whether the latter affirms a high or a low degree of probability. The fiduciary component must always be thought of as included in a prefixed affirmation sign, and never in the explicit statement itself.

The assumption that the truth we seek to discover exists by itself, hidden to us only by our misguided approach to it, represents correctly the feeling of an investigator pursuing a discovery which keeps eluding him. It may also express the ineradicable tension between our conviction that we know something and the realization that we may conceivably be mistaken. But in neither case can an outside observer of this relation compare another person's knowledge of the truth with the truth itself. He can only compare the observed person's knowledge of the truth with his own knowledge of it.

According to the logic of commitment, *truth is something that can be thought of only by believing it*. It is then improper to speak of another person's mental operation as leading to a true proposition in any other sense than that it leads him to something the speaker himself believes to be true. Let me illustrate the illegitimate use of the supposition that something is true in itself by an argument of R. B. Braithwaite concerning

305

induction.[1] He argues that if the policy of induction is true, a person B who believes it to be true would reasonably arrive, by the inductive method, at the conclusion that it is true. Three propositions are said to be involved as follows: $p$, the assertion of the evidence for the inductive hypothesis, i.e. the successful past applications of induction; $r$, the assertion of the effectiveness of the 'inferential policy' leading from $p$ to the assertion of the inductive hypothesis; and $q$, the inductive hypothesis itself. Now, Braithwaite argues, if B reasonably believes $p$, and either believes (but not reasonably) $r$, or if $r$ is true (though B does not believe it), or if B both (non-reasonably) believes $r$ and $r$ is true, then B may validly and without circularity infer from these premises a reasonable belief in $q$. Thus (1) if B reasonably believes $p$, and subjectively believes $r$, his consequent reasonable belief in $q$ establishes the 'subjective validity' of the inductive hypothesis. Or (2) if B reasonably believes $p$ and $r$ is true (whether he believes it or not), this establishes, Braithwaite argues, the 'objective validity' of the inductive hypothesis.

What *is* demonstrated here is that if we believe in the method of induction, $q$, we also believe that the past applications of this method, $p$, offer public evidence of its effectiveness, $r$, when examined in the very light of this method. This shows that belief in the method of induction is both self-consistent and implies a belief in its own self-consistency; but it shows nothing about the truth of this belief. If we do not believe in the method of induction nothing follows at all. The illusion that some progress has been achieved towards the establishment of $q$ as true is due once again to the illegitimate dismemberment of a commitment. 'B believes $r$' is shown to issue in a 'subjective validity' of $q$, and '$r$ is true' is used to establish $q$ as objectively valid. In the first case the conclusion is empty, unless the writer by his own commitment to $q$ transforms 'subjective validity' into possession of the truth by B. In the second case '$r$ is true' is used as a presupposition for deriving $q$, though it cannot be asserted by the author unless he has first endorsed $q$. In both cases the author's anterior commitment to $q$ reduces the process of inference which he attributes to B into a mere illustration of his own commitment.

We see it confirmed that we cannot compare subjective knowledge (in B) with objective knowledge, except in the sense of judging B's beliefs from the point of view of our own beliefs. The only proper comparison between imperfect knowledge and perfect knowledge remains the sense of risk and intimation of achievement in a heuristic endeavour towards knowledge, within a commitment situation.

## 4. EVASION OF COMMITMENT

Kant tried to salvage the justification of mechanics and geometry from the objectivist dilemma by deducing their basic concepts as *a priori* cat-

[1] R. B. Braithwaite, *Scientific Explanation*, Cambridge, 1953, pp. 278 ff.

egories or forms of experience. But since the end of the nineteenth century this has proved less and less tenable and an alternative teaching of Kant, represented by his regulative principles, gained predominance instead.

By regulative principles, in the general sense in which the term is employed here, I mean all manner of recommendations to act on a belief while denying, disguising, or otherwise minimizing the fact that we are holding this belief. Originally, Kant had recommended that certain generalizations (as for example the teleological aptness of living organisms) should be entertained *as if* they were true, without assuming that they were true. However, Kant does not say that we should entertain these generalizations as if they were true, even though we knew them to be false. His recommendation to entertain them *as if* they were true is thus seen to be based on the tacit assumption that they are in fact true. By conveying this assumption without asserting it, he avoids any formulation which would require to be upheld as his own personal judgment.

Modern descriptions of scientific truths as mere working hypotheses or interpretative policies are generalizations of the Kantian regulative principles to the whole of science.[1] For we would never use a hypothesis which we believe to be false, nor a policy which we believe to be wrong.[2] The suggestion which usually accompanies these regulative formulations of science, that all scientific theories are merely tentative, since scientists are ready to modify their conclusions in the face of new evidence, is irrelevant, for it does not affect the fiduciary content of a hypothesis or policy. There

---

[1] The following examples may illustrate the vagueness of this position. F. Waismann, 'Verifiability', in A. Flew, *Logic and Language*, I (Oxford, 1951), pp. 142–3: 'The way we single out one particular law from infinitely many possible ones shows that in our theoretical construction of reality we are guided by certain principles—*regulative principles* as we may call them. If I were asked what these principles are, I should tentatively list the following: (1) Simplicity or economy—the demand that the laws should be as simple as possible. (2) Demands suggested by the requirements of the symbolism we use—for instance, that the graph should represent an analytic function so as to lend itself readily to the carrying out of certain mathematical operations such as differentiation. (3) Aesthetic principles ("mathematical harmony" as envisaged by Pythagoras, Kepler, Einstein) though it is difficult to say what they are. (4) A principle which so regulates the formation of our concepts that as many alternatives as possible become decidable. This tendency is embodied in the whole structure of Aristotelian logic, especially in the law of excluded middle. (5) There is a further factor elusive and most difficult to pin down: a mere tone of thought which, though not explicitly stated, permeates the air of a historical period and inspires its leading figures. It is a sort of field organizing and directing the ideas of an age . . .' Or Cf. H. Feigl on 'Induction and Probability' (H. Feigl and W. Sellars, *Readings in Philosophical Analysis*, New York, 1949, p. 302); where it is argued that there is no problem of induction because the principle of induction is not a proposition at all, but 'a principle of procedure, a regulative maxim, an operational rule'. This is also the conclusion of Bertrand Russell's *Human Knowledge: Its Scope and Limits* (London, 1948, Part Six, ch. 2), where the premisses of science turn out to be a set of pre-suppositions neither empirical nor logically necessary. I have criticized this vagueness before (on p. 113). Explicit premisses of science are maxims, that can be acknowledged as such only as part of a commitment endorsing the scientist's vision of reality (pp. 160–70 above).

[2] The medieval principle of saving the phemonena by a theory without commitment to its truth was, as we have seen, rejected by Kepler on these very grounds (p. 146).

are admittedly various degrees of belief and our beliefs are changing. But belief does not cease to exist merely because it is weak or because it is variable. Zeno foolishly denied that physical motion was possible, because an object had to be at some place at every moment of time; it is equally foolish to argue, in reverse to Zeno, that we are never committed because our commitments are changing.

The purpose of endorsing one's belief in science without asserting it, can be achieved also by understating the claims of science to the point of insignificance and recommending science on these insufficient grounds. When science is said to be merely the simplest description of the facts, or a convenient shorthand, we rely on it that the reader will use the term 'simple' and 'convenient' in the sense of 'scientifically simple' and 'scientifically convenient'. We then accept science because it is scientific and not for being simple or convenient in the ordinary sense, which it is not. This procedure was described in earlier chapters as the bowdlerization of science. It results in a pseudo-substitution which refers to the mainsprings of scientific conviction in an emasculated language, for the sake of avoiding offence to a philosophy which cannot face up to our actual intellectual commitments.

## 5. THE STRUCTURE OF COMMITMENT: I

We have seen that the thought of truth implies a desire for it, and is to that extent personal. But since such a desire is for something impersonal, this personal motive has an impersonal intention. We avoid these seeming contradictions by accepting the framework of commitment, in which the personal and the universal mutually require each other. Here the personal comes into existence by asserting universal intent, and the universal is constituted by being accepted as the impersonal term of this personal commitment.

Such a commitment enacts the paradox of dedication. In it a person asserts his rational independence by obeying the dictates of his own conscience, that is, of obligations laid down for himself by himself. Luther defined the situation by declaring, 'Here I stand and cannot otherwise'. These words could have been uttered by a Galileo, a Harvey or an Elliotson, and they are equally implied in the stand made by any pioneer of art, thought, action or faith. Any devotion entails an act of self-compulsion.

We can watch the mechanism of commitment operating on a minor scale, and yet revealing all its characteristic features, in the way a judge decides a novel case. His discretion extends over the possible alternatives left open to him by the existing explicit framework of the law, and within this area he must exercise his personal judgment. But the law does not admit that it fails to cover any conceivable case.[1] By seeking the right de-

---

[1] In France, during the revolutionary period, judges were obliged to refer back to the legislature all matters not covered by statute, but this practice was abolished in 1804. Subsequently, the doctrine gained general acceptance that courts of law are competent

cision the judge must *find* the law, supposed to be existing—though as yet unknown. This is why eventually his decision becomes binding as law. The judge's discretion is thus narrowed down to zero by the stranglehold of his own universal intent—by the power of his responsibility over himself. This is his independence. It consists in keeping himself wholly responsible to the interests of justice, excluding any subjectivity, whether of fear or favour. Judicial independence has been secured, where it exists, by centuries of passionate resistance to intimidation and corruption; for justice is an intellectual passion seeking satisfaction of itself, by inspiring and ruling men's lives.

While compulsion by force or by neurotic obsession excludes responsibility, compulsion by universal intent establishes responsibility. The strain of this responsibility is the greater—other things being equal—the wider the range of alternatives left open to choice and the more conscientious the person responsible for the decision. While the choices in question are open to arbitrary egocentric decisions, a craving for the universal sustains a constructive effort and narrows down this discretion to the point where the agent making the decision finds that he cannot do otherwise. *The freedom of the subjective person to do as he pleases is overruled by the freedom of the responsible person to act as he must.*[1]

The course of scientific discovery resembles the process of reaching a difficult judicial decision—and the analogy throws light on a crucial issue of the theory of knowledge. Discovery stands in the same contrast to a routine survey, as does a novel court decision to the routine administration of law. In both cases the innovator has a wide discretion of choice, because he has no fixed rules to rely on, and the range of his discretion determines the measure of his personal responsibility. In both cases a passionate search for a solution that is regarded as potentially pre-existing, narrows down discretion to zero and issues at the same time in an innovation claiming universal acceptance. In both cases the original mind takes a decision on grounds which are insufficient to minds lacking similar powers of creative judgment. The active scientific investigator stakes bit by bit his

to decide every case brought before them, by applying the Code in conjunction with the legal principles embodied in the Code. (Cf. J. W. Jones, *Historical Introduction to the Theory of the Law*, Oxford, 1940.)

[1] In Freud's terms the subjective person is the appetitive id, controlled by the prudent ego. The responsible person is explained away by Freud as the outcome of interiorized social pressures, actuated from inside by the super-ego. This interpretation overlooks the fact that a responsible personhood, which curbs both the id and the ego, may at the same time rebel against the ruling orthodoxy, and that this is precisely where its presence is most impressively manifested. To accept moral conscience as the interiorization of social pressure renders nonsensical the very idea of respect being paid, or even due, by society to the conscience of its members. A super-ego cannot be free, and to demand liberty for it would be farcical. As to the Freudian interpretation of intellectual passions as a sublimation of appetite drives, it leaves unaccounted for everything that distinguishes science and art from the instincts of which these are supposed to be sublimates. 'Sublimation' is a circumlocution which relies for its meaning entirely on our previous understanding of the things which it is supposed to explain.

whole professional life on a series of such decisions and this day-to-day gamble represents his most responsible activity. The same is true of the judge, with the difference, of course, that the risk is borne here mainly by the parties to the case and by the society which has entrusted itself to the interpretation of its laws by the courts.

I have described before the principle which determines heuristic choices in the process of scientific research as a sense of growing proximity to a hidden truth, like that which guides us in groping for a forgotten name. Within the framework of commitment this determining force reappears now as a sense of responsibility exercised with universal intent. Scientific intuition is evoked by a strenuous groping towards an unknown achievement, believed to be hidden and yet accessible. Therefore, though every choice in a heuristic process is indeterminate in the sense of being an entirely personal judgment, in those who exercise such judgment competently it is completely determined by their responsibility in respect to the situation confronting them. In so far as they are acting responsibly, their personal participation in drawing their own conclusions is completely compensated for by the fact that they are submitting to the universal status of the hidden reality which they are trying to approach. Accidents may sometimes bring about—or prevent—discovery, but research does not rely on accident: the continuously renewed risks of failure normally incurred at every heuristic step are taken without ever acting at random. Responsible action excludes randomness, even as it suppresses egocentric arbitrariness.

Yet the explorer gambles for indefinite stakes. Columbus sailed out to find a Western route to the Indies; he failed and after repeating his voyage three times to prove that he had reached the Indies, he died in shame. Still, Columbus did not merely blunder into America. He was wrong in accepting on the evidence of the prophecies of Esdras and, presumably, of Toscanelli's map, that the westward distance of the Indies from Spain was only about twice that of the Azores, but he was right in concluding that the East could be reached from the West.[1] He staked his life and reputation on what appear now to be insufficient grounds for an unattainable prize, but he won another prize instead, far greater than he was ever to realize. He had committed himself to a belief which we now recognize as a small distorted fragment of the truth, but which impelled him to make a move in the right direction. Such wide uncertainties of its aims are attached to every great scientific enquiry. They are implicit in the looseness of the hold which a daring anticipation of reality has upon it. I have described before how the

---

[1] Salvador de Madariaga, in his *Christopher Columbus* (London, 1939), argues the case for Columbus' having seen, and surreptitiously copied, Toscanelli's map during his stay in Portugal. He also describes Columbus' reliance on the Apocryphal writer Esdras, who believed that the world is 'six parts dry land and one part sea'. 'Toscanelli, for Colon,' Madariaga writes, 'was on the way to truth, but as he had not read Esdras his plan still required mariners, not used to losing sight of land, to navigate 130° of 62½ miles, i.e. 8125 miles over unknown seas. Colon, through his study of Esdras, "knew" that the distance was only 2550 miles' (p. 101).

scientist must strike a balance between the opposite hazards of caution and daring—either of which might waste his gifts—so as to make the best use of these gifts. Scientists who rely on themselves to decide on this in a responsible manner—and their supporters who in their turn rely on them—believe that this is feasible, and I agree with them. I have implied this at the opening of this chapter by declaring my belief that, in spite of the appalling hazards, I am called upon to search for the truth and state my findings. My confident description of the heuristic commitment of scientists endorses here a similar belief, held by men of science pursuing their researches.

The science of today serves as a heuristic guide for its own further development. It conveys a conception about the nature of things which suggests to the enquiring mind an inexhaustible range of surmises. The experience of Columbus, who so fatefully misjudged his own discovery, is inherent to some extent in all discovery. The implications of new knowledge can never be known at its birth. For it speaks of something real, and to attribute reality to something is to express the belief that its presence will yet show up in an indefinite number of unpredictable ways.

An empirical statement is true to the extent to which it reveals an aspect of reality, a reality largely hidden to us, and *existing therefore independently of our knowing it.* By trying to say something that is true about a reality believed to be existing independently of our knowing it, all assertions of fact necessarily carry *universal intent. Our claim to speak of reality serves thus as the external anchoring of our commitment in making a factual statement.*

The framework of commitment is now established in outline for this particular case. The enquiring scientist's intimations of a hidden reality are personal. They are his own beliefs, which—owing to his originality—as yet he alone holds. Yet they are not a subjective state of mind, but convictions held with universal intent, and heavy with arduous projects. It was he who decided what to believe, yet there is no arbitrariness in his decision. For he arrived at his conclusions by the utmost exercise of responsibility. He has reached responsible beliefs, born of necessity, and not changeable at will. In a heuristic commitment, affirmation, surrender and legislation are fused into a single thought, bearing on a hidden reality.

To accept commitment as the only relation in which we can believe something to be true, is to abandon all efforts to find strict criteria of truth and strict procedures for arriving at the truth. A result obtained by applying strict rules mechanically, without committing anyone personally, can mean nothing to anybody. Desisting henceforth from the vain pursuit of a formalized scientific method, commitment accepts in its place the person of the scientist as the agent responsible for conducting and accrediting scientific discoveries. The scientist's procedure is of course methodical. But his methods are but the maxims of an art which he applies in his own original way to the problem of his own choice. Discovery forms part of the art of knowing; it can be studied by precept and example, but its higher performances require peculiar native gifts appropriate to particular subjects. Every factual statement embodies some measure of responsible judgment as the personal pole of the commitment in which it is affirmed.

We meet here once more the position which the Logic of Affirmation assigned to the intelligent person, who was defined there as the centre of unspecifiable intelligent operations. I shall show in Part Four that this is actually what we know as the mind of a person, when meeting and conversing with somebody. His mind is the focus which we look at by attending subsidiarily to the utterances and actions unspecifiably co-ordinated by his mind. Since the structure of commitment includes the logic of assent, it necessarily confirms this logic; yet it is worth noting that by relying on this logic my fundamental belief implies a belief in the existence of minds as centres of unspecifiable intelligent operations.

While the logic of assent merely showed that assent is an a-critical act, 'commitment' was introduced from the start as a framework in which assent can be responsible, as distinct from merely egocentric or random. The centre of tacit assent was elevated to the seat of responsible judgment. It was granted thereby the faculty of exercising discretion, subject to obligations accepted and fulfilled by itself with universal intent. A responsible decision is reached, then, in the knowledge that we have overruled by it conceivable alternatives, for reasons that are not fully specifiable. Hence to accept the framework of commitment as the only situation in which sincere affirmations can be made, is to accredit in advance (if anything is ever to be affirmed) affirmations against which objections can be raised that cannot be refuted. It allows us to commit ourselves on evidence which, but for the weight of our own personal judgment, would admit of other conclusions. We may firmly believe what we might conceivably doubt; and may hold to be true what might conceivably be false.

We reach here the decisive issue of the theory of knowledge. Throughout this book I have persistently followed one single endeavour. I have tried to demonstrate that into every act of knowing there enters a tacit and passionate contribution of the person knowing what is being known, and that this coefficient is no mere imperfection, but a necessary component of all knowledge. All this evidence turns into a demonstration of the utter baselessness of all alleged knowledge, unless we can wholeheartedly uphold our own convictions, even when we know that we might withhold our assent from them. I must yet face this issue more fully.

## 6. The Structure of Commitment; II

Let me return to fundamentals. In the theory of commitment the main division lies between experiences that are merely suffered or enjoyed and others that are actively entered upon. Convulsions or other incoherent motions are not activities, but everything that tends towards an achievement, whether it involves bodily motions or merely thought, is to be classed as an activity. Only an activity can go wrong, and all activity incurs the risk of failure. To believe something is a mental act: you can neither believe nor disbelieve a passive experience. It follows that you can only believe something that might be false. This is my argument in a nutshell; I shall now elaborate it in some detail.

In a widest sense every process of life, even in plants, is an activity that might miscarry. But as I am concerned here only with the way truth is found, I shall limit myself to the conscious achievement of knowledge. Even so, I shall have to supplement now what I have said in the previous section about scientific discovery, by describing how knowledge is acquired at lower levels, namely by perception and inarticulate learning. This will include all active 'epicritical' knowing, but exclude purely passive, 'proto-pathic' awareness, which I am classing as subjective.

Any act of factual knowing presupposes somebody who believes he knows what is being believed to be known. This person is taking a risk in asserting something, at least tacitly, about something believed to be real outside himself. Any presumed contact with reality inevitably claims universality. If I, left alone in the world, and knowing myself to be alone, should believe in a fact, I would still claim universal acceptance for it. Every act of factual knowing has the structure of a commitment.

Since the two poles of commitment, the personal and the universal, are correlative, we may expect them to arise simultaneously from an antecedent state of selfless subjectivity. That is in fact how the child's early intellectual development has been described by psychologists. The early behaviour of children suggests that they cannot distinguish between fact and fiction, or between their own person and another. They live in a world of their own making, which they believe to be shared by everybody else. This stage of infancy has been called 'autistic' by Bleuler and 'ego-centric' by Piaget; but the blurred distinction between self and non-self, which underlies the child's state of mind here, might as well be described as 'selfless'. So long, or in so far, as the external and internal worlds of a person do not interfere with each other, there can be no conflict between them and hence no attempt can be made to avoid such a conflict by discovering a correct interpretation of the world. Nor can any risk be taken in the pursuit of such a discovery. Only as we become divided from the world, can we achieve a personhood capable of committing itself consciously to beliefs concerning the world, and incurring thereby a fiduciary hazard. Autistic day-dreaming can then give way to acts of deliberate judgment.

The person that emerges at this level of commitment is only the ego, exercising discrimination, though lacking as yet responsible judgment. But we shall see later that even at this level an individual can be puzzled by a problem, to the point of suffering a nervous breakdown. His whole person is involved in his commitment; the effort of reaching out to reality involves, even here, compulsion of oneself to make oneself conform to reality.

313

Perception usually goes on automatically, but sometimes situations may present themselves in which all the senses are strained to the utmost in order to discriminate between two or more alternatives. If we then decide to see things in one particular way, we blot out any alternative vision of them for the moment. Experimental psychology provides examples of ambiguities between which our perception can decide at will. A flight of steps can be seen alternatively as an overhanging cornice. We may see two people facing each other on either side of a picture, or alternatively, a vase standing in the middle of the picture.[1] The eye may be able to switch at will from one way of seeing such a picture to the other, but cannot keep its interpretation suspended between the two. The only way to avoid being committed in either way, is to close one's eyes. This corresponds to the conclusion reached before in my critique of doubt; to avoid believing one must stop thinking.

So we see that even a primitive tacit act like perception may operate deliberately in search of the truth over an area of discretion, within which it overrules even more primitive, i.e. less discriminating, mental propensities. There is, indeed, complete continuity between a perceptive judgment and the process by which we establish responsible convictions in the course of scientific research. The assent which shapes knowledge is fully determined in both cases by competent mental efforts overruling arbitrariness. The result may be erroneous, but it is the best that can be done in the circumstances. Since every factual assertion is conceivably mistaken, it is also conceivably corrigible, but a competent judgment cannot be improved by the person who is making it at the moment of making it, since he is already doing his best in making it.

We cannot evade this logical necessity by suggesting that the mental act should be postponed until its grounds have been more fully considered. For every deliberate mental act has to decide its own timing. The risks of further hesitation must be weighed against the risks of acting hastily. The balance of the two must be left to be derived from the circumstances, as known to the person making up his mind. Hence an agent exercising a mental act competently with a view to the existing circumstances cannot, at the moment of acting, correct it as to its timing any more than as to its content.[2] To postpone mental decisions on account of their conceivable fallibility would necessarily block all decisions for ever, and pile up the

[1] Cf. e.g. E. S. and F. R. Robinson, *General Psychology*, Chicago, 1926, p. 242 (cornice and steps). R. H. Wheeler, *Science of Psychology*, 1929, London, p. 358 ('the falling cube' and 'vase and faces').
[2] This aspect of decision making was first formalized by A. Wald, *Annals of Mathem. Statistics*, 16 (1945), p. 117. The simplest case is that a null hypothesis $H_0$ is to be tested by collecting random samples. At each new test a threefold decision is to be made, namely (1) to accept $H_0$ or (2) to reject $H_0$ or (3) to continue the experiment. You fix a value $\alpha$ for the maximum tolerable probability of rejecting $H_0$ though $H_0$ is true. You go on testing till the actual probability of committing this error falls below $\alpha$. If you pre-assign an unreasonably small value for $\alpha$ you are incurring unreasonable costs in time and effort; if you fix $\alpha$ at zero you maximize unreason.

hazards of hesitation to infinity. It would amount to voluntary mental stupor. Stupor alone can eliminate both belief and error.

Strict scepticism should deny itself the possibility of advocating its own doctrine, since its consistent practice would preclude the use of language, the meaning of which is subject to all the notorious pitfalls of inductive reasoning. But strict scepticism could yet teach an ideal which it admits to be unattainable. Or the sceptic may excuse the imperfections of his scepticism by invoking the protection of regulative principles, which he professes to follow without accepting them to be true. He may thus retain his sense of intellectual superiority over others who—like myself—profess their fiduciary commitments without pretending that these are only temporary imperfections.

I shall not argue with the sceptic. It would not be consistent with my own views if I expected him to abandon a complete system of beliefs on account of any particular series of difficulties. Besides, by this time it should be clear how far-reaching are in my own opinion the changes in outlook that are required in order to establish a stable alternative to the objectivist position. I cannot hope to do more in this book than to exhibit a possibility which like-minded people may wish to explore.

I shall go on, therefore, to repeat my fundamental belief that, in spite of the hazards involved, I am called upon to search for the truth and state my findings. To accept commitment as the framework within which we may believe something to be true, is to circumscribe the hazards of belief. It is to establish the conception of competence which authorizes a fiduciary choice made and timed, to the best of the acting person's ability, as a deliberate and yet necessary choice. The paradox of self-set standards is eliminated, for in a competent mental act the agent does not do as he pleases, but compels himself forcibly to act as he believes he must. He can do no more, and he would evade his calling by doing less. The possibility of error is a necessary element of any belief bearing on reality, and to withhold belief on the grounds of such a hazard is to break off all contact with reality. The outcome of a competent fiduciary act may, admittedly, vary from one person to another, but since the differences are not due to any arbitrariness on the part of the individuals, each retains justifiably his universal intent. As each hopes to capture an aspect of reality, they may all hope that their findings will eventually coincide or supplement each other.

Therefore, though every person may believe something different to be true, there is only one truth. This can be substantiated as follows. The function of the word 'true' is to complete such utterances as ' "$p$" is true', which are equivalent to an act of assent of the form 'I believe $p$'. The question, whether a particular fact is true, e.g. whether Dreyfus did write the *bordereau*, challenges a person to such an act. Unless such a challenge is addressed to me—be it by other people or by myself—the question whether this fact is true does not arise for me. Questions and answers exchanged between other people on this matter are to me merely facts about these people, and not about the matter in question. The only sense in which I can speak of the facts of the matter is by making up my own mind about them. In doing so I may rely on an existing consensus, as a clue to the truth, or else may dissent from it, for my own reasons. In either case my answer will be made with universal intent, saying what I believe to be the truth, and what the consensus ought therefore to be. This is the only sense in which I can speak of the truth, and though I am the only person who can speak of it in this sense, this is what I mean by the truth. To ask what I would believe to be the true facts of a matter, if I were somebody else, means simply to ask what somebody else would believe them to be. This kind of question is interesting, and will yet be discussed later, but it is clearly not a question concerning the facts of the matter.

This position is not solipsistic, since it is based on a belief in an external reality and implies the existence of other persons who can likewise approach the same reality. Nor is it relativistic. This is already apparent from the previous paragraph, but may be stated in more formal terms as follows. The concept of commitment postulates that there is no difference, except in emphasis, between saying 'I believe $p$' or ' "$p$" is true'. Both utterances emphatically put into words that I am confidently asserting $p$, as a fact. This is something I am doing in the act of uttering the words in question, and is quite different from my reporting that I have done this in the past, or that somebody else has either done this or is doing it now. If I report 'I believed $p$', or 'X believes $p$', I am not committing myself in respect to $p$ and hence no utterance linking '$p$' to 'true' corresponds to these reports; they issue in no assertion of the sentence $p$ as true, be it in relation to my own past or to other people's beliefs. There remains therefore only one truth to speak about.

This is as far as I can take now this question of relativism.

## 7. INDETERMINACY AND SELF-RELIANCE

We have seen that the progress of scientific discovery depends on heuristic commitments which establish contacts with reality, and that the hazards incurred in entering on such a commitment are twofold: namely (1) that it may be mistaken and (2) that even if it is right, its future scope and significance is largely indeterminate. The foregoing section has followed up the hazards of error in the domain of the most primitive deliberate assertions, such as are made by the eye in deciding how to see an ambiguous set of objects. I shall supplement this now by briefly recalling the hazards incurred at the same level, owing to the indeterminate nature of real facts. This unspecifiable fund of implications can be revealed in a fact that is simple enough to be discovered by an earthworm.

An earthworm was taught in a famous experiment by Yerkes' to take the right turn at the branching of a T-shaped tube (through the stem of which it was made to crawl) by inflicting on it an electric shock whenever it tried to turn left. It took about a hundred trials to establish this habit.[1] A later investigator, L. Heck, confirmed this experiment and carried it a stage further.[2] After the earthworm's training was completed, he inverted the conditions between right and left. The new problem set to the worm was the reverse of that which it had first solved, but it still resembled it in the fact that one branch of the tube was painful to enter, while the other had no sting in it. The earthworm's behaviour in this second test was determined both by the contrast and the similarity of the two consecutive problems. At first the worm consistently turned to the right where it was now met with an electric shock. During this phase its previous training may be said to have exercised a misleading effect on it. But presently (after about thirty runs) the worm began to turn with increasing frequency into the branch now free from obstruction, and it eventually acquired the habit of turning in the direction opposite to that which it had been trained to take in the first place, within a much *shorter* series of trials than it had taken to establish the original reverse habit. The primary training which had taught the worm to turn in one direction thus proved a powerful help in training it subsequently to turn in the opposite direction. Here it is the similarity which the second problem bears to the first that asserts itself.

Since the range of problems which may arise in the future is unlimited and totally unpredictable, the bias which we adopt in respect to these problems by committing ourselves to any particular belief today is equally inexhaustible and unpredictable. Confrontation with these problems may bring to light, therefore, an indeterminate range of hidden implications that are inherent in any of our present beliefs.

I have mentioned this before in Part Two, ch. 5, when showing that all learning, even in animals, establishes a latent knowledge, the range of which is indeterminate. In the same chapter I enlarged upon this further by suggesting that in all our thoughts—whether tacit or articulate—we rely jointly on two faculties, namely (1) on the power of our conceptual framework, based on reality, to assimilate new experiences and (2) on our capacity to adapt this framework in the very act of applying it, so that it may increase its hold on reality. We can see this now in the perspective of commitment. The intellectual daring which impels our acts of commitment retains its dynamic character within the state of commitment, in relying on its own resourcefulness to deal with the unspecifiable implications of the knowledge acquired by the act of commitment. In this self-reliance lies our ultimate power for keeping our heads in the face of a changing world. It makes us feel at home in a universe presenting us with a succession of unprecedented situations and even makes us enjoy life best precisely on these occasions, which force us to respond to novelty by re-interpreting our accepted knowledge.

[1] See p. 122.
[2] L. Heck, 'Uber die Bildung einer Assoziation beim Regenwurm auf Grund von Dressurversuchen', *Lotos*, **67/8** (1916-20), p. 168.

## 8. EXISTENTIAL ASPECTS OF COMMITMENT

The enactment of commitment consists in self-compulsion with universal intent through the interaction of two levels: a higher self, which claims to be more judicious, taking control over a less judicious lower self. Self-reliance, which supports us to meet the indeterminate contingencies of commitment, has a similar structure; it makes us ready to suppress a routine operation of the mind in favour of a novel impulse. Self-compulsion and self-reliance both issue in acts of assent by which we definitively dispose of ourselves. The change may be large or small: a comprehensive conversion, or no more than a slight modification of our interpretative framework. The depth of the cognitive commitment may be measured in either case by the ensuing change of our outlook.

The hazards of such existential changes cannot be probed or delimited. If we believe—as I do—that it is incumbent on us to take these chances, we do so in the hope that the universe is sufficiently intelligible to justify this undertaking. The company of great scientists whom we acknowledge lends us courage. We draw confidence from the splendour of a thousand minds to which we pay homage. Yet this confidence will be cheap and vain if it keeps its eyes fixed on stories of success—though they be the success of martyrs. For the normal outcome of a daring commitment is failure. Or worse still, it may be the success of a vast error; the kind of error which, like the Big Lie, is irresistibly persuasive, since it sweeps away all existing criteria of validity and resets them in its own support, exactly as a great truth does when overthrowing big lies and great errors. A fiduciary philosophy does not eliminate doubt, but (like Christianity) says that we should hold on to what we truly believe, even when realizing the absurdly remote chances of this enterprise, trusting the unfathomable intimations that call upon us to do so.

But if an active mental process, aiming at universality, can turn out to have been altogether mistaken, can we still say that in it the subject has risen to the level of the personal by reaching out to reality? Though a Zande witch doctor arguing in terms of the poison-oracle is clearly a rational person, his rationality is altogether deluded. His intellectual system may gain a limited justification within a society which it supplies with a form of leadership and the means for deciding disputes, however unjustly. But as an interpretation of natural experience it is false.

To this I shall reply by distinguishing between a competent line of thought, which may be erroneous, and mental processes that are altogether illusory and incompetent. The latter I would class for the moment with passive mental states, as purely subjective.[1] Admittedly, the range within which I acknowledge mental activity as competent and beyond which I reject it as superstition, fatuity, extravagance, madness, or mere twaddle,

---

[1] This class is subdivided in Part Four, by separating mere incoherence from systematically pursued misinterpretations.

is determined by my own interpretative framework. And different systems of acknowledged competence are separated by a logical gap, across which they threaten each other by their persuasive passions. They are contesting each other's mental existence.

Such conflicts will take place within ourselves when we hesitate at the brink of being converted from one such system to another. This happens on a minor scale when we discredit the irresistible testimony of our eyes by classing something seen as an optical illusion. We see two adjacent circular segments as different in size, and we know that even animals see them the same way, yet we keep rejecting this universally compelling observation on the grounds that the segments are geometrically congruent. However, in other instances the conflict may be decided in favour of our perceptive faculties, as when the impressionist painters decided to accept the testimony of the eye, which sees shadows coloured merely by contrast to their coloured neighbourhood. For some time the public refused to recognize this manner of representation and rejected their paintings as shocking and absurd; but after a while they agreed to see as the impressionists saw and accepted their colouring as correct. Or consider the change in musical sensibility resulting from the introduction of the equal-temperament scale. It is first known to have been used in Hamburg about 1690; and Bach used it in his compositions for the clavichord. In 1852 Helmholtz could still write of the 'hellish row' resulting from its use in organ-building; whereas forty years later Planck, as he tells us in his autobiography, discovered that the tempered scale was 'positively more pleasing to the human ear, under all circumstances, than the "natural", untempered scale'.[1] In all such cases sense experience is found to conform to, or deviate from, certain norms. The person himself may be conscious of such a deficiency, and strive for a more correct experience.[2]

The case of the impressionists has numerous counterparts in the field of appetites. Peter the Great had to force his courtiers on peril of their lives to smoke cigars, by which he hoped to Westernize their outlook; and many of us have to pass through a similar ordeal under compulsion by fashion,

---

[1] Sir James Jeans, *Science and Music*, Cambridge, 1937, pp. 184–5, quotes Helmholtz as saying, 'When I go from my justly intoned harmonium to a grand pianoforte, every note of the latter sounds false and disturbing. . . . On the organ, it is considered inevitable that, when the mixture stops are played in full chords, a hellish row must ensue, and organists have submitted to their fate. Now this is mainly due to equal temperament, because every chord furnishes at once both equally-tempered and justly-intoned fifths and thirds, and the result is a restless blurred confusion of sounds.' Cf. Max Planck, *Scientific Autobiography and Other Papers*, London, 1950, pp. 26–7: '. . . Even in a harmonic major triad, the natural third sounds feeble and inexpressive in comparison with the tempered third. Indubitably, this fact can be ascribed ultimately to a habituation through years and generations.'

[2] For a survey and critique of the experimental work on the training of perceptual judgment, see Eleanor J. Gibson, 'Improvement in Perceptual Judgments as a Function of Controlled Practice or Training', *Psychological Bulletin*, **1** (1953), pp. 401–31. Mrs. Gibson's report recounts ample experimental evidence of the improvement of sensuous discrimination by training.

before we acquire some new taste to which we then become wholly addicted. In a conflict between our appetitive and our intelligent person we may side with one side or the other. Desire and emotion may educate our intelligence, as they do when we grow up to sexual maturity and parenthood; and the reverse may happen when we control and refashion our appetites in conformity to social custom. As we identify ourselves in turn with one level of our person or another, we feel passively subjected to the activities of the one which we do not acknowledge for the time being. When Penfield stimulates electrically in the brain the motion of a limb or the evocation of an image, the patient does not feel that he is carrying out the motion or recollecting the image.[1] The same sense of passivity accompanies the splitting of personality when a hypnotized subject carries out a post-hypnotic injunction. From his own experience under hypnosis Bleuler has likened post-hypnotic compulsion to the way we yield to reflex urges, like sneezing or coughing.[2] Each person within an individual may become a liability to another and may mould it to its commitments or be moulded by it in reverse. We may prefer to identify ourselves with the person on the higher level, but this is not invariably the case, and our choice between the levels is part of our ultimate commitment at any particular moment.

## 9. VARIETIES OF COMMITMENT

Within the framework of commitment, to say that a sentence is true is to authorize its assertion. Truth becomes the rightness of an action; and the verification of a statement is transposed into giving reasons for deciding to accept it, though these reasons will never be wholly specifiable. We must commit each moment of our lives irrevocably on grounds which, if time could be suspended, would invariably prove inadequate; but our total responsibility for disposing of ourselves makes these objectively inadequate grounds compelling.

Truth conceived as the rightness of an action allows for any degree of personal participation in knowing what is being known. Remember the panorama of these participations. Our heuristic self-giving is invariably impassioned: its guide to reality is intellectual beauty. Mathematical physics assimilates experience to beautiful systems of indeterminate bearing. Its application to experience may be strictly predictive within certain not strictly definable conditions. Alternatively, it may merely express a numerically graded expectation of chances; or provide only—as in crystallography—a system of perfect order by which objects can be illuminatingly classified and appraised. Pure mathematics attenuates

[1] Dr. Wilder Penfield, 'Evidence of Brain Operations', *Listener*, **41** (Jan.–June, 1949), p. 1063.
[2] C. L. Hull, *Hypnosis and Suggestibility*, New York, 1933, pp. 38–40 (quoting from E. Bleuler, 'Die Psychologie der Hypnose', *Munch. Med. Woch.*, 1889, No. 5).

empirical references to mere hints within a system of conceptions and operations constructed in the light of the intellectual beauty of the system. The act of acceptance becomes here entirely dedicatory. The joy of grasping mathematics induces the mind to expand into an ever deeper understanding of it and to live henceforth in active preoccupation with its problems.

Moving further in this direction we enter on the domain of the arts. Once truth is equated with the rightness of mental acceptance, the transition from science to the arts is gradual. Authentic feeling and authentic experience jointly guide all intellectual achievements; so that from observing scientific facts within a rigid theoretical framework we can move by degrees towards dwelling within a harmonious framework of colours, of sounds or imagery, which merely recall objects and echo emotions experienced before. As we pass thus from verification to validation and rely increasingly on internal rather than external evidence, the structure of commitment remains unchanged but its depth becomes greater. The existential changes accepted by acquiring familiarity with new forms of art are more comprehensive than those involved in getting to know a new scientific theory.

A parallel movement takes place (as we shall see) in passing from the relatively impersonal observation of inanimate objects to the understanding of living beings and the appreciation of originality and responsibility in other persons. These two movements are combined in the transition from the relatively objective study of things to the writing of history and the critical study of art.

The growth of the modern mind within these great articulate systems is secured by the cultural institutions of society. A complex social lore can be transmitted and developed only by a vast array of specialists. Their leadership evokes some measure of participation in their thought and feeling by all members of society. The civic culture of society is even more tightly woven into the structure of society. The laws and the morality of a society compel its members to live within their framework. A society which accepts this position in relation to thought is committed as a whole to the standards by which thought is currently accepted in it as valid. My analysis of commitment is itself a profession of faith addressed to such a society by one of its members, who wishes to safeguard its continued existence, by making it realize and resolutely sustain its own commitment, with all its hopes and infinite hazards.

## 10. Acceptance of Calling

We meet here the powers which call us into being: into our particular form of existence. Involuntary co-efficients of commitment, of which I have given instances before, become paramount here. Every deliberate act of our own relies on the involuntary functions of our body. Our thoughts are limited by our innate capabilities. Our senses and emotions can be enhanced by education, but these ramifications remain dependent on their native roots. Moreover, since our intellectual judgment ever relies on the services of an automatic sensory apparatus, it may always be misguided by it. The moral control of our drives, which would harness them to the service of a reasonably satisfying life, is ever in danger of being swamped and disorganized by them. Worse still, we are creatures of circumstance. Every mental process by which man surpasses the animals is rooted in the early apprenticeship by which the child acquires the idiom of its native community and eventually absorbs the whole cultural heritage to which it succeeds. Great pioneers may modify this idiom by their own efforts, but even their outlook will remain predominantly determined by the time and place of their origin. Our believing is conditioned at its source by our belonging. And this reliance on the cultural machinery of our society continues through life. We go on accepting the information given out by its leading centres of publicity and relying on its recognized authorities for most of our judgments of value. Nor are we merely passive participants in the social framework, sustaining the orthodoxy to which we adhere. For every society allocates powers and profits, to which the adherents of the intellectual *status quo* lend a measure of support. Respect for tradition inevitably shields also some iniquitous social relations.

How can we claim to arrive at a responsible judgment with universal intent, if the conceptual framework in which we operate is borrowed from a local culture and our motives are mixed up with the forces holding on to social privilege?

From the point of view of a critical philosophy, this fact would reduce all our convictions to the mere products of a particular location and interest. But I do not accept this conclusion. Believing as I do in the justification of deliberate intellectual commitments, I accept these accidents of personal existence as the concrete opportunities for exercising our personal responsibility. *This acceptance is the sense of my calling.*

This sense of calling may acknowledge many of the environmental antecedents of my thought as fulfilling a more primitive commitment. For the child which is prompted to listen and acquire the speech and conceptions of adults is an active intellectual centre. Though his efforts are not conducted deliberately, they are yet the intellectual gropings of a person in search of a valid result. Within the circumstances of his upbringing, this process fulfils, in my belief, the purpose of the child's mental powers and should be relied upon to operate within this framework, just as we rely on our own personal judgment for the solution of consciously envisaged problems. As I acknowledge, in reflecting on the process of discovery, the gap between the evidence and the conclusions which I draw from them, and account for my bridging of this gap in terms of my personal responsibility, so also will I acknowledge that in childhood I have formed my most fundamental beliefs by exercising my native intelligence within the social milieu of a particular place and time. I shall submit to this fact as defining the conditions within which I am called upon to exercise my responsibility.

I accept these limits, for it is impossible to hold myself responsible beyond such limits. To ask how I would think if I were brought up outside any particular society, is as meaningless as to ask how I would think if I were born in no particular body, relying on no particular sensory and nervous organs. I believe, therefore, that as I am called upon to live and die in this body, struggling to satisfy its desires, recording my impressions by aid of such sense organs as it is equipped with, and acting through the puny machinery of my brain, my nerves and my muscles, so I am called upon also to acquire the instruments of intelligence from my early surroundings and to use these particular instruments to fulfil the universal obligations to which I am subject. A sense of responsibility within situations requiring deliberate decisions demands as its logical complement a sense of calling with respect to the processes of intellectual growth which are its necessary logical antecedents.

The extension of commitment from deliberate judgments to innate intelligent impulses points towards further generalizations to include the whole process of life. My body may be said to be alive to the extent to which its parts are functioning as elements in a joint operation and these operational principles are rational. Life is a stratagem, in which each element must rely on it that the other elements will support it, and each consecutive step in the sequence is taken in the expectation that the next will suitably continue it. The higher the organism, the more involved is its plan of action and the more completely does each section of it rely in its location and timing, on the location and timing of all the others for its usefulness to the whole; the more useless does each become in itself. And consequently, the more fully is the living body committed to comprehensive governing principles of universal standing.

Thus we can see the same fundamental structure of personal commitment revealed on extremely different levels, with a correspondingly wide variation of its internal balance. The most strictly universalized processes of inference are shown to rely ultimately on their inarticulate interpretation by a person accepting them, and life pursuing its self-centred primitive urges is shown to rely on universal technical principles; while between the two we meet man's momentous acts of responsible commitment, made by accepting his own starting-point in space and time, as the condition of his own calling.

Within its commitments the mind is warranted to exercise much ampler powers than those by which it is supposed to operate under objectivism; but by the very fact of assuming this new freedom it submits to a higher power to which it had hitherto refused recognition. Objectivism seeks to relieve us from all responsibility for the holding of our beliefs. That is why it can be logically expanded to systems of thought in which the responsibility of the human person is eliminated from the life and society of man.

In recoiling from objectivism, we would acquire a nihilistic freedom of action but for the fact that our protest is made in the name of higher allegiances. We cast off the limitations of objectivism in order to fulfil our calling, which bids us to make up our minds about the whole range of matters with which man is properly concerned.

Those who are satisfied by hoping that their intellectual commitments fulfil their calling, will not find their hopes discouraged when realizing on reflection that they are only hopes. I have said that my belief in commitment is a commitment of the very kind that it authorizes: therefore, if its justification be questioned, it finds confirmation in itself. Moreover, any such confirmation will likewise prove stable towards renewed critical reflection, and so on, indefinitely. Thus, by contrast to a statement of fact claiming to be impersonal, an affirmation made in terms of a commitment gives rise to no insatiable sequence of subsequent justifications. Instead of indefinitely shifting an ever open problem within the regress of the objectivist criticism of objectivist claims, our reflections now move from an original state of intellectual hopes to a succession of equally hopeful positions; so that by rising above this movement and reflecting on it as a whole we find the continuance of this regress unnecessary.

Commitment offers to those who accept it legitimate grounds for the affirmation of personal convictions with universal intent. Standing on these grounds, we claim that our participation is personal, not subjective, except in so far as it is compulsive. While it then lies beyond our responsibility, it is yet transformed by our sense of responsibility into part of our calling. Our subjective condition may be taken to include the historical setting in which we have grown up. We accept these as the assignment of our particular problem. Our personhood is assured by our simultaneous contact with universal aspirations which place us in a transcendent perspective.

The stage on which we thus resume our full intellectual powers is borrowed from the Christian scheme of Fall and Redemption. Fallen Man is equated to the historically given and subjective condition of our mind, from which we may be saved by the grace of the spirit. The technique of our redemption is to lose ourselves in the performance of an obligation which we accept, in spite of its appearing on reflection impossible of achievement. We undertake the task of attaining the universal in spite of our admitted infirmity, which should render the task hopeless, because we hope to be visited by powers for which we cannot account in terms of our specifiable capabilities. This hope is a clue to God, which I shall trace further in my last chapter, by reflecting on the course of evolution.

# PART FOUR
# KNOWING AND BEING

# II

# THE LOGIC OF ACHIEVEMENT

## 1. INTRODUCTION

IN the rest of this book I shall outline some views on the nature of
living beings, including man, which clearly follow from the acceptance
of my commitment to personal knowledge. Having decided that I must
understand the world from my point of view, as a person claiming originality and exercising his personal judgment responsibly with universal intent, I must now develop a conceptual framework which both recognizes
the existence of other such persons and envisages the fact that they have
come into existence by evolution from primordial inanimate beginnings.

I shall use the following key-argument in a number of variants and
elaborations. Our comprehension of a living individual entails a subsidiary
awareness of its parts which is not wholly specifiable in more detached
terms. This understanding acknowledges a particular comprehensive—i.e.
'molar'—achievement of the individual itself. Since our knowledge of this
molar function is not specifiable in 'molecular' terms, the function itself is
not reducible to molecular particulars; it must be acknowledged therefore
as a higher form of being, not determined by these particulars. We can
reach this conclusion directly by recalling that the understanding of a
whole appreciates the coherence of its subject matter and thus acknowledges the existence of a value that is absent from the constituent particulars.

Arrived at this point, we can proceed further in two directions. One
leads to the contemplation by one person—the writer—of another person
in the process of acquiring knowledge. This relation will turn out to
duplicate in respect of the second person my reflections on my own
knowledge, which terminated in the acknowledgment of my intellectual
commitment. The new variant of this situation will establish a partnership
and a rivalry of commitments between the first person and the second
person, which will fall into the framework of individual culture. At the
same time we shall envisage the second person acquiring a knowledge of

the first person, appreciating both him and his knowledge and establishing thereby a whole range of interpersonal exchanges which, when extended to a group, form the civic culture and public order of society. Since both individual and interpersonal commitments are related socially and established institutionally, the perspective of commitment widens here to the whole of humanity pursuing its course towards an unknown destination.

## 2. RULES OF RIGHTNESS

We have seen that animals can learn (1) to perform tricks, (2) to read signs, (3) to know their way about. These activities were taken to prefigure primordially the three faculties of contriving, observing and reasoning, which are elaborated on the articulate level to the three domains of engineering, natural sciences and mathematics. I shall have to amend this scheme now to allow for the fact that only the physical sciences are predominantly observational, while biology and the study of mind and man have a more complex structure, in which observation plays but a subsidiary role.

I shall have, first of all, to claim a proper place for the logic of contriving. The logic of deductive reasoning has been systematically studied for two millennia, and the logic of empirical inference has been a major preoccupation of philosophy for centuries, but the logic of contriving has found its way only into scattered hints. One might think that pragmatism, operationalism or cybernetics had contributed to it in their attempt to explain thought as a process of contriving. But the endeavour to reduce all knowledge to strictly impersonal terms prevented this philosophic movement from attending to our knowledge of contriving which itself can never be impersonal.

We have seen that a tool, a machine or a technical process is characterized by an operational principle, which differs altogether from an observational statement. The former, if it is new, represents an invention and can be covered by a patent; the latter, if it is new, is a discovery, which cannot be patented. Contrivances are classes of objects which embody a particular operational principle. For the moment I shall deal only with mechanical contrivances and concentrate on things complicated enough to be called machines. Clocks, typewriters, boats, telephones, locomotives, cameras are the type of 'machines' I have in mind.

A patent formulates the operational principle of a machine by specifying how its characteristic parts—its organs—fulfil their special function in combining to an overall operation which achieves the purpose of the machine. It describes how each organ acts on another organ within this context. The inventor of a machine will always try to obtain a patent in the widest possible terms; he will therefore try to cover all conceivable embodiments of its operational principle by avoiding the mention of the physical or chemical particulars of any actually constructed machine, unless these

328

are strictly indispensable to the operations claimed for the machine. This will extend the conception of the machine to objects constructed from the most varied materials and differing altogether in shape and size. Just as the rules of algebra will operate for any set of numbers for which the algebraic constants may stand, so an operational principle applies to any collection of parts which are functioning jointly according to this principle.

It follows that the class of objects which could conceivably represent any particular machine would form, in the light of pure science—which ignores their operational principle—an altogether chaotic ensemble. In other words, *the class of things defined by a common operational principle cannot be even approximately specified in terms of physics and chemistry.*

Unless I believe a purpose to be reasonable or at least conceivably reasonable, I cannot endorse an operational principle which teaches how to achieve this purpose. Technology comprises all acknowledged operational principles and endorses the purposes which they serve. This endorsement also *appreciates the value of the machine as a rational means of securing the advantage in question.* The operational principle of the machine now functions as an ideal: the ideal of the machine in good working order. It sets a standard of perfection. Any 'clock', 'typewriter', 'locomotive', etc. can be judged by this kind of standard to be a more or less perfect clock, typewriter or locomotive. The conception of a machine exercises its evaluative function further, by setting off against itself the conception of a machine that is out of order. When a boiler bursts, a crankshaft snaps, or a train is derailed, these things behave against the rules laid down for them within the conception of the machine. So while this conception accredits certain events as orderly performances, it condemns others as failures.

But it can say nothing else about these failures. The conceptions of machines in good working order form a system which ignores the particulars of failures—in the same way as geometrical crystallography ignores the imperfections of crystals. The operational principles of machines are therefore *rules of rightness*, which account only for the successful working of machines but leave their failures entirely unexplained.

Can we turn to natural science to supplement this unrealistic approach? Yes and no. An engineer trained in physics and chemistry may be able to explain failures. He might observe strains under which the machine will break down, or corrosive effects which whittle its substance away. But it would be false to conclude that the physicist or chemist can replace the conception of the machine—as defined by its operational principles—by a more comprehensive understanding which accounts both for the correct functioning and the failures of a machine. A physical and chemical investigation cannot convey the understanding of a machine as expressed by its operational principles. In fact, it can say nothing at all about the way the machine works or ought to work.

This point is fundamental for the general understanding of different levels of reality and will therefore be followed up here, in respect of the machine, more closely than this subject matter would justify by itself.

The first thing to realize is that a knowledge of physics and chemistry would in itself not enable us to recognize a machine. Suppose you are faced with a problematic object and try to explore its nature by a meticulous physical or chemical analysis of all its parts. You may thus obtain a complete physico-chemical map of it. At what point would you discover that it is a machine (if it is one), and if so, how it operates? Never. For you cannot even put this question, let alone answer it, though you have all physics and chemistry at your finger-tips, unless you already know how machines work. Only if you know how clocks, typewriters, boats, telephones, cameras, etc. are constructed and operated, can you even enquire whether what you have in front of you *is* a clock, typewriter, boat, telephone, etc. The questions: 'Does the thing serve any purpose, and if so, what purpose, and how does it achieve it?' can be answered *only by testing the object practically as a possible instance of known, or conceivable, machines.* The physico-chemical topography of the object may in some cases serve as a clue to its technical interpretation, but by itself it would leave us completely in the dark in this respect.

We could extend the physico-chemical topography of the problematical object to include all possible future transformations which the object would undergo under the impact of all conceivable circumstances. Yet even this ensemble of all possible future configurations would still not tell us anything from the technical point of view.

This result is crucial. I shall repeat it therefore once more in concrete terms. We have a solid tangible inanimate object before us—let us say a grandfather clock. But we do not know what it is. Then let a team of physicists and chemists inspect the object. Let them be equipped with all the physics and chemistry ever to be known, but let their technological outlook be that of the stone age. Or, if we cannot disregard the practical incompatibility of these two assumptions, let us agree that in their investigations they shall not refer to any operational principles. They will describe the clock precisely in every particular, and in addition, will predict all its possible future configurations. Yet they will never be able to tell us that it is a clock. *The complete knowledge of a machine as an object tells us nothing about it as a machine.*

I have said that rules of rightness account only for the success, never for the failure, of things constructed and operated in accordance with them. We now see, on the other hand, that physics and chemistry are blind both to success and failure, since they ignore the operational principles by which success and failure are defined. We identify a machine by understanding it technically; that is, by a participation in its purpose and an endorsement of its operational principles. We do not exercise such a participation within a physical or chemical investigation. Indeed, the understanding of the structure and operation of a machine require as a rule very little knowledge of physics and chemistry. Hence the two kinds of knowledge, the technical and the scientific, largely by-pass each other.

But the relation of the two kinds of knowledge is not symmetrical. If any object—such as for example a machine—is essentially characterized by a comprehensive feature, then our understanding of this feature will grant us a true knowledge of what the object is. It will reveal a machine as a machine. But the observation of the same object in terms of physics and chemistry will spell complete ignorance of what it is. Indeed, the more detailed knowledge we acquire of such a thing, the more our attention is distracted from seeing what it is.

This unsymmetrical relationship prevails also in the manner in which the two kinds of knowledge can be fruitfully combined. We can make good use of physical or chemical observations in order to deepen our understanding of a machine, for example a clock. Having guessed that the clock is a time-keeping instrument and gained some intimation of the functions performed by its various parts—as of the weights which drive it, the pendulum which controls its speed by releasing the escape, and the hands which indicate the passage of time—we can proceed to examine the physical processes underlying these several operations. Thus we will establish the material *conditions* under which the parts can *fulfil* their functions and which will *explain* their occasional *failures*. As a result we shall be able to suggest improvements to avoid failures and perhaps even invent entirely new principles of clock-making. But no physical or chemical observations of clocks will be of any use to a clockmaker, unless such observations are related to the operational principles of a clock, as conditions for their success or causes of their breakdown. And we may conclude quite generally that in our knowledge of a comprehensive entity, embodying a rule of rightness, any information supplied by physics and chemistry can play only a subsidiary part.[1]

Some physical and chemical characteristics of a machine, such as its weight, size and shape, or its fragility, its susceptibility to corrosion or to damage by sunlight, will be of interest in themselves on certain occasions, for example to a carter undertaking the transport of the machine. But this is about as much as the scientific study of a machine can achieve when pursued in itself, without reference to the principles by which the machine performs its purpose.

### 3. Causes and Reasons

Technology, embodied in rules of rightness, teaches a rational way to achieve an acknowledged purpose. Such rules devise a stratagem consisting

[1] In the case of pharmacology and other areas where science and technology overlap (see Part Two, ch. 6, p. 179), the operational principles in question overlap with the laws of nature which are the conditions of their practicability. Even so, the action of an operational principle can always be distinguished from a natural law by its instrumental context. It is an action; which as such *can succeed or fail.*

of several steps, each of which performs a function of its own within a coherent, economic, and in this sense rational, procedure. The procedure may include the contriving of a machine, built of a number of parts, each of which has a function of its own within a coherent, rational operation. There is a specifiable *reason* for every step of the procedure and every part of the machine, as well as for the way the several steps and the various parts are linked together to serve their joint purpose. This chain of reasons is set out in the operational principles of the process or of the machine.

Since physics and chemistry ignore operational principles, they are blind also to the reasons which justify the successive steps of an operation. Yet a physico-chemical investigation can throw light on the rationality of a process or machine by establishing the physico-chemical conditions on which they have to rely, and warning that, if these conditions are not maintained, the contrivance will break down.

As an extreme case, natural science may declare that an alleged practical process is impracticable. It may say of a machine that it cannot work. I do say, for example, that the wheel of perpetual motion described by the Marquis of Worcester in 1663 could not be kept circling around by the succeeding descents of the weights attached to its rim and therefore it could not and cannot work. I can show in this case that the alleged operational principles of such a machine are incompatible with the law of conservation of energy, which I believe to be true. Hence I conclude that there can be *no* conditions in which these principles can be relied upon.

Since rules of rightness cannot account for failures, and reasons for doing something can only be given within the context of rules of rightness, it follows that there can be no reasons (in this sense) for a failure. It is best, therefore, to avoid the use of the word 'reason' in this context and to describe the origins of failures invariably as their *causes*. We can say then that physico-chemical investigations of a machine, carried out with a bearing on its operational principles, can elucidate both the conditions for their success and the causes of their failure. It would be wrong to speak of establishing the physical and chemical 'causes' of success, for the success of a machine is defined by its operational principles, which are not specifiable in physico-chemical terms. If a stratagem succeeds, it does so in accordance with its own premeditated internal reasons; if it fails, this is due to unforeseen external causes.

#### 4. Logic and Psychology

In so far as the deductive sciences consist in formalized operations, they can be embodied in the operational principles of computing machines; and the body of an animal also functions to a certain extent as a machine. Our enquiry into the logic of machines is, therefore, capable of generalization over a domain extending from mathematics to physiology. And we may add to this domain, as further rules of rightness, the principles of ethics and law. My enquiry can attend to these codes of human behaviour only in passing; while logic and mathematics as forms of knowledge, and the machine-like structure of animals—both as an object of natural science and as an instrument by which animals may achieve knowledge—will retain our full attention.[1]

Everything that I have said concerning the relation of machines to the laws of physics and chemistry, applies also to a digital computer that is operated as a machine of logical inference.[2] Here are some of the relevant points. The operational principles of logical machines are rules of rightness which can account only for success. These principles are dissolved altogether, and with them the difference between success and failure, in a physico-chemical topography of a machine. But physics and chemistry are highly illuminating when used subsidiarily, with a bearing on the previously established operational structure of logical machines, so as to define the material conditions under which they can work and to account for their occasional failures.

We can now define in similar terms the relation between logic and psychology, though for dealing with this subject our terms will have to be somewhat expanded. Thought proceeds largely by an irreversible process of comprehension and not according to specifiable rules. Only the latter will be called logical thinking, in which I shall include mathematics. Logic, thus defined, is a rule of rightness: it tells us how we must reason in order to derive correct and ample conclusions from given premises. When I listen to another person's argument I appraise it in relation to the standards of correctness which I have set up by acknowledging certain rules as the rules of logic. Piaget has made a systematic series of such appraisals at successive stages in the mental development of children. His *epistémologie génétique* shows that the child's reasoning fulfils year by year ever higher standards of logical performance.[3] We may say that in his efforts to reason rightly the child strives towards the fulfilment of these logical standards. And we may add that by accepting these as obligatory he lays them down as universally valid. The fact that he will argue in terms of these rules shows that he holds them with persuasive intent. Thus the child's growing logical coherence appears as a progressive elaboration of his commitment to logical rules of rightness. He discovers, submits to, relies

---

[1] I have pointed out before (Part Two, ch. 6, p. 161) that the rules of scientific procedure on which we rely, and the scientific beliefs and valuations which we hold, are mutually determined, and also (Part Two, ch. 6, p. 184) that mathematics can be equally affiliated to natural science or technology. We have seen later that the truth of a factual sentence is equivalent to the rightness of its assertion. But these interrelations, which could be traced back to the structure of perception and the exploratory activities of the lower animals, are intrinsic to the process of acquiring knowledge. They represent the universal interpenetration of sensing and groping (see Part One, ch. 4, pp. 55-6), and not the duality of an operational principle contrasted with the medium in which it is incorporated.

[2] Logical machines were discussed earlier in Part Three, ch. 8, p. 261.

[3] Cf. e.g. Jean Piaget, *Logic and Psychology*, Manchester, 1953.

on, and declares to be valid universally, ever higher standards of logical excellence.

The operation of these rules of rightness takes place within the flow of conscious and unconscious awareness which is the subject matter of psychology. It may be set in motion by appetite or by intellectual passion, and operates on a fund of memories. It may work by the powers of the visual imagination, by the aid of verbal or other symbolism, or altogether conceptually. It will be profoundly influenced by the language and the conceptual framework in which the child was brought up. But no study of this medium of thought can reveal whether a particular deduction—for example a proof of the binomial theorem—is correct or not. The correctness of such a proof can be supported only by logical reasons, not by psychological observations. Psychology cannot distinguish by itself between true and false inferences, and hence is blind to logical principles; but it can throw light on the conditions under which the understanding and operation of correct logico-mathematical reasoning may develop, and it may supply an explanation for errors in reasoning. Indeed, an error in reasoning can never be the subject of a logical demonstration; it can be understood only by psychological observations which reveal its causes. On the other hand, it is meaningless to speak of the causes of a mathematical theorem. We may enquire into the conditions which favoured its discovery, but the validity of a theorem can only be justified by reasons, not accounted for by causes.

The relation between the rules of logic and the subject matter of psychology which I have just described is the same as that between the operational principles of machines and the subject matters of physics and chemistry, except for one additional feature—I have introduced a *second person* striving to reason rightly, according to the rules of logic. This active, responsible personal centre will emerge more fully later. But we may anticipate this here: whatever rules of rightness a person tries to fulfil and establish—be they moral, aesthetic or legal—he commits himself to an ideal; and again, he can do so only within a medium that is blind to this ideal. The ideal determines the standards to which a person holds himself responsible; but the ideal-blind medium both grants the possibility for striving for this ideal and limits this possibility. It determines his calling.

At the other end of our line of generalization (which will be fully set out in the next chapter) we have placed *the organism functioning as a machine*. We may conclude straight away that in this respect an organism is represented by operational principles of the kind which define machines. Physiology is the technology of healthy achievements: of wholesome feeding, good digestion, effective locomotion, sharp perception, fertile copulation, etc. The argument could be pursued further on now familiar lines. But I shall prefer to postpone this until I have defined the machine-like features of an animal more precisely, by contrast to its 'organismic' functions, which cannot be aptly formulated in terms of definite operational principles.

## 5. Originality in Animals

The first time I mentioned the inventive powers of animals I used Köhler's examples, in which a chimpanzee achieves a set purpose by suitably reorganizing his field of vision. Next I identified in a rat the latent knowledge of a maze that it had learned to run, by its capacity for suitably reorganizing its knowledge to deal with the emergency of blocked paths.[1] The Yerkes-Heck experiment showed how earthworms can reorganize latent knowledge indefinitely in the face of new circumstances.[2] On a similarly inarticulate level, men grope their way towards a skilful performance, unconsciously readjusting the co-ordination of their muscles in the direction of success.[3] The way we strain our eyes in order to 'make out' what it is that we see, gave us an example of a discovery pursued by conjointly reorganizing a set of unconscious muscular actions with our simultaneous interpretation of the impressions shaped by these actions.[4] This urge to reorganize our experiences and capabilities in a manner more satisfying to ourselves was traced upwards through the entire range of human inventiveness. It defined man the innovator and explorer, passionately pouring himself into an existence closer to reality.

All this was surveyed with a view to the reinterpretation of truth carried out in Part Three. I have accepted there a fiduciary commitment which authorizes me to choose my fundamental beliefs in harmony with my total situation; I must try to decide now, on such grounds, whether the originality of animals and men can be accounted for by some ingenious automatic machinery, or should be acknowledged as an independent force operating through the body in combination with the existing machinery of the body.[5]

---

[1] P. 74.　　[2] Pp. 316-17.　　[3] P. 62.　　[4] Pp. 96-7.

[5] I have given ample evidence that such comprehensive questions are decided by our vision of the general nature of things, and that the guidance of such a vision is indispensable to science. (see p. 135). But since the fiduciary character of such general conceptions is rarely acknowledged, I shall quote here two statements by K. S. Lashley in respect to my present subject matter. Speaking for the assembled members of the Hixon Symposium of 1948 on cerebral mechanisms in behaviour, Professor Lashley declared: (1) 'Our common meeting ground is the faith to which we all subscribe, I believe, that the phenomena of behavior and of mind are ultimately describable in the concepts of the mathematical and physical sciences' (p. 112), and later, speaking only for himself: (2) 'I am coming more and more to the conviction that the rudiments of every human behavioral mechanism will be found far down in the evolutionary scale and also represented even in primitive activities of the nervous system. If there exist, in human cerebral action, processes which seem fundamentally different or inexplicable in terms of our present construct of the elementary physiology of integration, then it is probable that that construct is incomplete or mistaken, even for the levels of behavior to which it is applied' (*Cerebral Mechanisms in Behavior: The Hixon Symposium*, ed. Ll. A. Jeffress, New York and London, 1951, p. 135). While I fully endorse the necessity of such fiduciary statements and also accept the principle of continuity (2) as propounded by K. S. Lashley, I do not share his belief (1) that mind can be represented in terms of physics and chemistry. Not even a machine can be thus represented.

The machine-like conception of living beings can be extended to account in principle for their adaptive capacities. An automatically piloted aeroplane approximates the skills of an air pilot. Its mechanical self-regulation co-ordinates its activities in the service of a steady purpose, and it may even appear to show a measure of resourcefulness in responding to ever new, not exactly foreseeable situations. There is a school of thought today which passionately pursues this mechanical conception of all vital adaptive functions, including the activities of human intelligence. I shall briefly enumerate the clues which suggest to me that this endeavour is misguided.

Our existing knowledge of physics and chemistry can certainly not suffice to account for our experience of active, resourceful living beings, for their activities are often accompanied by conscious efforts and feelings of which our physics and chemistry know nothing. But let us assume for the sake of argument that physics and chemistry could be expanded to account for the sentience of certain physico-chemical systems. It might not be inconceivable that a machine of sufficient complexity would develop conscious thinking, without losing its machine-like character. However, conceived in this sense, conscious thoughts would be the mere accompaniment of automatic operations, on the outcome of which they could exercise no influence. We should have to imagine, for example, that Shakespeare's conscious thoughts had no effect on the writing of his plays; that the plays were subsequently performed by actors whose thoughts had no effect on their acting; while successive generations of audiences flocked to see the plays without being impelled by the fact that they enjoyed them.

None of this is *strictly* inconceivable; it forms a closed interpretative system. Though nobody can believe in it in practice, one might regard this as a failing due to primitive habits of mind which a perfect scientific knowledge would eliminate. I am committed to a different belief. I accept the responsibility for drawing an ever indeterminate knowledge from unspecifiable clues, with an aim to universal validity; and this belief includes the acknowledgment of other persons as responsible centres of equally unspecifiable operations, aiming likewise at universal validity.[1] To me, therefore, the works of Shakespeare offer a massive demonstration of a creativity which cannot be explained in terms of an automatic mechanism. In my view, whenever we are confronted by a work of genius and submit to the leadership of its author, we emphatically acknowledge originality as a performance, the procedure of which we cannot specify.[2]

This conceptual framework strongly suggests to me the presence of an active centre operating unspecifiably in all animals. In 'The Logic of Affirmation' (Part Three, ch. 8), I have linked the unformalizable powers of originality to the whole range of tacit, often passionate, coefficients which account for all the powers of an articulate intelligence. I have said that this tacit urge sustains throughout our culture the coherence and fertility of fixed symbolic operations that were initially contrived by this urge itself. I believe that, on grounds of continuity, we should acknowledge that the same urge operates also primordially throughout the animal kingdom. There are then two principles at work in animals: namely, (1) the use of machine-like contrivances and (2) the inventive powers of animal life. Accordingly, while the animal's machinery embodies fixed operational principles, this machinery would be impelled, guided and readapted by the animal's unspecifiable inventive urge—even as rigid symbolic operations are accredited and steadily reinterpreted by the tacit powers that affirm them.

For the sake of brevity, I shall present here only a few characteristic pieces of evidence for the existence of such generalized creative powers. Lashley[1] has observed that mutilated rats which had learned a maze continued to find their way through it, though the neural paths used in learning had been cut. Naturally, the manner of their progression was completely altered: 'One drags himself through with his forepaws (writes Lashley); another falls at every step but gets through by a series of lunges; a third rolls over completely in making each turn, yet manages to avoid rolling into a cul-de-sac and makes an errorless run. . . .' He concludes that 'If the customary sequence of movements employed in reaching the food is rendered impossible, another set, not previously used, and constituting an entirely different motor pattern, may be directly and efficiently reconstituted, without any random activity. . . .' The operated rats retain a memory and a purpose that evokes in each of them a different set of operational principles for achieving the same persistent aim. These instantly improvised alternative combinations of organs may be said to be *equipotential* in achieving the same overall action. They offer a series of solutions for the same technical problems.[2]

Similar instances of equipotentiality may be found at levels far above and far below the maze-running rat. With advancing age Renoir became crippled with arthritis. He lost the use both of his feet and hands; his fingers were immobilized in perpetual cramped rigidity. Yet Renoir went on painting for another twenty years until his death, with a brush fixed to his forearm. In this manner he produced a great number of pictures hardly distinguishable in quality or style from those he had painted before. The skill and the vision which he had developed and mastered by the use of his fingers, was no longer in his fingers. It had become a knowledge and purpose of a highly abstract, totally unspecifiable kind: a purpose which could evoke from his mutilated body a set of implementations that were equipotential to his previous performance.

[1] K. S. Lashley, *Brain Mechanisms and Intelligence*, Chicago, 1929, pp. 136 ff. See also *ibid.*, p. 99.
[2] At this level equipotentiality is invariably a discovery of the means to a predetermined end. Its generalization to intellectual comprehension, and to the whole field of heuristics beyond that, transcends this limitation on grounds of the continuity between groping and sensing.

At the other extreme of the evolutionary scale Buddenbrock[1] and Bethe[2] have shown that insects, spiders, centipedes and water-beetles, instantly adapt their mode of locomotion to the amputation of a leg or indeed of any particular combination of legs. Bethe argued that these improvised equipotential co-ordinations are so varied that they cannot be due to the action of predetermined anatomic paths.[3] He thought they manifested a capacity of the nervous system to reorganize itself adaptively.

This process of spontaneous adaptive reorganization, by which a pre-determined end is achieved under profoundly modified conditions, finds an important parallel in the process of embryonic development. The fragments detached from embryos of certain lower animals have the capacity of regenerating the whole embryo and of producing normal individuals. This ontogenetic principle was first discovered by H. Driesch in the embryo of the sea urchin. Throughout its cleavage stage any cell or combination of cells detached from the embryo will develop into a normal sea urchin. Driesch characterized these regenerative powers of an embryo by describing it as an 'harmonious equipotential' system. The capacity of the germ to build up a normal embryo in spite of severe amputations is more widely referred to today as 'morphogenetic regulation'.

I shall speak later of another principle of ontogenesis, operating by locally fixed potentialities, and of the combination of this mosaic principle with the equipotential reorganization of embryonic fragments. For the moment it is enough to observe that the powers of improvisation, discovered by Driesch in embryonic fragments, have proved up to this day just as inexplicable in terms of fixed anatomical structures as have the powers of functional regeneration manifested throughout the animal kingdom—from the amputated centipede to the paralysed Renoir.[4]

---

[1] W. v. Buddenbrock, *Biologisches Zentralblatt*, 1921, **41**, 41–8.

[2] A. Bethe, *Handbuch der normalen und pathologischen Physiologie*, 1931, **15** (Zweite H.), 1175–1220.

[3] K. S. Lashley, in the Hixon Symposium of 1951 (p. 124), concurs in Bethe's argument. Bethe's work has been followed up extensively by E. v. Holst (see particularly *Die Naturwissenschaften*, **37** (1950), 464–76) in an attempt to lend greater precision to Bethe's dynamic conception of muscular co-ordination. Paul Weiss has proved the absence of anatomically fixed co-ordinative paths in the central nervous system by showing that the co-ordination of muscles remains unaffected when these are attached to nerve fibres belonging to an arbitrary assortment of neurons. For this fact and other supporting evidence see *Analysis of Development*, ed. B. H. Willier, P. A. Weiss and V. Hamburger, Philadelphia and London, 1955, 'Neurogenesis', by Paul Weiss, p. 388.

[4] Recent experiments carried out in the laboratory of Paul Weiss (*Proc. Nat. Acad. U.S.A.*, **42** (1956), 819) have strikingly expanded the domain of equipotential reorganization. Embryonic skin-, cartilage- or kidney-tissue was subjected to complete dissociation into separate freely floating cells. It was found that cells from any one of these tissues, thrown together at random, proliferated in cultures to form higher stages of the tissue in question; thus producing respectively feather germs, kidney tubules or a cartilage of a characteristic type.

We may add also that the powers of morphogenetic integration have long since been acknowledged by some investigators as essentially akin to the powers of comprehension, to which Gestalt psychology has directed our attention. The concluding paragraph of the Silliman Lectures delivered in 1938 by the great master of experimental embryology,

The continuity between heuristics and morphogenetic equipotentiality can now be outlined in more specific terms. We start from the fact that no material process governed by the laws of matter as known today can conceivably account for the presence of consciousness in material bodies. I have refused to assume that if we succeeded in revising the laws of physics and chemistry, so as to account for the sentience of animals and men, these would still appear to us as automata—with the super-added absurdity of a totally ineffectual mental life accompanying their automatic performances. To represent living men as insentient is empirically false, but to regard them as thoughtful automata is logical nonsense. For we are aware of a man's thoughts only by listening to him, i.e. by attending subsidiarily to certain bodily actions in the assumption that they are impelled by his thoughts, which are in fact known to us only as the effective centre of his meaningful actions. Nor can we speak therefore of thinking which totally lacks originality and responsibility; or indeed, envisage another person's considered judgment without acknowledging its universal intent, which challenges us to follow or contradict it. These features are essential to our pre-scientific conception of thought, and neurology cannot be said to account for thinking unless it represents it as something in which these features are still recognizable.

A big step towards the generalization of the powers of thought downwards in the direction of morphogenetic originality is made by acknowledging the originative powers of unconscious thought. The unconscious exercise of originality is usually still prompted by a conscious effort and a judgment of a high order, as in the case cf the heuristic efforts which induce discovery during a subsequent period of latency. An effort will usually be also at work in causing the reorganization of available means for a pre-determined end.

Ultimately, by dropping also the element of effort, the capacity for coherent and resourceful action can be generalized to a process of growth. The principle of equipotentiality is thus equated to the acknowledgment that we cannot identify the phenomenon discovered by Driesch, except

Hans Spemann, expresses this eloquently. 'There still remains an explanation (he writes) which I believe I owe the reader. Again and again terms have been used which point not to physical but to mental analogies. This was meant to be more than a poetical metaphor. It was meant to express my conviction that the suitable reaction of a germ fragment, endowed with the most diverse potencies, in an embryonic "field", its behavior in a definite "situation", is not a common chemical reaction, but that these processes of development, like all vital processes, are comparable, in the way they are connected, to nothing we know in such a degree as to those vital processes of which we have the most intimate knowledge, viz. the mental ones. It was to express my opinion that, even laying aside all philosophical conclusions, merely for the interest of exact research, we ought not to miss the chance given to us by our position between the two worlds. Here and there this intuition is dawning at present. On the way to the high new goal I hope to have made a few steps with these experiments' (Hans Spemann, *Embryonic Development and Induction*, New Haven, 1933, p. 371. I have taken the liberty of replacing in ll. 3 and 13 the word 'psychical' by 'mental', the equivalent of the German 'seelisch' which, I believe, Spemann had in mind).

by crediting any fragment whatever of the early sea-urchin embryo with the capacity to grow into a complete individual. This is a capacity for utilizing indeterminate means for achieving a comprehensive feature that we deem to be right, and it can be envisaged only by acknowledging this instrumental relationship in these terms. It foreshadows, therefore, to this extent the kind of faculty which enabled Renoir to continue painting after he was paralysed, and beyond that, the whole range of personal judgment and originality which we acknowledge when properly attending to a thoughtful person. The morphogenetic principle discovered by Driesch thus reveals itself as *the primordial member of an ascending series of homologous processes, which cannot be understood except as the resourceful achievement of a comprehensive rightness, and every one of which dissolves altogether in the light of any more impersonal examination.*

## 6. EXPLANATIONS OF EQUIPOTENTIALITY

Many contemporary scientists insist that all intelligent behaviour is based on a machinery which, in organisms possessing a nervous system, operates on the principles of digital computers. This is the McCulloch-Pitts theory of neural network. It shows that a suitable linkage of neural circuits can account for the responses given by an intelligent person to the stimuli impinging on his sensory organs. Adherents of this theory go so far as to assert that even the discoveries of Kepler and Darwin are but the output of a computing machine capable of solving a very great number of simultaneous equations. To represent Kepler and Darwin (and presumably also Shakespeare and Beethoven) as automata is, according to K. Z. Lorenz who puts forward this view, imperative for 'the inductive research worker who does not believe in miracles'.[1] I have dealt with this theory in the previous section.

Others have criticized the digital computer model on the grounds that it does not account for the great resilience of the nervous system under the effects of widespread injuries. One would, indeed, hardly expect such a delicate machinery so promptly to resume its functions—often in a novel manner—when large parts of it are excised, its peripheral network cut through at essential points, or some of its effector organs amputated.[2]

A radical alternative to the digital computer as the operational principle of the neural network has long been advanced by W. Köhler in the form of his principle of 'isomorphism'. Köhler points to certain orderly physical

---

[1] K. Z. Lorenz in *Aspects of Form*, ed. L. L. Whyte, London, 1951, pp. 176–8.

[2] Examples of such functional stability despite anatomical lesions were given in the foregoing section. The discovery that even extensive brain ablations produce little reduction in the intelligent performances of animals is due to K. S. Lashley (*Brain Mechanisms and Intelligence*, Chicago, 1929). Objections against the digital computer model were raised at the Hixon Symposium (1951) on these and partly similar grounds, by K. S. Lashley, Paul Weiss, Ralph Gerard and Lorente de Nó. (But no one dissented from the fundamental belief, for which I quoted K. S. Lashley above on p. 335, that thinking beings are automata.)

systems which can be described in two alternative terms: one referring directly to their comprehensive orderly features, the other stating the dynamic conditions underlying this orderly state. Thus Kepler's laws describe directly certain comprehensive orderly features of the solar system, which can be shown to be but a manifestation of interactions based on Newtonian dynamics. The principle of isomorphism assumes that the neural traces of stimuli interact likewise in accordance with some dynamic laws of physics or chemistry, thus giving rise to an orderly configuration inside the nervous system: a configuration which has all the comprehensive features of the objects from which the stimuli originated. The excitation of this orderly condition within our central nervous system is supposed to make us aware of all the relations entering into the gestalt of the objects confronting us. Thus, for example, the cortical counterpart of a square is supposed to have all the structural properties of a square and enable us thereby to respond to any of these properties.[1]

From this principle it would follow that the whole of mathematics —whether known or yet to be discovered—is latent in the neural traces arising in a man's brain when he looks at the axioms of *Principia Mathematica*, and that the physico-chemical equilibration of these traces should be capable of producing a cerebral counterpart (a coded script) comprising this entire body of mathematics. However, if *per incredibile*, any such equilibration existed, it could certainly not be based on the physico-chemical interaction of neural traces.

Köhler's theory is also logically defective in failing to account for the external manifestations of thought. It does not tell us how the duplication of an external gestalt inside the brain will produce any overt responses corresponding to itself—and a moment's reflection shows that it is indeed just as difficult to account for the formation of an appropriate response *to a cerebral gestalt* as for a similar response *to the original gestalt* outside the body. In order to satisfy us in this respect, isomorphism would have to be supplemented by an effector mechanism, for which so far the only principle suggested is that of the computer which we have seen to be inadequate. Köhler's theory, therefore, leaves the problem of intelligent behaviour where it stood before.

But the idea of equilibration as an ordering principle has wider implications, presenting a more general problem. The predominant school of biologists regard the persistent achievement of the same typical form from a variety of embryonic cell-combinations as a process of equilibration. They assume that the physico-chemical interaction of every part with every other part of the developing embryonic fragment brings about each time the same overall configuration.

Though I have already shown that isomorphism is untenable as a theory of conceptual understanding or of intelligent behaviour, I shall neverthe-less momentarily accept it for the sake of the following argument as an

---

[1] W. Köhler (Hixon Symposium (1951) p. 68).

explanation for our sensory awareness of gestalt. The sensory shaping of gestalt is then on a par with morphogenesis, as an ordering process in which the comprehensive physico-chemical interactions of the parts are supposed to produce orderly entities. The question is: whether any such equipotential processes, leading to comprehensive achievements of rightness, can in fact ever be represented in terms of physico-chemical equilibration.

My answer to this at this stage, is as follows. (1) Where science and technology overlap, operational principles overlap with certain laws of nature (see p. 331 above), and in the same sense a physiological function may conceivably coincide with certain laws of physics and chemistry. Yet both in the case of technology and physiology something is being achieved which neither physics nor chemistry can define.

(2) The seeing of a pattern or a shape is such an achievement. A process of physico-chemical equilibration is indifferent to the success or failure of gestalt seeing and therefore cannot express the difference between illusion and knowledge, or represent the effort made by the subject to avoid error and achieve knowledge. Morphogenesis is the formation of right shapes, a process which may succeed or fail. A physical-chemical explanation would not account for these alternatives, it would merely shift the problem back to the rightness of the conditions from which the process started.

(3) All questions raised by psychology and morphogenesis are rooted in our interest in mental activities and embryonic development. Studies of physical-chemical processes can never *take the place* of these interests; they can belong to psychology or embryology only to the extent to which they *have a bearing on anterior interests arising within these sciences*. Physical and chemical knowledge can form part of biology only in its bearing *on previously established biological shapes and functions*: a complete physical and chemical topography of a frog would tell us nothing about it as a frog, unless we knew it previously *as a frog*. In this sense both psychology and morphogenesis would remain unspecifiable in terms of physics and chemistry even if the mechanistic assumptions, which I have admitted here for the sake of argument, were fulfilled. I shall illustrate this in the next chapter in respect of morphogenesis.

For the moment we establish the following lessons. Living beings function according to two always interwoven principles, namely as machines and by 'regulation'. Machine-like functions operate ideally by fixed structures; the ideal case of regulation is an equipotential integration of all parts in a joint performance. Both kinds of performances are defined by rules of rightness and these refer in either case to a comprehensive biotic entity. But there is this difference. Machine-like functions are ideally defined by precise operational principles, while the rightness of a regulative achievement can be expressed only in gestalt-like terms. One's comprehension of a machine is, accordingly, analytical, while one's appraisal of regulation is a purely skilful knowing, a connoisseurship. Yet both kinds of performances have it in common that their rightness cannot be specified in the more impersonal terms of physics and chemistry.

The suggestion made in the foregoing Section 5, that equipotential processes are a primordial form of originality, will be taken up later in the chapter dealing with Evolution.

### 7. LOGICAL LEVELS

To the extent to which our personal participation in knowing a fact contributes to making it what it is, we may call it a *personal fact*. To the extent to which our personal knowing of a thing is unspecifiable, the thing itself cannot be represented exhaustively in terms of its less personal particulars. This is true in respect of inanimate things like a piece of information, an accident, a noise or a pattern, and was implied, and sometimes also expressly stated, in my discussion of these personal facts in Part One. I have also spoken of this aspect of personal knowledge at some length in the present chapter, with reference to machines. But its major importance emerges only when we turn to living beings, where an important additional feature is added to it: this feature is our recognition of individuals.

There is life in tissue cultures and viruses which are not segregated in the form of individuals, and the germ plasm transmitting heredity has a continuously extended life, which transcends the individuals through which it passes. Fragments of plants and lower animals may be viable in themselves. Yet the bulk of living matter is found embodied in a finite set of individuals, circumscribed in space and of limited duration in time. Each has come into existence at a definite moment, to remain alive for a certain period, after which it will die.

Our acknowledgment of an individual is an act of personal knowing which is clearly foreshadowed, as follows, in the claim I have made to hold my personal knowledge responsibly, with universal intent. (1) I am myself an individual living being. Therefore as I gave instances of my personal knowledge and analysed my essential participation in it, I was already describing a living being, and crediting it with certain arts of doing and knowing of which I believe myself possessed. (2) Having such confidence in myself, I went on to recognize the companionship of other people, thus using my powers of personal knowing for crediting others with the exercise of similar powers.[1] (3) By the same token I can now generalize this fiduciary act to a recognition of all kinds of individual living beings.

Individuality is, accordingly, a personal fact, and to that extent unspecifiable. To this I shall return in the next chapter; for the time being, I shall deal with other peculiarities of our knowledge concerning individuals. In the first place, a living individual strikes us as a personal fact,

[1] Pp. 263–4.

having a much more tangible and active being than any other personal fact that we have yet encountered. Of course, every comprehension of a whole acknowledges the reality of it; and whatever we comprehend both means something to us and to a certain extent at least means something also existentially, in itself. We experience this meaning by pouring ourselves out for the sake of achieving a focal awareness of the whole. By dwelling in a harmonious sequence of sounds, we acknowledge their joint meaning as a tune: a meaning they have in themselves, existentially. Up to a point our acknowledgment of the existence of a living individual proceeds on quite similar lines. We appraise in it a significant orderliness, which as such means something in itself. But a living individual is altogether different from any of the inanimate things, like tunes, words, poems, theories, cultures, to which we have ascribed meaning before this. Its meaning is different, perhaps richer, and above all, it has a *centre*. The focus of our comprehension is now something active, that grows, produces meaningful shapes, survives by the rational functioning of its organs; something that can behave and acquire knowledge, and at a human level, can even think and affirm its own convictions.

The acknowledgment of such a centre is a logical novelty. This becomes strikingly apparent at the human level. When we know somebody who himself knows something, his knowledge is part of our subject matter. We must decide whether it is in fact knowledge. A man's illusions are not the same as his knowledge. We must undertake therefore to discriminate between the two and to understand the ground on which knowledge was acquired. So presently we find ourselves examining knowledge or alleged knowledge, in the same way as when we reflect on what we ourselves know, or believe we know.

This is very peculiar. For logicians discriminate sharply between our *knowledge of things* and our *reflections on our knowledge of things*. Natural science is regarded as a knowledge of things, while knowledge *about* science is held to be quite distinct from science, and is called 'meta-science'. We have then three logical levels: a first floor for the objects of science, a second for science itself and a third for meta-science, which includes the logic and epistemology of science. But since we have seen that sign-learning is logically equivalent to the establishment of truth in the natural sciences, it follows that the process of sign-learning takes place on two logical levels: learning occupying the higher, and the discrimination box, etc., the lower level. And hence to the *study* of sign-learning we now ascribe the three-storied structure of a meta-science, the animal psychologist occupying its topmost, third level. I have anticipated this already up to a point on p. 262 by defining the tripartite situation in which the neurologist investigates the functions of the brain.

A science dealing with living persons appears now logically different from a science dealing with inanimate things. In contrast to the two-storied logical structure of inanimate science, biological science, or at least some parts of biology, seem to possess a three-storied structure, similar to that of logic and epistemology. This conclusion presents us with the following paradox. The evolutionary process forms a continuous transition from the inanimate stage to that of living and knowing persons; how can it then bring forth an additional logical level—two in place of one, three in place of two?

Let us look first at the stage where the three-storied structure is fully established. Once we have before us the deliberate behaviour of an animal, by which it commits itself to a mode of action which can be right or wrong, and which thus implies assumptions about external things that can be true or false, the understanding of such a commitment is a theory of rightness and knowledge. It is clearly three-storied. But some aspects of the three-storied structure emerge much earlier, at the very first appearance of individual living beings. Any such individual may be said to be normal or abnormal; healthy or sick; it may be mutilated, malformed, or else intact and well-shaped. The three-storied structure manifests itself at this stage only in the fact that any distinction between a physiological and pathological shape or process must necessarily be based on standards of rightness *that are proper to the individual in question.* These standards, which are common to the species to which the individual belongs, acknowledge our interest in the existence of normal specimens of the species and endorse their normal functions as proper to them. The observer's judgment of rightness already operates here on two consecutive levels. On the upper level, it establishes the physiological characteristics of the species, as opposed to its pathological anomalies, while, on the lower level, it applies these criteria for the appraisal of a single individual, on the assumption that they tell us what is good for him. The most primitive stage at which a third, lowest level becomes apparent, is when animals operate externally, though without deliberation, e.g. by co-ordinating their limbs for locomotion, guiding their migration, etc. The animal may then be said to be doing something that can be right or wrong, though in a weaker sense than when it is acting deliberately. Where no action is involved no lower level exists, and the judgments of morphology and physiology then merely imply that the normal existence of the animal is right in being what it is, as it is.

Biology is therefore three-storied in so far as the individual under observation is doing or knowing something, and two-storied when it observes an individual existing by himself, without bearing on things outside it. This reduction in the number of logical levels is similar to the transition from technology and the natural sciences, which both operate on *two levels*, to pure mathematics and music which—having no bearing on things outside themselves—are experienced by indwelling, on *one level.* Life lived for itself is equated here logically with artistic experience. Since passive existence awakens gradually to active performances, there is no discontinuity in the transition from the two-storied biology of plants and the lowest animals, to the three-storied biology of the more active and more knowledgeable animals. This resolves our paradox.

The appraisal of an animal's behaviour (as well as of its shape and the functioning of its organs) on two consecutive levels is an important generalization of a principle that we have met before, when we acknowledged the greatness of a scientific work of the past, even when its results had been largely mistaken. We did so because we judged the merits of the work within the framework of the means available to the author at the time. It was the same principle, once more, which defined our own calling in the light of the particular conditions to which we were born and brought up. All the applications of this principle were made with universal intent, by distinguishing between the subjective or contingent in us, which is part of our calling, and the personal in us which acts within this setting.

This brings me to a second logical novelty arising from the acknowledgment of a centre of individuality. For it shows another way in which logical levels can be effaced, this time particularly on the human level. Another person can judge us as we can judge him, and his judgment may affect our judgment of ourselves. Our relation to him may, indeed, be predominantly passive, as when we acknowledge the person's authority. For to the extent to which we accept a statement on trust, we forgo enquiring into its justification and cannot be said to be examining it from our own superior logical level. A measure of companionship prevails even between the animal psychologist and a rat on which he is experimenting, but interpersonal relations become ampler as we deal with higher animals, and even more as we reach the inter-human level. Mutuality prevails to such an extent here that the logical category of an observer facing an object placed on a lower logical level becomes altogether inapplicable. The I-It situation has been gradually transformed into an I-Thou relation. This suggests the possibility of a continuous transition from statements of fact to affirmations of moral and civic commands. We shall see this confirmed at the close of the next chapter.

# 12

## KNOWING LIFE

### 1. INTRODUCTION

FACTS about living things are more highly personal than the facts of the inanimate world. Moreover, as we ascend to higher manifestations of life, we have to exercise ever more personal faculties—involving a more far-reaching participation of the knower—in order to understand life. For whether an organism operates more as a machine or more by a process of equipotential integration, our knowledge of its achievements must rely on a comprehensive appreciation of it which cannot be specified in terms of more impersonal facts, and the logical gap between our comprehension and the specification of our comprehension goes on deepening as we ascend the evolutionary ladder. I shall demonstrate this in the present chapter. But before entering on this enquiry, I want to anticipate yet another point; namely, that as we proceed to survey the ascending stages of life, our subject matter will tend to include more and more of the very faculties on which we rely for understanding it. We realize then that what we observe about the capacities of living beings must be consonant with our reliance on the same kind of capacities for observing it. Biology is life reflecting on itself, and the findings of biology must prove consistent with the claims made by biology for its own findings.[1]

And as we shall find ourselves accrediting living beings with a wide range of faculties, similar to those which we have claimed for ourselves in the foregoing enquiry into the nature and justification of knowledge, we shall see that biology is an expansion of the theory of knowledge into a theory of all kinds of biotic achievements, among which the acquisition of knowledge is one. These will all be comprised by a generalized conception of commitment. The critique of biology will then turn out to be an analysis of the biologist's commitment, by which he accredits the realities on which living beings rely in the stratagem of living. And while these

---

[1] Cf. p. 142 above on self-confirmatory progression.

347

realities will fall into line with the realities to which our knowledge of inanimate things commits us, another line of generalization, ascending from the I-It to the I-Thou and beyond it to the study of human greatness, will transform the biologist's relation to his subject matter to that between man and the abiding firmament which he is committed to serve.

## 2. TRUENESS TO TYPE

The lowest manifestation of individual life—but not the least wonderful —is its manner of appearance, in shapely forms, ruled by specific standards. The meaning of such harmonious being and our appreciation of its significance are two allied forms of life. For the appreciation of harmonious beings is, just like the enjoyment of a work of art, itself a harmonious being. Our contemplation of living beings finds a justification in itself—a justification derived from the significance which it accords to the living beings that it contemplates, as beings in themselves.

There is a science—a descriptive science—which undertakes the classification of living beings according to shapes. This most ancient form of botany and zoology goes today by the technical name of taxonomy.[1] The basic performance of the taxonomist is actually practised every day without any scientific aid, whenever we identify a cat, a primrose or a man. Even animals have this capacity, and can exercise it even in respect of a species which has normally no vital interest for them, whether as a menace or as a quarry. Lorenz found that young birds who fix their filial sentiments on a human being will show the same attitude towards all members of the human race.[2]

Common Law makes the crime of murder, and punishment for murder, dependent on the human shape of the individual whose death has been caused. It demands that through all its variations—caused by differences of age and race, by malformations and mutilations, or by ravaging disease —we should always identify the presence of the human shape. Nor does this demand seem excessive, since no case is known in which an accused has pleaded failure to recognize the human shape of an individual he had killed.

Yet it would seem impossible to devise a definition which would unambiguously specify the range over which human shape may, and beyond which it may not, vary; and it is certain that those who recognize this shape are not in possession of any such explicit definition. Instead, they have exercised their art of knowing by forming a conception of the human shape. They have trusted themselves to identify noticeably different instances of what—in spite of these differences—they judge to be the same features, and to discriminate in other cases between things which, in spite

[1] The expression was first used in 1813 by de Candolle in his *Theorie Elémentaire de la Botanique*.
[2] Cf. K. Z. Lorenz in *Aspects of Form*, ed. L. L. Whyte, London, 1951, p. 169.

of some similarities, they judge to be instances of different features. Sustained by the belief that a human type exists, they have continued to build up their conception of it by noticing human beings as instances of this type. In doing this, they have practised the kind of power used for generating a focal awareness of a comprehensive entity from a subsidiary awareness of its parts.

I have already acknowledged my belief in the competence of this power of personal knowing, and said that it will be found to predominate in the descriptive sciences. I have endorsed especially our competence for classifying things—and living beings in particular—according to criteria which we believe to be rational, and to expect that the classes thus formed will prove real in the future, by revealing an indefinite range of uncovenanted common properties.[1]

Let me repeat now also that, by acknowledging that a specimen is normal, the biologist appraises an achievement on a scale of merit that he had set up for the specimen in question. This process is similar to the appreciation of individual crystals as specimens of the crystallographic class to which they are thought to belong. But even apart from the important fact that its objects have a centre of individuality, biology differs from crystallography in that the biologist's standards are empirical. They are not deduced from a single highly generalized assumption, based on a summary experience of the specimens in question, but are shaped piecemeal by a series of conceptual decisions, made by a close observation of every new specimen to be subsumed under the species to which it is thought to belong.[2] Thus every time a specimen is appraised, the standards of normality are somewhat modified so as to make them approximate more closely to what is truly normal for the species.[3] These standards are themselves subject to appraisal by the biologist. He will regard some species as well-defined, others as uncertain or altogether spurious. He will apply similar standards also to supra-specific groupings, such as genera, families, orders, classes; and he will apply them to the entire classificatory systems of which these form part.

The most important difference of value lies here between an artificial and a natural classification. The Linnean classification of plants according to the number and arrangement of stamens and carpels was excellent for the practical purpose of discriminating species and filing away specimens. But it was an artificial system, elegant, but lacking in real scientific beauty. Linnaeus knew that this system was not natural, and laboured relentlessly to replace it by one that would reveal the true kinship between species

[1] Cf. p. 112.  [2] Cf. Part Two, ch, 5, pp. 114–17.

[3] This process is not a statistical observation. Statistics can refer only to measurable parameters varying within a given population. Taxonomy judges non-measurable combinations of qualities within a population selected by the taxonomist himself with a view to the presence of these qualities. Nor is it even true that what is widespread is considered normal. Perfectly normal—as distinct from malformed or mutilated—specimens of a species may be the rarest.

according to their nature. Linnaeus believed species to be immutably fixed;[1] yet he clearly appreciated the deeper significance of a Natural Classification, the discovery of which he thought the Alpha and Omega of Systematic Botany. He said that while artificial systems serve to distinguish one plant from another, natural systems serve to teach the nature of plants.[2]

The effort made by Linnaeus himself to establish natural classifications both for plants and animals was successfully resumed about half a century later by A. P. de Candolle for plants, and by Lamarck and Cuvier for animals. Subsequent work has enormously expanded, but not changed fundamentally, the principles of these natural classifications; indeed, the publication of Darwin's *Origin of Species* in 1859 revealed a deeper meaning of this system than its authors had ever clearly envisaged. The hierarchy of the two kingdoms of plants and animals, with their subordinate Classes, Orders, Families, Genera and Species, was reinterpreted here as the branches of a family tree, the successive stages of which could be verified by paleontology.

Some idea of the size and complexity of this system may be gleaned from the estimates of the number of species into which our contemporary fauna and flora are divided, and the number of classes formed by these species. A standard textbook published in 1953 estimates that there are 1,120,000 known species of animals,[3] forming 30 phyla and 68 classes;[4] while G. N. Jones estimated in 1951 the number of known plant species at 350,000.[5]

One might expect to find this grandiose achievement celebrated wherever biology—the science of animals and plants—is taught and cherished. But no; classical taxonomy has almost ceased to count as a science. The explanation seems to lie in a change in the valuation of knowledge. It is due to a steadily mounting distaste for certain forms of knowing and being; a growing reluctance to credit ourselves with the capacity for personal knowing, and a corresponding unwillingness to recognize the reality of the unspecifiable entities established by such knowing.[6]

[1] J. Ramsbottom, *Linnaeus and the Conception of Species*, Presidential Address to the Linnean Society, 1938, shows that while Linnaeus rigidly adhered to the fixity of species till 1751, he later suggested a kind of evolutionary scheme. His first outline of a natural system of plants was, however, already published in 1751.

[2] A. J. Wilmott, 'From Linnaeus to Darwin', in *Lectures on the Development of Taxonomy*, delivered in the Linnean Society, 1948–9, published London, 1950, p. 35.

[3] Mayr, Linsley and Unsinger, *Methods and Principles of Systematic Zoology*, New York, 1953, p. 4.          [4] *ibid.*, p. 53.

[5] G. N. Jones, *Scientific Monthly*, 72 (1951), p. 293. The number is increasing rapidly, and not through subdivision of known species but by new discoveries, particularly in the New World tropics. Jones thinks it probable that we have not yet made the acquaintance of half the existing species of plants.

[6] See for example the account of the discussion held in London in December, 1950, by the *Systematic Association* on the subject 'Phylogeny in Relation to Classification' (*Nature*, 167 (1951), p. 503). The tendency, approvingly stated by one of the participants, to regard 'any attempt to make a classification without a motive as a waste of time' was the main concern of this meeting of taxonomists.

For taxonomy is based on connoisseurship. The nature of this faculty can be best recognized in a great naturalist who displayed it to a high degree. Take Sir Joseph Hooker. In 1859 he brought together and published evidence of nearly 8000 species of flowering plants in Australia, more than 7000 of which he himself had collected, seen and catalogued.[1] The 8000 generic entities which Hooker derived from the individual specimens coming under his notice, have been recognized as valid in the vast majority of cases by subsequent observations of botanists. Of Hooker's special gifts it was said: 'Few, if indeed any, have ever known or will ever know plants as he did. . . . He knew his plants personally.'[2]

More recently, C. F. A. Pantin has described how a new species of worm is discovered

> by a peculiar feeling of discomfort that something is not quite right, followed by a sudden detection of the error and simultaneous realization that it is highly significant—'It is a *Rhynchodemus* all right, but it is *not bilineatus*— it is an entirely new species!'

Pantin calls this mode of identification an 'aesthetic recognition' by contrast to a systematic recognition based on key features. He shows that the former predominates in field work.[3]

Once a species is established it is usually defined by the presence of certain distinctive key features. But these key features themselves are variable in shape, and hence reference to them represents once more a claim to the identification of a typical shape in its variable instances. This was made clear at the Fifth International Botanical Congress, held in Cambridge in 1930, partly for the purpose of finding a definition for a species. The features of plants are characterized by different authors as 'ovate, oval, patent, hirsute, ciliate . . .' but these authors may have quite different attributes in mind, said A. J. Wilmott. 'The lanceolate of Linnaeus (he continued) is very different from that of Lindley. . . . No two of my colleagues draw the same form of lanceolate.'[4] The knowledge of key features is invaluable as a maxim for the identification of specimens, but like all maxims it is useful only to those who possess the art of applying it.[5]

But the exercise of such exceptional skill weakens the position of the scientist today in the eyes of scientific opinion and tends to depreciate both his knowledge and its subject matter. The exceptionally high degree of connoisseurship required for establishing a species, combined with the

---

[1] Sir Joseph Hooker, *Introductory Essay to the 'Flora of Tasmania'*, London, 1859, p. iii.
[2] Leonard Huxley, *Life and Letters of Sir J. D. Hooker*, London, 1918, p. 412.
[3] C. F. A. Pantin, 'The Recognition of Species' in *Science Progress*, 42 (1954), p. 587.
[4] Fifth International Botanical Congress, August, 1930, *Report of Proceedings*, Cambridge, 1931, p. 542.
[5] This is why the British Museum has compiled a collection of 15 million insects against which they can match any new specimen submitted to them. Even so it requires the unique experience of the Museum's staff to carry out this feat successfully.

enormous extension of the domain over which it is to be exercised and the comparative shallowness of the knowledge thus acquired, lays the taxonomist open to the charge of indulging in merely subjective imaginings. When members of the Fifth International Botanical Congress declared that 'the concept of most species must rest on the judgment and experience of the individual taxonomist',[1] they invited this criticism. Reflecting on this discussion on the definition of a species, S. C. Harland recalled how in *Fanny's First Play*, by Bernard Shaw, the dramatic critic replies to the question whether the play was a good play, that if the play was by a good author, then it was a good play. 'The situation would appear to be somewhat similar', writes Harland, 'in regard to what constitutes a species.'[2]

I would suggest that the answer lies behind the Shavian joke itself. Just as plays written by good writers are, as a rule (though, of course, not always) good plays, so species described by good systematists are as a rule good species. In other words: owing to their acknowledged skill, good playwrights and good systematists alike enjoy considerable authority. This is conspicuous for both kinds of authors, because the rules by which they work and by which their work is judged are extremely delicate and altogether unspecifiable. Only if you refuse to accept any such highly unspecifiable knowledge, so that you renounce altogether the possibility of knowing a good play or a good species—and wipe out thereby the very conception of a good playwright or a good systematist—can you repudiate also the authority of any such persons.

Of course, the widespread distaste for the inexactitude of systematic morphology, which Professor Harland expresses here, issues in no demand for denying the existence of different animals and plants of typical shapes and structures. Professor Harland (and other scientists expressing similar tendencies) would only wish to recast the concept of species in the more impersonal terms of genetics. Thus defined, a species (a 'geno-species') would be formed by a world population of an organism, where there is—or is at least believed to be—a potentiality for an exchange of chromosomal material throughout the entire population.[3] However, the genetic investigation of a population presupposes its morphological distinctness.

[1] Statement by A. S. Hitchcock in *Report of Proceedings*, Cambridge, 1931, p. 228. Another member, Professor C. H. Ostenfeld, said (*ibid.*, p. 114) that a species consisted of all the individuals the character of which is in all main points the same, so far as the characters 'which we consider essential' are concerned.

[2] S. C. Harland, 'The Genetical Conception of the Species', *Cambridge Biol. Review*, **11** (1936), pp. 83–112. A number of years later we see the taxonomist still embarrassed by this charge. In *Lectures on the Development of Taxonomy* (London, 1950, p. 81), John Smart describes the increasingly delicate work of the modern taxonomist as follows: 'ultimately, the systematist could do little more than say that a species was such a segregate of organisms as he decided to designate as a species.' This, he says, 'sounds absurd', even though there is evidence that 'the really competent systematist had pretty shrewd opinions in the matter'. And in 1954 Pantin (*loc. cit.*) defends taxonomy against the appearance of thoroughly begging the question by referring to the 'competent taxonomer' in the definition of a species. [3] John Smart, *op. cit.*, p. 82.

The task of observing the process and the outcome of interbreeding within a given population is difficult enough and often quite impracticable for the vast majority of known morphological species. To undertake genetic experiments irrespective of morphological differences, with the intention of establishing from these alone the boundaries of specificity, would be absurd. It has certainly never been contemplated.

The same is true of all other tests that have been recently suggested with a view to placing taxonomy on more objective foundations. Thanks to the work of I. Manton, cytology has yielded very interesting corrections and amplifications of the morphological system of ferns.[1] But, once more, the range of such tests is comparatively limited, and above all, it has to rely on the existing morphological system for its guidance.

It all comes down to this. If you want to bring order into the multitude of animals and plants on earth, you must first of all look at them. Many thousands of millions of insects are crawling, swimming, burrowing and hopping all over the world and they fall into about 800,000 species. To apply any kind of test for identifying and discriminating these teeming multitudes, without paying attention to their characteristic shapes and markings, would be obviously impossible.

Of course, nobody has suggested this. Projects for the application of additional, and in particular more objective, taxonomic tests, have all set themselves their tasks *within* the existing morphological system. They propose to amend it or merely to understand it better, by bringing to bear on the existing system the methods of other branches of biology, whether the more objective tests of genetics and cytology, or the descriptive methods of anatomy, physiology, histology, ecology, phyto- and zoo-geography, etc. This would certainly not abolish Natural History in favour of a system based on objective tests. Yet the deprecation of the original conception of Natural History as a contemplative, rather than an analytical, achievement persist throughout modern biology.[2]

Not that the joy of seeing animals and plants and of entering into their forms of existence, by carefully studying their shapes and behaviour, is extinct among the naturalists of our time. Far from it; listen to K. Z. Lorenz:

> I confidently assert (he writes) that no man, even if he were endowed with a superhuman patience, could physically bring himself to stare at fishes, birds or mammals, as persistently as is necessary in order to take stock

[1] See C. Wardlaw, *Phylogeny and Morphogenesis*, London, 1952, pp. 99–102.

[2] Some biologists, consciously opposing this movement, have argued effectively that the art of recognizing kinds of living things is fundamental to their science. Thus A. Naef (*Handbuch der Vergleichenden Anatomie der Wirbeltiere*, Bd. 1, Berlin-Wien, 1921, pp. 77–118) has developed a pure typology of vertebrates. Agnes Arber (*Biol. Rev.*, **12** (1937), pp. 157–84) carried on the movement, initiated in Germany by Troll (1928), for 'returning to Goethe' and extensively developed Goethean morphology in *The Natural Philosophy of Plant Form* (Cambridge, 1950). J. Kälin (*Ganzheitliche Morphologie und Homologie*, Freiburg (Schweiz) and Leipzig, 1941), stresses the 'logical primacy of morphology', and O. Schindewolf (*Grundfragen der Paläontologie*, Stuttgart, 1950) insists on the priority of systematics to phylogeny.

of the behaviour patterns of a species, unless his eyes were bound to the object of his observation in that spellbound gaze which is not motivated by any conscious effort to gain knowledge, but by that mysterious charm that the beauty of living creatures works on some of us![1]

Indeed, biology remains the study of living beings, deriving its value ultimately from the intrinsic interest of living things—a general human interest which Natural History has immensely widened and deepened. Experimental studies made on animals and plants remain meaningless, except through their bearing on animals and plants as known to us by ordinary experience and through Natural History.

Of course, the scientific study of a subject matter may justifiably destroy our interest in it if it proves that the matter is in fact illusory. Astronomy, which started in Babylonian times as part of astrology, eventually proved astrology to be illusory; and the study of chemistry, originally initiated within the framework of alchemy, finally discredited and replaced alchemy. If experimental biology could discredit the existence of animals and plants— or at least prove that their alleged typical shapes and their systematic classifications are illusory—in the sense in which the shapes of constellations are illusory—then experimental biology might indeed supersede Natural History and be studied for its own sake, without bearing on Natural History. To aim at this would no doubt be foolish, but at least it would be consistent. Instead, we meet with the typical device of modern intellectual prevarication, first systematized by Kant in his regulative principles. Knowledge that we hold to be true and also vital to us, is made light of, because we cannot account for its acceptance in terms of a critical philosophy. We then feel entitled to continue using that knowledge, even while flattering our sense of intellectual superiority by disparaging it. And we actually go on, firmly relying on this despised knowledge to guide and lend meaning to our more exact enquiries, while pretending that these alone come up to our standards of scientific stringency.

If consistently carried out, the denial of contemplative value in science would cut off biology from the intellectual passions from which it takes its origin, and could not stop short of denying altogether scientific reality to the beings in which life manifests itself. Of course, biology may continue to flourish vigorously (as other branches of science have done) by wisely disregarding its own professed philosophy. But we shall see, as we proceed further, that this cannot be altogether relied upon.

### 3. MORPHOGENESIS

We ascend to the second level of biological achievement, by passing from the study of typical shapes to the science dealing with their coming into being: from the appraisal of living forms to the appraisal of the processes of regeneration and embryonic growth.

[1] Konrad Z. Lorenz in *Physiological Mechanisms in Animal Behaviour*, Symposia of the Society for Experimental Biology, No. 4, Cambridge, 1950, p. 235.

Regeneration is the restoration of a mutilated organism. Some lower animals, like hydra or planaria, have exceptional powers of regeneration, so that tiny pieces of their body will regenerate to complete individuals.[1] This is a manner of vegetative, a-sexual reproduction, commonly found in plants. It forms a transition from regeneration to sexual ontogenesis, which may be regarded as the regeneration of a complete individual from a fragment formed by the fusion of two parental gametes. The fact discovered by Driesch that any cell or cell-group, detached from the embryo of a sea urchin at the stage of segmentation, grows into a complete embryo is another extension of regeneration into ontogenesis.[2]

But complete regeneration is not universal, and at the limits set to regeneration we meet another principle of morphogenesis, which replaces the equipotentiality of all detachable fragments by a system of fixed potentialities. If the fertilized Ascidian egg in the two or four cell stage is cut in two, each half develops only into half an embryo.[3] Though this type of ontogenesis is never free of regulative tendencies, its principle can be clearly distinguished as a pattern of independently proceeding processes of growth. The organism is built up in sections which must fit together and be ready to function together when the moment arrives for it. Such a mosaic of independently proceeding interlocking sequences corresponds to the conception of ontogenesis which Roux and Weismann had formulated and made universally current before Driesch's observations on equipotentiality.

The regulative principle of Driesch and the mosaic principle of Roux-Weisman actually operate in combination. This is revealed in Spemann's principle of localized embryonic *organizers*. Spemann found that in the newt embryo at the gastrula stage there is a certain region adjoining the entrance of the primitive gut which dominates the further segmentation of the embryo. If the embryo is cut up, any part of it in which this dominant region is included—or in which it is engrafted—will proceed to develop further, while in the embryonic tissue from which it is eliminated,

[1] 1/200th of a hydra is able to regenerate the whole animal, while in the planaria 1/280th and even less has been shown to regenerate completely. A. E. Needham, *Regeneration and Wound-Healing*, London, 1952, p. 114.

[2] The pioneering experiments of Driesch have been greatly amplified by his successors, particularly Hörstadius (*Acta Zool.*, 9 (1928), p. 1; *Roux' Arch.*, 135 (1936), pp. 69–113) whose experiments are illuminatingly analysed by P. Weiss, *Principles of Development* (New York, 1939), pp. 249–88. Hörstadius observed that meridional halves of the sea urchin blastula regulate into normal, though small, pluteus larvae. Though in the course of this reorganization 'in general vegetative material was used to build the intestine and animal material to build the ectoderm, there is no detailed correspondence whatsoever between the actual use to which individual portions have been put in the experimental germ and their prospective fate in the normal germ' (*ibid.*, p. 261).

[3] Discovered by W. Roux in 1888 (see W. Roux, *Gesammelte Abhandlungen über die Entwicklungsmechanik der Organismen.* II, Leipzig, 1895, pp. 419–521). Later experiments by E. G. Conklin, greatly expanded by Dalcq and collaborators, have, however, established considerable powers of regulation at early stages of the Ascidian development, particularly in the virgin egg (see A. M. Dalcq, *Form and Causality in Early Development*, Cambridge, 1938, pp. 103–27).

individuation comes to a stop. Thus the dominant region, which is the seat of the organizer, moulds a whole region under its control into one complete embryo, irrespective of any previously differentiated character of its several component cells, which respond equipotentially to the organizer's stimulus. The effect of this stimulus on the area under its control is ascribed to the organizer's *morphogenetic field*.[1] The morphogenetic powers of the individual are localized at this stage in a single organizer; but presently this centre splits up into sub-centres of organization, each of which controls by its field the development of one section of the embryo. Later, these sub-organizers divide up in their turn by stages into secondary, and possibly tertiary, specialized sub-organizers, each of which controls the development of a limb, or part of a limb, or of some other organ or feature emerging from the progressive differentiation of the individual. A segregated area provided with its own organizer may be cut off with it and will then go on differentiating in isolation—producing, for example, an isolated limb. At this higher stage the development of the embryo may be regarded as a mosaic of interlocking independent sequences, each controlled by its organizer, while equipotentiality has been reduced within the confines of the several morphogenetic fields controlled by their separate organizers. This mosaic structure prefigures the fixed localization of specific regenerative powers found in the adult higher animal.

But to complete this picture of morphogenesis even in the crudest outline, we must yet add the fact that embryonic tissues do not always submit unconditionally to the field of an organizer. This preparedness was defined from embryological observations by Waddington and called the 'competence' of the tissue.[2] As the result of his grafting experiments, P. Weiss established more generally that 'a field cannot make any cell produce any specific response unless that cell is intrinsically prepared to do so'.[3] Owing to this condition, the part played by the organizer may be reduced to a mere evocation of the potentialities preformed in the tissue subjected to its influence. This opens up a wide range of rivalry between the morphogenetic potentialities that are proper to an embryonic tissue and those induced in it by the dominant influence of adjacent tissues.

All these principles of morphogenesis were discovered by the new experimental methods applied for the first time by Wilhelm Roux in 1885.[4] The work was based throughout on a previous knowledge of des-

---

[1] The field concept was first used by Spemann (1921), in describing the organizer; Paul Weiss (1923) introduced it for the study of regeneration and extended it (1926) to include ontogeny. Cf. Paul Weiss, *Principles of Development*, New York, 1939, p. 290. The most striking manifestations of morphogenetic fields were revealed by the cultivation of embryonic tissues, described by Paul Weiss (1956), see note on pp. 338-9.

[2] C. H. Waddington, *Phil. Trans. Roy. Soc.*, **B, 221** (1932), 179. See also C. H. Waddington, *The Epigenetics of Birds*, Cambridge, 1952, pp. 106 ff. Embryological competence was described earlier as *Reaktionsfähigkeit* by O. Mangold, *Roux' Arch.*, **47** (1929), 249.

[3] P. Weiss, *Principles of Development*, 1939, p. 359.

[4] See W. Roux, *op. cit.*, 'Einleitung' zu den *Beiträgen zur Entwicklungsmechanik des Embryos*. (1885). This paper contains the first definition of Entwicklungsmechanik.

criptive embryology, which relied in its turn on a previous knowledge of systematic morphology. It was these descriptive sciences, therefore, that jointly set the standards for appraising the achievement of normal shapes by normal embryonic stages, and experimental embryology was thus an attempt to analyse hitherto descriptively defined performances. The morphogenetic principles sketched out here may accordingly be taken to define operations for the achievement of ontogenetic success conceived in morphological terms. In the pursuit of this study several factors of form-achievement are tested in isolation and under variable conditions, and their operation observed also by experiments of regeneration, of trans-plantation, of the influence of toxic media, etc. While the investigation will thus extend to the production of abnormal forms, these processes derive their interest from their bearing on normal development.

In the previous chapter I have distinguished two kinds of biological achievements, namely, (1) achievements performed by the rational con-currence of several parts with fixed functions and (2) achievements per-formed by the equipotential interplay of all parts of a system. In morpho-genesis the first, machine-like, type is present in the stratagem of inde-pendent interlocking morphogenetic sequences, based on a mosaic of fixed potentialities; the second, integrative, type is found in the mor-phogenetic achievements induced by the field of an organizer, as well as in the autonomous morphogenetic responses of isolated tissues. Embryo-genesis appears to be a comprehensive achievement due to the rational combination of these two types of rational principles.

The analysis of the process by which living beings are formed cor-responds to the logic of achievement, as illustrated by the manner in which we find out how a machine works. We must start from some anterior knowledge of the system's total performance and take the system apart with a view to discovering how each part functions in conjunction with the other parts. The framework of any such analysis is logically fixed by the problem which evoked it. Its content may be extended indefinitely and it may penetrate thereby ever further into the physical and chemical mech-anism of morphogenesis; but its meaning will always lie in its bearing on living structures that are true to type, emerging from a mosaic of mor-phogenetic fields

The meaning of experimental embryology is thus doubly dependent on personal knowledge: both in respect of the unspecifiable knowledge of true shapes, and in respect of the appreciation of the process by which highly significant shapes and structures are brought into existence. This situation has caused uneasiness among scientists. 'Morphogenesis'—complains Paul Weiss—'is still in the transition phase from descriptive "natural history" to analytical science.' He finds that when modern physical and chemical tools of great precision are applied to problems formulated in such less precise terms, the results are equally imprecise and ambiguous.[1] This complaint

[1] Paul Weiss, *Quarterly Review of Biology*, **25**, 1950, p. 177.

was voiced even more radically by F. S. C. Northrop and H. S. Burr in a summary of their electro-dynamic theory of life, published in 1937.[1] Physico-chemical explanations, they suggest, correspond to a Democritean philosophy of science, while 'perceived organization' is an Aristotelian concept. But: 'The Aristotelian and Democritean philosophies of science do not combine.' Hence we are asked to replace in biology the visible appearance of organisms by an observation of the electrodynamic field produced by them. C. M. Child had insisted likewise that morphological differentiation must be defined in quantitative terms, since otherwise we are inevitably led to 'barren neo-vitalistic assumptions'.[2]

Throughout this literature the word 'vitalistic' is used as a term of condemnation, even as Wöhler and Liebig used it, for the purpose of discrediting evidence which threatened a more objectivistic framework.[3] However, in this case no such objectivistic framework exists. No one has yet seriously envisaged that we should study living beings without noticing them; yet as soon as we do notice them we are relying on those very features which a 'Democritean' science must ignore. Indeed, a complete 'Democritean' or Laplacean knowledge can tell us nothing without relying on our personal knowledge of these comprehensive features. Suppose we were given a complete topographical chart of all physico-chemical changes taking place in our surroundings. It would require a superhuman feat of insight to discover from this information the fact that there are somewhere such things as chickens, which are hatched from eggs. But suppose we could do this—that we could achieve this feat and became familiar through it with chickens and their hatching from eggs—we should still have gained only the same kind of comprehensive view of morphogenesis as our ordinary insight conveys to us.

But—we may be asked—could the shape of chickens, eggs, etc. not conceivably be determined in mathematical terms? Could we not give then an exact and strictly objective account of morphogenesis? No, we could not; for even on this fantastic assumption (which Northrop and Burr seem to have envisaged in their own way) we would be still ultimately relying on ordinary morphological observations. Normal shapes—as distinct from abnormal, malformed, stunted shapes—would have to be identified by our own standards of rightness *before* they could be defined in mathematical terms. Mathematical relations are, like the processes of physics and chemistry, neutral in respect to morphogenetic success or failure; these alternatives must therefore be identified by ourselves before we can analyse them in terms of mathematics or physics and chemistry.[4]

[1] F. S. C. Northrop and H. S. Burr, *Growth*, **1**, 1937, p. 78.
[2] C. M. Child, *Individuality in Organisms*, Chicago, 1915, pp. 183–4.
[3] See Part Two, ch. 6, p. 157.
[4] Recent experiments by Holtfreter (1951) and others, suggesting that only living tissues act as organizers in the full sense of the term, have been taken by some embryologists as a warning that causal embryology must continue to rely on morphological knowledge. See Clifford Grobstein in *Aspects of Synthesis and Order in Growth*, ed. by Dorothea Rudnik, Princeton, N.J., 1954, p. 233.

We may conclude that the insights by which we recognize life in individual plants and animals, and distinguish their several kinds—and by which we appraise them as normal or abnormal, establishing thereby the success or failure of the process by which they come into existence—that these insights reveal a reality to which we have access by no other channels, and that the mechanism of morphogenesis can therefore never amount to anything but the observation and understanding of patterns and processes expressly bearing on that reality.

### 4. LIVING MACHINERY

Both plants and animals have a great many ingenious devices within their bodies which are used for the benefit of the organism. Animals, operating more energetically than plants, are much richer in them. Claims for hundreds of patents could be found by describing the rational interaction of the animal's organs in the service of its various interests. The operational principles defined by such patents would be the principles of animal physiology.

I have analysed in the previous chapter our knowledge of machines. They can be recognized as such only by first guessing, at least approximately, what they are for and how they work. Their operational principles can then be specified further by technological investigations. Physics and chemistry can establish the conditions for their successful operation and account for possible failures, but a complete specification of a machine in physico-chemical terms would dissolve altogether our knowledge of the machine.[1]

The logic of engineering applies also to physiology, but with some modifications. The organs of the body are more complex and variable than the parts of a machine, and their functions are also less clearly specifiable. While considerable connoisseurship may be required to judge the shape of a good air-screw and to diagnose any possible defect of it, the skill needed for judging the shape of a heart and its possible malformations is more delicate. A knowledge of the shape and location of the organs in the whole variety of animals known to zoology forms a vast body of morphological information; it is a descriptive science. Moreover, any particular function can be performed in a great many ways. In breathing, for example, the two sides of the chest, the diaphragm, and the muscles of the neck may be used in variable co-ordinations. This reduces the specifiability of a living performance as compared with that of a machine and contributes once more to the descriptive character of physiology.

Thus in physiology, the twofold unspecifiability of organized shapes

---

[1] Except that to the extent to which laws of nature overlap with an operational principle it suffices to transpose the natural laws in question into the form of an instrument serving a purpose acknowledged by the observer. Cf. Part Four, ch. 11, pp. 331 and 342 above.

and of the processes occurring within them is added to the inherent un-specifiability attached to operational principles in general, and to this extent the logic of physiology differs from that of engineering. Otherwise we have the same set of relations in both cases. The study of an organ must begin with an attempt to guess what it is for and how it works. Only then can it proceed further by combined physiological and physico-chemical enquiries, both being conducted with a bearing on the purposive physio-logical framework which they help to elucidate. Any attempt to conduct physico-chemical investigations of a living being irrespective of physio-logical assumptions will lead as a rule to meaningless results; and any attempt to replace physiology altogether by a physico-chemical chart of the living organism would completely dissolve any understanding of the organism.[1]

Of course, living machinery has a purpose only in the interest of the living individual as appraised by the observer. But it must possess this purpose. Organs and their functions exist only in their bearing on the presumed interest of the living individual. All physiology is teleological, and in this sense we may speak here also of reasons and causes. We say that the *reason* for having valves in the circulatory system is to prevent the regurgitation of the blood; while we ascribe the *causes* of any regurgita-tion, occurring in spite of these, to an insufficiency of the valves owing to malformation or disease. Physiology is a system of rules of rightness, and as such can account only for health. Accordingly, we do not enquire into the causes of health—any more than into the causes of a mathematical proof; but we do enquire into the causes of disease, as we do into the causes of a mathematical error.

Once more, as in morphology and morphogenesis, the existence of every living being is acknowledged as an aim in itself; however nasty a flea or liver fluke may be to us, we recognize the rational functioning of its organs in their own interest. The purely scientific interest of physiology depends, therefore, ultimately on the passions which make us pursue Natural History. It relies on the passions which account for the impor-tance which we attribute to a living being in itself; on its intrinsic interest, and on our contemplation of it as it *is*—and as it ought to be.

[1] Comparative physiology shows that widely different mechanisms are used by organisms for a single purpose, e.g. for digestion, breathing, etc.; such mechanisms are defined, therefore, by their common operational principle, not by their physical and chemical structure. Cf., for example, the chapters in J. T. Bonner, *Cells and Societies*, London, 1955, pp. 116–21, on 'Feeding in Animals', 'Breathing in Animals', 'Circulation in Animals', 'Excretion in Animals', 'Development and Reproduction in Animals', 'Coordination in Animals'. H. Graham Cannon (*Linnean Soc. J.*, **43** (1956), p. 9) gives ample evidence that the same operational principle—the filter mechanism of shrimps —is realized in different species by quite different constituent members. Moreover, corresponding (i.e. homologous) characters are produced in different mutants by different genes and the same genes may produce different characters (A. C. Hardy, *Proc. Psych. Res.*, **50** (1953), 96). The earliest notice of this fact, to which Professor Hardy called my attention, is by G. R. de Beer, 'Embryology and Evolution' in *Evolu-tion*, ed. G. R. de Beer, Oxford, 1938, pp. 65–6.

According to the current theory of evolution all living machinery has come into existence by accident and is found in existence only because it has conferred on the individual living beings, of which it forms part, competitive advantages which have secured the survival of their kind. This conception of evolution (to which I shall yet return in detail) would eliminate any true achievement from the phylogenesis of living beings; but even so this would not affect the teleological character of their machine-like equipment, which is logically inherent in the conception of jointly functioning organs.

## 5. ACTION AND PERCEPTION

I have surveyed so far the being, the growing and the functioning of organisms as subject matters of knowledge. These vegetative manners of living are common to plants and animals; yet they are usually more striking and better known in animals, and hence I took my examples from zoology. The examination of biology as a knowledge of action and perception, to which I shall now proceed, will apply exclusively to the study of animals. Action will be taken here to differ, by being deliberate, from the mere functioning of organs. This assumes the prompting of a conscious motive which I shall call a drive.[1] The term 'perception' will be applied here in its usual sense, to designate the process of getting to know an external object by the impression made by it on our senses. Disregarding thus the earlier stage of protopathic sentience, I shall envisage perception from the start as consciously discriminating, even though not yet capable of strenuous deliberation. This stage corresponds about to the transition by which human beings emerge from childish autism and recognize the world outside as a field of hazardous doing and knowing.

At the level of being, of growing, or of functioning, an individual could fail through being abnormal, malformed or diseased. The active, perceiving person has two more possibilities for going wrong, namely, *subjectivity* and *error*; and again, it falls to the observer to appreciate the rightness which is impaired by these shortcomings. You cannot observe deliberate action or perception except by legislating for it in these respects.

Take the feeding of a higher animal as an example of conscious action. This may be defined as the ingestion of food. But since we recognize as 'food' only materials which we believe to be nutritive, or at least not deleterious to the animal, it falls to us to this extent to determine what is *right feeding*. This is often far from obvious. When a sheep eats the wool off the back of another sheep, or cattle eat bones, the uninstructed may object to this as an aberration, but physiologists approve of it as a compensation for certain mineral deficiencies in the animal's diet. Yet not everything that animals eat is nutritive or even wholesome. It is easy to

---

[1] This term, to which R. S. Woodworth first gave currency in his *Dynamic Psychology* (New York, 1918), in place of 'instinct', is not always used in this sense in psychology.

poison animals by arsenic or strychnine, or to deceive them as the angler does by his fly. Rats will drink saccharine solution, which has no nutritive value, and captive apes eat their faeces, which seem to be quite useless as food. In all these cases it is the observer's judgment which appraises what is right and what is wrong feeding.

The nature of this judgment is qualified by the fact that feeding is normally actuated by a drive. The dog-lover is painfully conscious of his dog's hunger when it whines for food; the rat's enjoyment of a sweet taste is the only reason that can be found for its feeding on a solution of saccharine. In recognizing this we acknowledge the presence of a rational centre in the animal, to which we attribute both its correct and its mistaken decisions. In this sense we shall deprecate the drinking of a saccharine solution in a rat, as offering a purely *subjective satisfaction*, and class the swallowing of an angler's fly by a fish as a *reasonable error* in an otherwise altogether rational way of feeding. On the other hand, we shall deny to a maniac devouring paper or sand any degree of rationality; this kind of false feeding will be classed as a *meaningless act*. It is a compulsive pathological process, endured by the diseased mind and as such to be classed with purely passive bodily malformations.

The process of perception has a similar logical structure. An object approaching the eye is seen as constant, so long as a certain relationship prevails between the effort of accommodation and the size of the retinal image. More precisely, we are jointly aware of the retinal image and of the adaptive effort, as well as of certain relations of the two, while both are undergoing a change, in terms of the constant size of an object seen at variable distances. The observer of this process of perception will regard it as a *correct performance* if he himself endorses the affirmations implied in it, namely that the object did in fact remain of constant size. But it may happen, as in the experiments of Ames already mentioned, that unnoticed by the subject, the observer alters the size of the object—a rubber-ball—by inflating it. We have seen that the subject may then increase his accommodation as if the object were approaching him and become aware of this increased effort in its conjunction with an increased retinal image, by seeing the swelling object as coming nearer at constant size. In this case the seeing of a constant size may be regarded as a *reasonable error*. On the other hand, if the effort required for a certain measure of accommodation is increased by atropin poisoning, an approaching object will be seen shrinking to a tiny size and the reduction of its size will make it appear farther off; but owing to our knowledge that this cannot be true we shall know this anomalous appearance to be deceptive.[1] We shall regard it then as a *subjective experience* of the perceiving person which is *rational from his own point of view*, but not otherwise.[2] And again,

[1] William James, *Principles of Psychology*, 2, London, 1910, p. 93.
[2] To the extent to which a false perception is corrigible it is an error; to the extent to which it is compelling, it is an illusion.

we know of hallucinations, the falsity of which cannot be accounted for either by subjective rationality or reasonable error; they are *devoid of reason*.

We have met here some primitive forms of commitment, and biology has been revealed as an appreciation of commitment. To swallow something in the hope that it may be wholesome is clearly a commitment, and so is every act of seeing things in one particular way. I have suggested before that in a generalized sense commitment may be acknowledged even at the vegetative level, since it is of the essence of a living organism that each part relies for its function, and for its very meaning as part of the organism, on the presence and proper functioning of a number of other parts.[1] In this sense our knowledge of the normal growth, functioning and being of the organism is an appraisal of its primordial commitments which accredits their success. Commitment may then be graded by steps of increasing consciousness; namely, from *primordial*, vegetative commitment of a centre of being, function and growth, to *primitive* commitment of the active-perceptive centre, and hence further again, to *responsible* commitments of the consciously deliberating person. The aphorism that biology is life reflecting on itself now acquires a fuller meaning. Biology is a responsible commitment which appraises other commitments. In its usual narrower sense, biology is a responsible commitment which appraises primordial and primitive commitments. But I shall break through the limitation implied in this formulation and proceed presently to consider the appraisal of responsible commitments (which includes the justification of my own convictions) as the extension of an ascending series of biological observations beyond biology, into a domain that may be called 'ultra-biology'.

I have dealt before with the molar features that characterize the vegetative level; let me now sum up the new features that are added to these on the active-perceptive level. They are *sentience* of motive and knowledge; an effort to *do right* and *know truly*; a belief that there exists an *independent reality* which makes these endeavours meaningful, and a sense for the consequent *hazards*.

On the morphological and vegetative level we had only two classes of appraisal: namely normality and abnormality: health or disease. The intervention of sentience enlarges our scale to four significant classes:

(1) a correct satisfaction of normal standards,
(2) a mistaken satisfaction of normal standards,
(3) action or perception satisfying subjective, illusory standards,
(4) mental derangement issuing in meaningless reactions.

The first three kinds of appraisals are those of a normal individual, the fourth case is pathological. This classification shows that the presence in a living being of sentience, purposive action and the knowing of external things elevates our knowledge of the living being into a *critical meeting* of it.

[1] See p. 323.

By including a critique of the handling and the knowing of things by its subjects, biology becomes three-storied. Our personal knowing becomes then the perceiving of an actively intended meaning, which we are trying both to understand and to judge with a view to the facts on which it bears. It is, in fact, the reception of a convivial communication, subject to its critical appraisal by ourselves.

Our understanding of the hungry animal choosing its food, or of an animal on the alert listening, watching and reacting to what it notices, is an act of personal knowing similar in its structure to the animal's own personal act which our knowing of it appraises. And accordingly, our knowledge of the active-perceptive animal would dissolve altogether if we replaced it by our focal knowledge of its several manifestations. Only by being aware of these particulars subsidiarily, in relation to a focal awareness of the animal as an individual, can we know what the animal is doing and knowing. Besides, when the subsidiary particulars of a comprehensive entity are as highly complex and variable as in these cases, attempts to specify them can do no more than highlight some features, the meaning of which will continue to depend on an unspecifiable background that we only know within our understanding of the entity in question. In other words, the meaning of an animal's actions can be understood only by *reading* the particulars of its actions (or by reading its mind in terms of these actions) and not by observing the actions themselves as we may observe inanimate processes.

Behaviourists teach that in observing an animal we must refrain above all from trying to imagine what we would do if placed in the animal's position. I suggest, on the contrary, that nothing at all could be known about an animal that would be of the slightest interest to physiology, and still less to psychology, except by following the opposite maxim of identifying ourselves with a centre of action in the animal and criticizing its performance by standards set up for it by ourselves.

## 6. LEARNING

In this rapid survey of the ascending stages of biological knowledge I must disregard many aspects of the subject. I shall proceed now to reflect on our knowledge of learning, without taking into account the component due to learning in the animal's capacity for primitive action and perception. I shall also leave out altogether the domain of ethology, and concentrate on the psychology of learning, based on animal experiments. I shall take advantage of my earlier treatment of this subject by using mainly examples already mentioned there.

Take first experiments with the discrimination box. Here the psychologist places the animal in a situation which constitutes a problem for the satisfaction of some of its major drives, usually hunger. A process of learning will originate from this arrangement only if (1) the animal recognizes the problem and responds to it, and if (2) this problem demands an appreciable measure of ingenuity, but not more than the animal in fact possesses. The limited alternatives offered to the animal force it to respond (if it responds at all) in a manner that can be classed as correct or false. Moreover, the experiment is so devised that the animal's choice between a correct and a false response has to be made at a particular point in time and space. The narrowness of the experimental situation tends to key up the animal's state of perplexity at a choice point, to a tension which is not likely to be reached in the wider circumstances of nature. Thus the laboratory both intensifies and spotlights the moments of heuristic effort by which the animal's active centre rises to the performance of an intelligent judgment.

I have given evidence for this before, which I shall presently expand further. Meanwhile let me recall that we have identified sign-learning with a process of inductive inference.[1] The question: How does an animal learn to recognize a sign? (or if the reflex language is preferred, How is an animal conditioned to a particular stimulus?) is therefore essentially akin to the epistemological question: how can correct generalizations be drawn from experience? The fact that the animal is generalizing about events engineered by ourselves does not distinguish it from us in this respect, since as subjects both the animal and we ourselves are faced with events beyond our control.[2]

But there are certain differences. Epistemology reflects on knowledge which we ourselves believe we possess; the psychologist studies knowledge which he believes to have been acquired by another individual and studies also the shortcomings of such knowledge. No knowledge, whether our own or that of a rat, is fully specifiable; but the fact that we must rely on recognizing the rat's knowledge, or ignorance, from our own knowledge of the rat's behaviour, involves an additional enquiry and an additional unspecifiability. Let me add also that (in view of the fact that in animal experiments the achievement of learning must always manifest itself in appropriate behaviour) I shall subsume here, for brevity, trick-learning under sign-learning, except when the trick is manifestly contrived from known elements, as in Köhler's experiments on apes.[3] I shall also admit fully, from the start, the presence of latent learning in all types of learning, though in some cases it is found almost lacking.

We can now run through the various alternative outcomes of learning experiments.

(1) We consider that learning has been fully achieved only if, judging by the animal's behaviour, we believe that it has formed a generalization which we consider to be the correct solution to its problem. In experiments

---

[1] Part Two, ch. 5, p. 76.

[2] Of course, learning corresponds to the drawing of empirical inferences concerning natural regularities only if we assume that the initial conditions of the learning experiment are maintained unchanged, indefinitely.

[3] This corresponds to classing *empirical* technology with natural science.

of the type used by Guthrie, in which learning is achieved by pure accident, the ensuing generalization will usually contain many irrelevant elements. (The animal may be said to be mistaken then in the sense in which primitive man is, who does not know clearly whether it is his axe, or the incantation by which he accompanies its strokes, that fells the tree.) A *correct generalization* must be free of such errors. It should offer a sufficient understanding of the problematic situation for establishing the *necessary* conditions of success.

(2) Clever Hans, faced with a blackboard which meant nothing to him, found a solution to the problem of obtaining the reward offered by the experimenter, by watching the man's behaviour while he, Hans, was tapping the ground. This generalization may be regarded as *subjectively correct*, as it was the most reasonable that could be established within the range of the animal's competence. We may similarly regard as subjectively correct the generalization by which red-green colour blind people distinguish by means of secondary signs the two kinds of colouring. The forming of false 'initial hypotheses' ('turn always to the right', or 'always to the left', or 'alternately right and left') may be also classed in this category.[1]

(3) Lashley and Franz (1917), experimenting with a 'problem box', observed that a rat which accidentally opened the box by a fall from the roof of the restraining cage, attempted to repeat this feat fifty times in vain before the method was abandoned.[2] This rat had formed a *mistaken generalization*.

(4) Rats with extensive cerebral destruction never learn anything. In a maze they move about at random. Rats suffering from experimental neurosis behave obsessively.[3] These animals form *no generalizations*.

The four grades according to which we classified reasonable action and perception reappear here in the classification of empirical inferences. We have (1) objectively reasonable inference, (2) reasonable error, (3) subjectively reasonable inference, and (4) unreason, i.e. no inference. And once more, each of these grades assesses the subject's performances by standards set for it by the observer, from his own understanding of the problem he has set to it.

But we notice also, proceeding from (4) to (1), a hardening of the *claims to universality*, combined with a quickening of the *heuristic impulse*, and as a joint result of these a more emphatic act of *commitment*. This threefold shift of emphasis could already be noticed when passing from consummatory action to perception. It has been actually anticipated already in Parts Two and Three by the linking of learning to problem-solving. For a problem is the intimation of a hidden rational relationship which is felt to be accessible by an heuristic effort, and the discovery of which may

[1] Cf. p. 73.
[2] Lashley, K. S., *Brain Mechanisms and Intelligence*, Chicago, 1929, p. 133.
[3] N. P. F. Maier, *Frustration, The Study of Behavior without a Goal*, New York, Toronto, London, 1949, pp. 25–76.

be accompanied, even in animals, by the lively enjoyment of their own ingenuity. By searching for such a hidden relation and by its eventual joyful acceptance, the animal reaches out towards something more objectively satisfying than food or sex, and in this sense the ensuing commitment becomes more radical. Egocentric desire gives way to personal assertion; the corruptible puts on incorruption.

I have given evidence before of the emotional upheaval which accompanies the mental reorganization necessary for crossing the logical gap that separates a problem from its solution. I have pointed out that the depth of this upheaval corresponds to the force of personal judgment required to supplement the inadequate clues on which a decision is being based. Experiments producing a nervous breakdown in animals lay bare in the simplest possible terms both the tension of this choosing power, and the limits within which this tension is bearable.

In Pavlov's classical investigations a dog was first trained to accept a circle, or a nearly circular ellipse, as a sign for immediately forthcoming food and a flat ellipse as a sign for 'no food just now'.[1] The hungry animal watching the different signs was found committing itself—as the variations in the secretion of its saliva showed—to the two alternative expectations which these two signs justified. So long as the signs of opposite significance were widely different—the ellipse being either very flat or nearly circular—the dogs reacted to them without developing symptoms of nervous strain. But when the hungry animal was repeatedly shown intermediate shapes, its behaviour underwent a profound change. It turned wild and angrily strained and snapped to set itself free. At the same time it had lost all its powers of discrimination and was giving false reactions to signs to which it had been perfectly conditioned before. After a while the animal would fall into abnormal listlessness and refuse to react altogether to any of the formerly established signs.

Previously I derived the presence of intellectual passions in animals from the way they rejoice in performing a new trick, regardless of its material result. We may now observe, similarly, that Pavlov's dogs were affected by their incapacity to distinguish between the signs of Food or No Food, far more than their care for food would warrant. We may take this to prove that they were labouring under an effort to discriminate, and that, as the problem facing them was made increasingly difficult, this effort eventually exhausted, or temporarily overstrained and paralysed, their powers of rational control.

The extent of this damage shows the depth to which the animal's person is involved even in such an elementary heuristic effort. The animal disintegrates emotionally, as well as intellectually. The neurotic dog which can only snarl or sulk ceases to be a companion to us. And we realize then, if we had not done so before, that the intelligence of the animal and

[1] I. Pavlov, *Conditioned Reflexes*, Oxford, 1927, pp. 290–1; also *Selected Works*, Moscow, 1955, pp. 235 f. (from *Skand. Arch. Physiol.*, **47** (1926), 1–14).

our appreciation of it was *convivial*: it formed a link between his person and ours.

Pavlov has observed that the experimental neurosis of dogs can be healed by presenting the animals for a while with signs of a clearly distinguishable kind and accompanying these consistently with the offer of food, or the reverse. The successful solution of these simple problems seems to restore an animal's self-confidence, much as occupational therapy helps to restore the shattered personality of the neurotic.[1]

A manifest proof that an animal's capacity for straining its powers of rational inference is linked to the very core of its emotional and intellectual personality, was discovered by Jacobsen in 1934.[2] He found that chimpanzees who were liable to nervous breakdown when subjected to excessive mental strain, were rendered safe against such ill-effects if their frontal lobes were severed or eliminated. Though the animal's ability to solve problems is noticeably impaired, its intellectual frustrations cease to worry it and no longer endanger its balance of mind.[3] Soon after this discovery, Edgar Moniz showed that a similar operation, when performed on patients suffering from melancholia, may relieve their depression, and that it markedly reduces at the same time the depth of their personality, rendering them crude, improvident and grossly inconsiderate. The chimpanzee's capacity intensely to worry about a problem is thus seen to be akin to man's capacity for self-control, guided by a sense of responsibility.

---

[1] Numerous cases of experimental neurosis were reported since 1938 from psychological laboratories in America, and some from Britain. Though the pathogenic situations are often described simply as conflicts, it appears, particularly from the comprehensive studies of N. R. F. Maier (*op. cit.*) that only such conflicts are effective from which the animal endeavours vainly to escape by problem solving. This author points out that, when subjected to opposite impulses which it can clearly envisage as such, the animal may simply do nothing. '*Frustration' develops, therefore, only when the hidden promises of a puzzling situation keep stimulating its vain efforts to gain intellectual control over the possibilities assailing him.*

[2] C. F. Jacobsen, *Res. Publ. Ass. nerv. ment. Dis.*, **13** (1934), 225; cf. J. F. Fulton, *Act. med. scand. suppl.*, **196** (1947), 617.

[3] Fulton (*op. cit.*, p. 621) describes the behaviour of the chimpanzee after lobotomy: 'The chimpanzee offered the usual friendly greeting, and eagerly ran from its living quarters to the transfer cage, and in turn went promptly to the experimental cage. The usual procedure of baiting the cup and lowering the opaque screen was followed. The chimpanzee did not, however, show its usual excitement, but rather quietly knelt before the cage or walked around. Given an opportunity, it chose between the cups with its customary eagerness and alacrity. However, whenever the animal made a mistake it showed no emotional disturbance, but quietly awaited the loading of the cup for the next trial. The opaque door was again lowered, but without untoward effect, and if the animal failed again it merely continued to play quietly or to pick over its fur. Thus, while the animal repeatedly failed and made a far greater number of errors than it had previously, it was quite impossible to evoke even a suggestion of an experimental neurosis. It was as if the animal had joined the "happiness cult of the Elder Michaeux", and had placed its burdens on the Lord.'

### 7. LEARNING AND INDUCTION

A learning experiment is a teaching experiment. We must start it by judging the animal to be ignorant in certain respects and by trusting that after certain experiences, for which we shall offer it an opportunity, its behaviour will reveal that it has—or has not—acquired the knowledge which it should properly derive from these experiences. If we eventually come to believe that it has acquired this knowledge and that it has acquired it by the experience in question, we shall call this 'learning'; while if we deny this achievement, we shall say that the animal has failed or that our teaching was inadequate.

Psychologists would almost unanimously reject such a definition today; in the first place, I think, because it is teleological. But that this is unjustified can be shown easily, in respect of strict behaviourism which describes its subjects as machines, and *a fortiori* for other schools of psychology. A machine is defined by operational principles which achieve an acknowledged purpose.[1] That is why the McCullough-Pitts model of the nervous system, or C. L. Hull's robot representing the process of learning, are machines, while the solar system is not. And this is why psychology differs from astronomy; it does not describe events related to no purpose, but analyses a certain class of *achievements* believed to be mental. The result is a system of rightness, which depends on certain not normative elements for its success or failure.[2]

Since the success of learning consists in the acquisition of knowledge, a mechanical theory of learning can be represented as the operations of a machine drawing correct inferences from observable facts. Such machines have been devised in principle, and their mechanism closely resembles that which many behaviourists, following Thorndike, have attributed to the process of learning in animals. The machine is designed to produce a series of random responses to a given state of affairs until it finally hits on the right response, which it henceforth invariably reiterates on every similar occasion.

Any machine that is to represent learning presupposes a theory of acquiring knowledge and a theory of knowledge itself. The machine which I have just mentioned assumes that in spite of the incessant changes sweeping through the world, identifiable states of affairs keep recurring and can actually be recognized as such both by the animal and the observer; and that there are right responses to such identifiable occasions which can be reiterated, so that the responses too are identifiable. We have seen before (p. 81) that a belief in the existence of identifiable things, to which we can respond by identifiable actions, underlies the process of

---

[1] It is difficult to understand the customary condemnations of teleology by neurologists, psychologists, etc., for example, when W. R. Ashby, *Design for A Brain*, London, 1952, pp. 1–10, emphatically renounces all teleological explanations in the very act of constructing a machinery to explain the functions of the brain.

[2] For a more careful formulation see p. 370 below.

denotation and that it justifies the kind of induction which underlies the descriptive sciences. This justification can be readily extended further to other processes of inductive reasoning by interpreting a learned response 'If A then do X' as saying likewise 'If A, then expect B'. The learning machine is then seen to operate by the kind of random accumulation of observations which, according to the currently predominant conception of the scientific method, results by chance in the discovery of the constant conjunctions known to science.

But I must digress here briefly in order to keep clear the tracks along which this enquiry is proceeding. Granted that the study of learning is an appreciation of an animal's behaviour by the standards of inductive logic, the following question arises. I have separated earlier on our acknowledgment of a deductive inference from the study of the psychological process which embodies it and may interfere with it (pp. 332–4). Hence it may seem questionable whether, in the study of learning, the acknowledgment of rightness which accounts for the success of learning and accredits its achievements with universal intent, may be lumped together with the study of the conditions and shortcomings of learning. My answer is that the distinction in question is sharply pronounced only in the case of highly formalized logical operations. It becomes blurred and should be allowed to lapse altogether, when rightness is achieved according to vague maxims which are effective only when applied with exceptional skill and understanding. Such, I believe, is the case for inductive inferences. The analysis of such operational principles is so closely interwoven with a study of the conditions under which they can operate or fail to operate, that the two aspects of the subject must be treated jointly.[1] Thus in spite of the logical and epistemological affirmations contained in the theory of learning, we shall accept it wholly as a branch of psychology and authorize this branch to study—as all biology does—certain achievements ascribed to living beings.

Looking then at the psychology of learning as a study of empirical inference, we can assess its current methods and results by recalling our earlier critique of the philosophic theories of empirical inference. Since such inference can be formalized only superficially, any rules laid down for carrying out empirical inferences must be highly ambiguous. Hence a machine designed to carry out such inferences can present but a clumsy imitation of actual processes of inference. A psychology of learning which strives for objectivity by representing the process of learning in terms of a formalized inductive logic can likewise achieve, therefore, only a semblance of its

---

[1] The following parallel may illustrate this situation. We speak of 'applied mathematics' when a technology can be mathematically formulated (as in electro-technics), and such an abstract technology is then sharply separated from the study of the materials by which it can be implemented. By contrast, chemical technology has no theory that could be developed irrespective of the chemical properties of the materials on which it relies. Hence the technical chemist's subject matter represents a fusion of certain operational principles with the material conditions of their success or failure.

aim. It will have (1) to curtail its subject matter to the crudest forms of learning and (2) to exploit at the same time the ambiguity of its supposedly impersonal terms, so that they will appear to apply to the performances of a living being which are covertly kept in mind.

I shall illustrate both these points from the otherwise distinguished work of C. L. Hull, whose method has exercised a profound influence on psychologists since the publication of *The Principles of Behavior* in 1943.[1] The treatise opens with the definition of a stimulus by the example of a light ray entering the eye. This (it says) is the stimulus $S$. But later—halfway through the work—it is admitted that there is always an indefinite variety of stimuli impinging on the animal's sense organs, and hence there is no such thing as *the* stimulus $S$. At this stage 'attention' is mentioned in quotation marks, as a loan from introspection which we must ignore, and the role of 'attention' is replaced forthwith by a previously established habit which is supposed to have been acquired by the very process—involving the alleged objective predominance of a single stimulus—which had just been abandoned as fallacious. Actually, the animal's role in directing its own attention remains unacknowledged and unaccounted for; as it is found missing also from all formalized theories of induction. This deficiency reappears with even more far-reaching consequences in Hull's analysis of discrimination. This opens by defining generalization as the capacity to respond in a similar manner to similar stimuli, and observes then that if the response to the similar stimulus remains repeatedly unrewarded it will cease to elicit the response, so that as a result of this, the animal discriminates between the two similar stimuli. This theory is an application of induction 'by agreement and difference' as laid down by J. S. Mill, and it suffers from the same shortcomings. Since no ingenuity is supposed to be involved on the part of the animal (which functions as an automaton) there is no limit set either to its powers of induction—provided that its sense organs are adequate to the task. Hence, if a dog were consistently offered food whenever it was shown the radiogram of diseased lungs and no food when shown the radiogram of healthy lungs, it should learn to diagnose pulmonary diseases. An objectivist theory of learning leads to the same absurdities as an objectivist theory of induction: since it has no place for heuristic powers it cannot account for their obvious limitations. And, of course, it likewise fails to account for such heuristic powers as even rats do manifest—as when Lashley's mutilated rats produced entirely new motor patterns for running a maze which they had learned before as intact animals.[2]

Yet all these oversimplifications fail to achieve their purpose. For even the most elaborate objectivist nomenclature cannot conceal the teleological

[1] Robert Leeper (*Amer. Journ. Psychol.*, **65** (1952), p. 478) describes Hull as in many respects *the* major figure in learning theory. Since then the rise of Cybernetics has increased even further the attractions of a behaviourism based on a strictly mechanistic model. E. R. Hilgard, *Theories of Learning*, 2nd edn., New York, 1956, p. 182, acknowledges that Hull has set 'the ideal . . . for a genuinely systematic and quantitative psychological system'. [2] See p. 337 above.

character of learning and the normative intention of its study. Its supposedly objective terms still do not refer to purposeless facts but to well functioning things. Something is a 'stimulus' only if it succeeds in stimulating. And though 'responses' may be meaningless in themselves, the state of affairs called 'reinforcement' functions as such by converting at least one particular response into a sign or a means to an end. Moreover, the result of a series of successful reinforcements is, by definition, a habit which the experimenter deems right. So even the most rigidly formalized theory of learning does lay down a system of rightness for the purpose of assessing and interpreting the rationality of the animal's behaviour.

Besides, the behaviourist vocabulary of learning, intelligence, etc., would be unintelligible to us but for our convivial understanding of the animals under observation. It is a mere pseudo-substitution, which relies for its meaning entirely on our familiarity with the conceptions it is trying to replace. This applies even to the most liberal and creative behaviourism, that of E. C. Tolman, and to the logical behaviourism of Gilbert Ryle which I now propose to criticize from this point of view.

Tolman assumes that what is commonly called the observation of a mental state is additional to an observation of its manifestations. Hence he declares (as others of this school have done before) that 'all that can ever actually be observed in fellow human beings . . . is behavior',[1] and concludes that any reference to mental states is unnecessary. Ryle argues, on the contrary, that there is no mind as distinct from its workings and that it is meaningless to refer to it as such.[2] Both conclusions fail to take into account that a focal observation of the particulars by which a person's mind manifests itself is something different from a subsidiary awareness of these particulars within a focal observation of the mind. Owing to the absence of this distinction, both psychological and logical behaviourism miscarry.

Take first the latter. The focal observation of the workings of someone's mind *dissolves* our knowledge of his mind, so that in this sense these workings are certainly *not* his mind. On the other hand, a *comprehensive awareness* of these workings *constitutes* an observation (or reading) of the mind, which may appear to vindicate Professor Ryle, but does not in fact do so. For Ryle does not have the conception of subsidiary awareness, and his identification of the mind with its workings can therefore only mean that the two are identical in the usual sense, i.e. as focally observed facts, which is false.

The fundamental postulate of Tolman's behaviourism collapses in either formulation. For it is not possible to keep track of a mind's workings except by comprehending them, so that a focal observation of the particulars of intelligent behaviour is impossible. And if, on the contrary, we

---

[1] E. C. Tolman, *Purposive Behavior in Animals and Men*, New York, 1932, p. 2.
[2] Gilbert Ryle, *The Concept of Mind*, London, 1949, p. 58: 'Overt intelligent performances are not clues to the workings of minds; they are those workings.'

observe these particulars comprehensively, we are in fact focussing not on the behaviour, but on the mind of which they are the workings. We are reading the mind at work in these particulars.

Such is the convivial relation which serves as the channel of all psychological observations, and within which all the terms of psychology must be interpreted. It is the same relation in which we have observed the active and perceptive centre of animals and have seen, at the level of learning, the animal committing its whole person to an effort of rational inference. It is the relation in which we take an interest in a fellow-being for its own sake and appreciate its achievements by standards set for it by ourselves. We shall see presently that this conviviality comprises at a further stage, when the other person rises above ourselves, an acceptance of another's iudgment of ourselves.

## 8. HUMAN KNOWLEDGE

But before arriving at this point we have yet to consider a previous stage, at which we achieve equality between ourselves and the person whose knowing we examine. This situation is of special interest, for it is here that my biological acknowledgment of rising levels of personhood comes to coincide—or nearly to coincide—with the position from which, at the opening of this enquiry, I first envisaged the unfathomable range of the scientific mind. Let me recall how this spectacle induced me to undertake a systematic survey of the tacit coefficient of knowing, as it ascends and broadens out from the primordial activity of the lowest animals to the whole edifice of human thought in human society. As a result of these reflections I have acknowledged my capacity, and my calling, to pursue knowledge and to declare it responsibly, within my own limited possibilities. Consider now how the critique of biology, ascending from morphology to psychology, has shown that the knowing of life entailed at all these levels an appreciation of biological achievements by standards set to the organism by ourselves; and has shown that a more detached manner of observing life would dissolve altogether our knowledge of life. We can see then how the extension of this progression to an examination of the knowledge of another person—of a standing *equal* to our own— places us in a situation virtually identical with that in which we reflected on our own knowledge in Parts One to Three of this book.[1] For if we agree with that which the other person claims to know and with the grounds on which he relies for this knowledge, the critical examination of this knowledge will become a critical reflection on our own knowledge. Biology then comes to include the accrediting of our own intellectual powers and the confirmation of our commitments within the framework

---

[1] Using here the symbolism of an earlier chapter, we are making here a transition from $\vdash . H/E$, which utters my own commitment, to $\vdash . P(/HE)$, by which I acknowledge another person's similar commitment (Part One, ch. 3, p. 32).

of our calling. It acknowledges, in particular, our capacity for continually discovering new interpretations of experience which reveal a deeper understanding of reality, and takes us eventually to the point where the whole panorama of science unfolds for a second time within a biology of man immersed in thought.

The significance of this confluence of an extended biology with the theory of knowledge will soon become more fully apparent. I shall pause here only in order to glance from this angle at some of the features of human knowledge, already identified in our critique of science and of other systematic interpretations of experience. We can identify in these the four grades according to which we have classified reasonable action and perception, as well as animal inference. We have

(1) *Correct* inferences reached within a *true system.*
(2) Erroneous conclusions arrived at within a true system (like an *error* committed by a *competent* scientist).
(3) Conclusions arrived at by the correct use of a fallacious system. This is an *incompetent* mode of reasoning, the results of which possess *subjective validity*.[1]
(4) *Incoherence* and *obsessiveness* as observed in the ideation of the insane, particularly in schizophrenia. The morbid reasoning of sufferers from systematic delusions should also be classed here, rather than under (3), since such delusions impair the very core of a person's rationality.

These alternatives correspond to the appraisal of a commitment in two stages, namely in respect of (*a*) its framework and (*b*) the application of this framework. If we accredit both (*a*) and (*b*), we have case (1); if (*a*) but not (*b*): case (2); if (*b*) and not (*a*): case (3). Our rejection of both (*a*) and (*b*) defines the unimportant case that I have not listed, when a false framework is applied erroneously, while case (4) is now seen to represent the absence of *any* interpretative framework, whether true or false.

These stages of knowing are of course all appraised by him who speaks of them, on the assumption that he can judge their truth content, critically. But we must take into account now, in addition to this critical relation, the possibility of exchanges between the speaker and the person whose claims he is assessing, i.e. the exchanges by which they mutually question, inform, criticize and persuade each other.

## 9. SUPERIOR KNOWLEDGE

Let me concentrate on the exchanges taking place between equals within the medium of a common complex culture. I must recall here briefly the social matrix of a complex culture, in order to look upon it now

---

[1] Cf. Part Three, ch. 9, pp. 286-8.

as an extension of the rising levels of life which form the subject matter of biology. Take two scientists discussing a problem of science on an equal footing. Each will rely on standards which he believes to be obligatory both for himself and the other. Every time either of them makes an assertion as to what is true and valuable in science, he relies blindly on a whole system of collateral facts and values accepted by science. And he relies also on it that his partner relies on the same system. Indeed, the bond of mutual trust thus formed between the two is but one link in the vast network of confidence between thousands of scientists of different specialities, through which—and through which alone—a consensus of science is established which may be said to accept certain facts and values as scientifically valid. I have described before how small a fragment of science is clearly visible to any one scientist. I have shown also that a system of scientific facts and standards can be said to exist only to the extent to which each scientist trusts all the others, to uphold his own special sector of the system in respect of his research, his teaching and his administrative actions. Though each may dissent (as I am myself dissenting) from some of the accepted standards of science, such heterodoxies must remain fragmentary if science is to survive as a *coherent system of superior knowledge, upheld by people mutually recognizing each other as scientists, and acknowledged by modern society as its guide.*

I have shown in some detail also how this mediated consensus operates in the pursuit and dissemination of scientific knowledge, and have outlined the analogous operations of such a consensus in the wider domains of a complex modern culture (pp. 216 ff.). Envisaging this consensus now as an extension of the biological achievements to which our ascending survey of living beings has led us, I shall regard the entire culture of a modern, highly articulate community as a form of superior knowledge. *This superior knowledge will be taken to include,* therefore, beside the systems of science and other factual truths, *all that is coherently believed to be right and excellent by men within their culture.* My own appreciation of any 'superior knowledge' within a foreign culture is subject, of course, to my acknowledgment of the superior knowledge of my own culture, and this will have to be allowed for.

Only a small fragment of his own culture is directly visible to any of its adherents. Large parts of it are altogether buried in books, paintings, musical scores, etc., which remain mostly unread, unseen, unperformed. The messages of these records live, even in the minds best informed about them, only in their awareness of having access to them and of being able to evoke their voices and understand them. And this leads us back to the fact, implied in describing science as superior knowledge, that all these immense systematic accumulations of articulate forms consist of the records of human affirmations. They are the utterances of prophets, poets, legislators, scientists and other masters, or the messages of men who, by their actions, recorded in history, have set a pattern for posterity; to which are added the living voices of contemporary cultural leaders, competing for the allegiance of the public. Thus we may regard, in the last analysis, the entire superior knowledge embodied in a modern highly articulate culture as the sum total of what its classics have uttered and its heroes and saints have done. If we belong to this culture then these are our great men: men to whose superiority we entrust ourselves, by trying to understand their works and to follow their teachings and examples. Our adherence to the common beliefs and standards on which intellectual exchanges within a culture depend, appears then equivalent to our adherence to the same masters as fountains of authority. They are our intellectual ancestry: 'the famous men and fathers who begat us' in whose heritage we enter.

It follows, therefore, that a dialogue between equals within a complex culture acknowledges a further (fifth) grade of knowledge, not appraised critically by those who recognize it, but accepted by them largely unseen, on the authority of those whom they believe to possess it. When referring to such superior knowledge we are not laying down standards for judging the persons to whom we attribute this knowledge; we are submitting, on the contrary, to the standards laid down by them for our own guidance.

In the chapter on Conviviality I have distinguished the following types of societies, characterized by their relation to thought. (1) Pre-modern static societies which recognize thought as an independent force, but only as embodied in a specific orthodoxy; we may call them authoritarian. (2) Modern dynamic societies which are either (*a*) free, if they acknowledge thought as an independent force or (*b*) totalitarian, if they deny this independence in principle. A free society differs from a static, authoritarian, society by accepting a wide range of rival thoughts for its guidance. Its members share the bulk of their heroes and masters, but may disagree in respect of some of them. A totalitarian society differs both from a free and an authoritarian society by inverting in principle the relation between power and thought; I have explained the principles of this inversion in the chapter on Conviviality.

What I have said so far about superior knowledge referred in the main to its position in a free society; what follows now shall acknowledge expressly my allegiance to such a society. I shall speak of its heroes and masters, who are also my own heroes and masters, and shall refer to the liberal orthodoxy established by them in terms that are consistent with the content of this orthodoxy, to which I myself subscribe.

Let me recall once more, for a start, that everything by which we mentally surpass the animals is first evoked in us by learning to speak. Mentally, we are called into being by accepting an idiom of thought. The child accepts it almost passively. Yet from the masters of a free society he learns a language which implicitly restricts the authority to which he is submitting—not because of its occasional admonitions to scepticism, but on the contrary, because it acknowledges the universality of truth and other forms of excellence. The language of these ideals, anchored in the works and lives of our masters, grants to each one of us the right to uphold these ideals against any particular utterance of these same masters. For it is not to their person, but to what we understand to be their teaching, that we pledge ourselves. It is indeed only by the lives of ordinary men within a free society that the principles to which it is dedicated acquire their effective meaning. The superior knowledge guiding a free society is formulated by its great men and embodied in its tradition.

Such is man's relation to his ideals: he can know them only by freely following them. This has been said before in the chapter on Commitment. Let me substantiate it once more here by recalling the various scattered references made in this book to the ideals of a free society. In the chapter on Intellectual Passions I have shown how the values of science are rooted in the work of great scientists, and how our aesthetic sensibilities are developed likewise by the masters of music and painting. In 'Conviviality' I have spoken of the moral passions which inspire our modern political dynamism, and in 'The Critique of Doubt' I gave some evidence of the deepening and purification of religious passions in our time. In 'Commitment' I spoke of the sustained passion for justice which has eventually secured the independence of law courts, and in the present chapter I have shown how our appreciation of living beings and their various achievements are sustained by biology.

These brief texts can serve only as pointers towards this limitless subject matter: the kinds of excellence to which our great men have testified is inexhaustible and no attempt can be made in this book even to classify them. Yet we must now try to bring this whole domain of superior knowledge into focus as an overlapping of our ascending biological survey with the extension of our previous epistemological enquiry. The framework of commitment, by which I have stabilized my personal knowledge of facts, must be capable of justifying also—by a suitable generalization of its terms—my adherence to the beliefs and standards which underlie the culture of a free society, and this result should fall into line with an extension of biology to the study of great men.

I shall arrive at this confluence of ultra-biology with the upholding of human ideals by a further pursuit of my ascending biological survey accompanied by a running critique of biology. Remember how biology rises from the appraisal of *primordial*, vegetative, commitments, to the appraisal of *primitive*, active-perceptive, commitments and then, by the study of animal learning, to the appraisal of commitments entered on *intelligently* and with *universal intent*. We start by observing a living body with a primordial centre of individuality, and are led on by a continuous progression to a situation in which we face a subject committing himself deliberately to the solution of an external problem. And as we rise stage by stage from morphology to animal psychology, our convivial participation in the living organism becomes increasingly richer, more intimate and less unequal. So, arriving finally at the study of human thought, conviviality becomes mutual. A conscious, responsible person—the biologist—is now appraising the achievements of another person of the same rank, whose thoughts can claim respect on the same grounds as his own. It is the reference to these grounds that inevitably expands the dialogue of two responsible human beings into an acknowledgment of a knowledge that is superior to their own: the superior knowledge of their culture, as mediated by the great men who are the founders and exemplifiers of that culture. A dialogue can be sustained only if both participants belong to a community accepting on the whole the same teaching and tradition for judging their own affirmations. A responsible encounter presupposes a common firmament of superior knowledge.

In the course of this progression our convivial passions undergo a fundamental development. Our love of harmonious being makes us study living shapes; our pleasure in the ingenuity of living functions upholds embryology and physiology; our love of animals sustains the study of their behaviour; and as we finally ascend to human companionship, we necessarily arrive also beyond it, by finding a spiritual home in the society on which this companionship is grounded. Thus the mental life developing between two equal human beings necessarily includes an emotional relation to the whole galaxy of their common superiors. The riches of mental companionship between two equals can be released only if they share a convivial passion for others greater than themselves, within a like-minded community—the partners must belong to each other by participating in a reverence for a common superior knowledge.

We can now appreciate also the changes in the logical structure of the convivial relations along the line of this progression. The feelings by which we appreciate the achievements of beings lower than ourselves, involve an extension of ourselves by which we participate in their achievements. But though the naturalist is inspired by the love of nature, and all biology derives its interest ultimately from the fascination exercised on us by living beings, even the most passionate animal lover receives no instruction from his pet. Only as the biologist's participation rises to the level of human companionship, does it become distinctly self-modifying and thus eventually loses altogether its observational character, to become a condition of pure indwelling. The decisive break occurs when we accept another person's superior knowledge. By applying his thoughts or deeds as our standards for judging the rightness of our own thoughts and deeds, we surrender our person for the sake of becoming more satisfying to ourselves in the light of these standards. This act is irreversible and also a-critical, since we cannot judge the rightness of our standards in the sense in which we judge other things in the light of these standards. At this point the three-levelled structure of biology proper gives way (as I foreshadowed at the close of the previous chapter) to a two-levelled structure. But these are not the same two levels on which the observation of inanimate nature takes place, with the observer occupying the higher plane. The three levels which had emerged from these original two levels by the extension of our attention to active beings centred on themselves, have been replaced now by two levels representing the outlook of man centring on things higher than himself. He may be said to stand on the lower level of this commitment. Alternatively, we may describe him as forming the personal pole of a commitment of which the ideals of man form the universal pole.

## 10. AT THE POINT OF CONFLUENCE

This completes the extrapolation of biology to the point where it coincides with our commitment to the intellectual standards of our culture. Looking back from this point of confluence on the two branches of the argument which it unites, we can see that each can be generalized to include the other, and that this brings out their joint ontological significance.

The enquiry into the nature and justification of personal knowledge, which fills Parts One, Two and Three of this book, has led to the acceptance of our calling—for which we are not responsible—as a condition for the exercise of a responsible judgment with universal intent. Our calling was seen to be determined by our innate faculties and our early upbringing within our own culture, and these conditions were made to subserve an act of commitment by relying on them for the fulfilment of standards believed to be universal. Calling; personal judgment involving responsibility; self-compulsion and independence of conscience; universal standards; all these were shown to exist only in their relation to each other within a commitment. They dissolve if looked upon non-committally. We may call this the ontology of commitment.

This ontology can be expanded by acknowledging the achievements of other living beings. This is biology. It is a participation of the biologist in various levels of commitment of other organisms, usually lower than himself. At these levels he acknowledges trueness to type, equipotentiality, operational principles, drives, perception and animal intelligence, according to standards accepted by him for the organisms in question. I have demonstrated that these achievements are personal facts which are dissolved by any attempt to specify them in impersonal (or not sufficiently personal) terms. The unspecifiability of such achievements can now be seen to represent a generalization of the theorem that the elements of a commitment cannot be defined in non-committal terms. The paradox of self-set standards and the solution of this paradox are thus generalized to include the standards which we set ourselves in appraising other organisms and attribute to them as proper to them. We may say that this generalization of the universal pole of commitment acknowledges the whole range of being which we attribute to organisms at ascending levels.

On the other hand, the extrapolation of biology to the acknowledgment of human greatness, by which we first reached the point of 'confluence', shows how a reverse generalization could be carried out, by which biology would come to include the whole ontology of commitment. For human greatness can be recognized only by submission to it and thus belongs to the family of things which exist only for those committed to them. All manner of excellence that we accept for our guidance, and all obligations to which we grant jurisdiction over us, can be defined by our respect for human greatness. And from these objects of our respect we can pass on continuously to purely cognitive targets, such as facts, knowledge, proof, reality, science—all of which can likewise be said to exist only as binding on ourselves. We can then work our way back from this point, by aid of reflection, to a recognition of ourselves as the persons deliberately entering on these commitments and can extend our recognition also to all the members of a society sharing similar beliefs and obligations. The whole ontology of commitment and of a free society dedicated to the cultivation of thought by responsible commitments of its members can in fact be built up, in this manner, as a generalization of biology followed by reflection on this generalized biology.

Thus, at the confluence of biology and philosophical self-accrediting, man stands rooted in his calling under a firmament of truth and greatness. Its teachings are the idiom of his thought: the voice by which he commands himself to satisfy his intellectual standards. Its commands harness his powers to the exercise of his responsibilities. It binds him to abiding purposes, and grants him power and freedom to defend them.

And we can establish it now as a matter of logic that man has no other power than this.

He is strong, noble and wonderful so long as he fears the voices of this firmament; but he dissolves their power over himself and his own powers gained through obeying them, if he turns back and examines what he respects in a detached manner. Then law is no more than what the courts will decide, art but an emollient of nerves, morality but a convention, tradition but an inertia, God but a psychological necessity. Then man dominates a world in which he himself does not exist. For with his obligations he has lost his voice and his hope, and been left behind meaningless to himself.

# I3

# THE RISE OF MAN

## 1. Introduction

I HAVE arrived at the opening of this last chapter without having
suggested any definite theory concerning the nature of things; and I
shall finish this chapter without having presented any such theory.
This book tries to serve a different and in a sense perhaps more ambitious
purpose. Its aim is to re-equip men with the faculties which centuries of
critical thought have taught them to distrust. The reader has been invited
to use these faculties and contemplate thus a picture of things restored to
their fairly obvious nature. This is all the book was meant to do. For once
men have been made to realize the crippling mutilations imposed by an
objectivist framework—once the veil of ambiguities covering up these
mutilations has been definitively dissolved—many fresh minds will turn to
the task of reinterpreting the world as it is, and as it then once more will
be seen to be.

There is one more move to be made towards reopening this vision.
I have shown in the last two chapters what I mean by the achievements
of living beings and have exhibited in these examples the logic of achievement. These were our results:

(1) Living beings can be known only in terms of success or failure.
They comprise ascending levels of successful existing and behaving.

(2) We can know a successful system only by understanding it as a
whole, while being subsidiarily aware of its particulars; and we cannot
meaningfully study these particulars except with a bearing on the whole.
Moreover, the higher the level of success we are contemplating, the more
far-reaching must be our participation in our subject matter.

(3) Therefore, to interpret systems that can succeed or fail in the more
detached terms, by which we know systems to which no distinction of
success or failure applies, is logically impossible. Systems that can succeed
or fail are properly characterized by operational principles, or more
generally, by certain rules of rightness; and our knowledge of any class of

things that is characterized by a rule of rightness disappears when we attempt to define it in terms that are neutral to this rightness.

(4) Accordingly, it is as meaningless to represent life in terms of physics and chemistry as it would be to interpret a grandfather clock or a Shakespeare sonnet in terms of physics and chemistry; and it is likewise meaningless to represent mind in terms of a machine or of a neural model. Lower levels do not lack a bearing on higher levels; *they define the conditions of their success and account for their failures, but they cannot account for their success, for they cannot even define it.*

The step that remains to be taken in this chapter is to confront this vision of an essentially stratified world with the facts of evolution. We must face the fact that life has actually arisen from inanimate matter, and that human beings—including the teachers of mankind who first shaped our knowledge of rightness—have evolved from tiny creatures resembling the parental zygote in which each of us had his individual origin. I shall meet this situation by re-establishing within the logic of achievement, the conception of emergence first postulated by Lloyd Morgan and Samuel Alexander. The heuristic act of leaping across a logical gap will prove paradigmatic in this respect. We shall find indications of such inherently unformalizable processes at a variety of levels and suggest that evolutionary achievements should be classed among them.

## 2. Is Evolution an Achievement?

The conception of evolution as a process of fundamental innovations, tending to produce ever higher biotic achievements, cannot be taken for granted. For it is sharply contested by the predominant school of scientific thought, a school which can claim to have produced most of the brilliant modern work on heredity and the modification of heredity, as well as some excellent studies of paleozoology. However, far from being discouraged by it, I find this array of distinguished opponents most heartening, for only a prejudice backed by genius can have obscured such elementary facts as I propose to state here.

I shall argue on two lines, marked A and B, both of which have already been indicated. In A, I shall try to establish an ordering principle of evolution, by distinguishing the *actions* of such a principle from the *conditions* which *release* and *sustain* its actions. This argument is too general to be carried out fully here, and I shall turn therefore to argument B by pointing out that the observed evolution of human consciousness plainly exemplified this kind of active emergence.

A. The predominant modern theory, usually described as Neo-Darwinism, which I shall criticize here, regards evolution as the sum total of successive accidental hereditary changes which have offered reproductive advantages to their bearers. The sequence of hereditary changes, leading to the replacement in succeeding generations of the original types by better equipped variants, is described as 'natural selection', and the 'force of natural selection' is supposed to have brought forth the successive forms of life that have eventually produced man.[1]

There is a fundamental vagueness inherent in this theory which tends to conceal its inadequacy. It consists in the fact that we lack any acceptable conception of the way in which genic changes modify ontogenesis—a deficiency which is due in its turn to the fact that we can have no clear conception of living beings, as long as we insist on defining life in terms of physics and chemistry.[2] My argument will be based on a different conception of life. I shall regard living beings as instances of morphological types and of operational principles subordinated to a centre of individuality and shall affirm at the same time that no types, no operational principles and no individualities can ever be defined in terms of physics and chemistry. From which it follows that the rise of new forms of life—as instances of *new* types and of *new* operational principles centred on *new* individualities —is likewise undefinable in terms of physics and chemistry.

To simplify the argument I shall concentrate here on the rise of novel modes of operation, which are as a rule the most striking advantage in the new forms of life arising from evolution. A theory of evolution must explain, then, the rise of novel individuals performing new biotic operations. But the question how instances of new biotic operations come into existence, leads obviously back to the coming into being of life itself from inanimate origins. It is clear that for such an event to take place two things must be assured: (1) Living beings must be possible, i.e. there must exist rational principles, the operation of which can sustain their carriers indefinitely; and (2) favourable conditions must arise for initiating these operations and sustaining them. In this sense I shall acknowledge that the *ordering principle* which *originated* life is the

[1] See R. A. Fisher, *The Genetical Theory of Natural Selection*, Oxford, 1930; J. Huxley, *Evolution: The Modern Synthesis*, London, 1942; G. G. Simpson, *The Major Features of Evolution*, New York, 1953. Replacement by better equipped variants is said frequently to take place by the migration of the new type into areas not accessible to the original type, so that the latter remains in undisturbed possession of its dwelling place.

[2] (a) Throughout the process of morphogenesis the chromosomes are reproduced at each successive cell division and thereby place a replica of themselves into every cell of the final organism. But the successive differentiations achieved in these consecutive cell divisions become ever more specialized. This progressive differentiation appears therefore unaffected by the chromosomes present in the cells in question; it is in fact determined instead by the 'morphogenetic field'. (b) Regeneration, which in the lower animals can reproduce whole organs, including the head of the animal, proceeds likewise under control of a morphogenetic field, while duplicating a set of chromosomes which cannot be seen to exercise any effect on this morphogenetic process. (c) Specialized tissues continue to proliferate in cultures, while duplicating again at each cell division sets of chromosomes which should reproduce the whole organism.

How can the duplication of the same chromosomes produce the most varied types of cells? If the chromosomes do not control the nature of the cells produced in the course of morphogenesis, what agent does exercise this control? And how can the chromosomes still be said to control morphogenesis as a whole? There is some fundamental principle missing here. Perhaps it will be supplied by accepting the morphogenetic field as the true ordering principle of ontogenesis. Of this more later.

*potentiality* of a stable open system; while the inanimate matter on which life feeds is merely a *condition* which *sustains* life, and the accidental configuration of matter from which life had started had merely *released* the operations of life. And evolution, like life itself, will then be said to have been *originated* by the *action* of an ordering principle, an action *released* by random fluctuations and *sustained* by fortunate *environmental conditions*. I shall now elaborate this analysis.

The stability of a living being has been strikingly compared by W. Ostwald to that of a flame. One speaks today more generally of 'open systems',[1] but a simple gas flame contains all that is relevant. It represents a phenomenon of constant shape, fed by a steady inflow of combustible material and releasing a continuous flow of waste products and of the energy produced by combustion. Once a flame has been started, its shape and chemical composition can be varied without extinguishing it. To this extent, its identity is not defined by its physical or chemical topography, but by the operational principles which sustain it. A particular collocation of atoms may accidentally fulfil the conditions for starting a flame, but this accident itself can be defined as such only by its bearing on the system of ordering principles which establishes the possibility of stable flames.

Thus the potentiality of a stable flame bears the same relation to any random fluctuation which ignited it as the ordering principle inherent in the potential energy of biassed dice does to the randomness of Brownian motion, as described in the third imaginary experiment in Chapter 3 of Part One.[2] But we must note the following important difference. The fluctuation which leads to the establishment of an open system does not vanish after the event, as does the Brownian impulse which made the dice tumble into stable positions. The atomic configuration which ignited a flame keeps renewing itself within the flame. It is a fundamental property of open systems, not described before now, that they stabilize any improbable event which serves to elicit them. R. A. Fisher's observation of the way in which natural selection makes the improbable probable[3] is but a particular application of this theorem. The first beginning of life must have likewise stabilized the highly improbable fluctuation of inanimate matter which initiated life.

Owing to the slowness of evolution, no complete functional innovations can be seen to occur within any observable period. But they undoubtedly do take place over longer periods. There is a cumulative trend of changes tending towards higher levels of organization, among which the deepening of sentience and the rise of thought are the most conspicuous. And in this sense we can acknowledge that certain lines of evolution have been more effective than others; for example, that the

---

[1] This goes back to Reiner and Spiegelman, *J. Phys. Chem.*, **49** (1945), 81 and to Prigogine and Wiame, *Experientia*, **2** (1946), 451.

[2] Part One, ch. 3, p. 39.

[3] Huxley, Hardy, Ford, *Evolution as a Process*, London, 1954, p. 91.

principle of an exoskeleton used by the arthropods offered less scope for evolution than the endoskeleton of the chordates. But these comprehensive operations of evolution are not observable within the short span of contemporary experience, the less so, since any indications of them are likely to be swamped by ephemeral genetic variations which are taking place, as it were, in the interstices of the dominant evolutionary trend. Hence the long-range operations of evolution will not be noticed by the experimental geneticist, nor even by the students of population genetics, and geneticists will have no difficulty in explaining all hereditary variations observed by them, without reference to the action of evolutionary trends.

Indeed, in so far as variations lack—or fail to reveal—any long-range evolutionary significance, they can be described only by the present-day theory of natural selection. They must appear as random mutations, establishing themselves merely by their reproductive advantage. And this explanation will in fact apply to a host of striking adaptive changes which actually form no part of any long-range evolutionary achievement; some intricate devices, such as those of protective coloration, will rightly be explained in this manner. But I deny that any entirely accidental advantages can ever add up to the evolution of a new set of operational principles, as it is not in their nature to do so.

The grounds for this assertion have been laid down in the previous argument and will yet be clarified later. Let me observe here only that the theory of natural selection, by subsuming all evolutionary progress under the heading of adaptation as defined by differential reproductive advantage, necessarily overlooks the fact that the *consecutive* steps of a long-range evolutionary progress—like the rise of human consciousness—cannot be determined *merely by their adaptive advantage*, since these advantages can form part of such progress only in so far as they prove *adaptive in a peculiar way, namely on the lines of a continuous ascending evolutionary achievement*. The action of the ordering principle underlying such a persistent creative trend is necessarily overlooked or denied by the theory of natural selection, since it cannot be accounted for in terms of accidental mutation plus natural selection. Its recognition would, indeed, reduce mutation and selection to their proper status of merely *releasing and sustaining the action of evolutionary principles* by which all major evolutionary achievements are defined.

B. I shall now substantiate this general argument by focussing it more fully on the rise of man.

Since we can know living beings only by appreciating their achievements, we can know their evolution only by appreciating the development of their achievements in the course of succeeding generations. We have seen that such appreciations are integral to biology. But the achievements of man's evolution are exceptionally high. While animals are acknowledged as centres of interest to themselves, we owe respect to our fellow men. Hence we know man to be the most precious fruit of creation—and hence

385

also the knowledge of this fact lies outside natural science. For we possess this knowledge only by our submission to a firmament of obligations to which we believe all human beings to be equally subject. On these grounds man's supreme position among all known creatures is safely established; but at the same time the study of man's rise extends thereby far beyond biology, into our acceptance of what we believe to be man's nature and destiny.

In order to contemplate clearly the process by which the rise of man was achieved I shall trace the ancestry of one single human being to its beginnings. Since each of the man's parents, grandparents and more remote forebears, have in their turn a definite set of parents, grandparents, etc., a man's genealogical tree comprises an unambiguously determinate set of individuals. As the ancestral series recede in time they descend to ever more primitive forms of life, and where sexual reproduction is eventually replaced by asexual propagation they cease to branch out and continue instead along single lines. They penetrate here to the realm of unicellular organisms and beyond that to the realm of submicroscopic, virus-like specks of living protoplasm.

I shall call this ancestral system an *anthropogenesis*. The bodies of the successive generations of metazoa comprised in one man's ancestry appear as mere carriers of a continuously surviving germ plasm. The carrier may possibly modify his charge; but even so, while he himself dies and passes away, the germ plasm lives on mixed with that of another parent in the body of their joint progeny. We may thus regard an anthropogenesis in its entirety as a continuous proliferation of germ plasm, from unicellular origins down to the germ plasm of the human couple of whom the man in question is born. Since throughout the range of sexual reproduction, the begetting of each new individual marks the confluence of two branches of germ plasm, this proliferation of germ plasm is accompanied by a steady reduction of the number of individuals carrying the germ plasm. And the whole process of convergent proliferation—extending over many millions of years—is brought eventually to its close by the fertilization of the maternal egg followed by the embryonic development and the birth and growing up, of the man whose genesis we are contemplating.

This entire evolutionary achievement can be localized within a circumscribed material system. Its operation must have taken place within this system while interacting with its environment. In the light of the logical analysis applied before under (A), the process must have been directed by an *orderly innovating principle*, the action of which could have only been *released* by the random effects of molecular agitations and photons coming from outside, and the operations of which could only have been *sustained* by a favourable environment. But in this case we need no such abstract analysis to recognize that an orderly transforming principle has been at work. We have direct evidence, anticipating the result of our logical analysis, in the manifest rise of human consciousness. From a seed of submicroscopic living particles—and from inanimate beginnings lying beyond these—we see emerging a race of sentient, responsible and creative beings. The spontaneous rise of such incomparably higher forms of being testifies directly to the operations of an orderly innovating principle.

In the previous chapter I have surveyed a series of ascending biotic levels and exhibited in terms of these the logic of successively rising achievements. This progression made me realize that biology can be extended by continuous stages into epistemology, and more generally, into the justification of my own fundamental commitments. And so this ultrabiology went on extending further into the acknowledgment of all my obligations. In the course of evolution this series should present itself as a series of successive existential achievements. It should show how in the course of anthropogenesis the descending lines of our ancestors have taken on by stages the full capacities of personhood and have inherited eventually all the hazardous aspirations of humanity. Let me outline this process briefly.

The first small and yet decisive step towards man's destiny was made when ultramicroscopic, virus-like specks of living matter gained standard shapes and sizes, presumably with a correspondingly integrated internal organization. The bacillus which thus emerged carried the stamp of individuality. Its self-controlled shape and structure, and the physiological functions serving its survival, set up a centre of self-interest against the world-wide drift of meaningless happenings.

The next stage on the way towards personhood was reached by the protozoa. The appearance of a nucleus within a bed of protoplasm indicates an increased complexity of internal organization, underlying an external behaviour of immensely augmented self-control. Protozoa move about of their own accord and engage in a variety of deliberate purposive activities. A floating amoeba emits exploratory pseudopodia in all directions, which will catch food or else attach themselves to solid ground and then drag the whole mass of protoplasm with the nucleus in it towards this foothold. All these manoeuvres are co-ordinated: the amoeba hunts for food.[1] Thus it grows fatter until it reaches the size at which its personal life ends by fission.

A further great step was achieved by the aggregation of protozoan-like creatures to multicellular organisms. This enabled animals to evolve a more complex physiology based on sexual reproduction, a manner of propagation which greatly strengthened their personhood. The story of the Fall presents a strangely apt symbol of this event. For as one part of the body took over procreation and the animal ceased to survive in its progeny, lust and death were jointly invented. And as the achievement of metazoic existence established the rudiments of this tragic combination, a finite

---

[1] For a vivid description of this pursuit see H. S. Jennings, *Behavior of the Lower Organisms*, New York, 1906, p. 15.

personal destiny arose to challenge the surrounding deserts of deathless inanimate matter.

We do not know at what stage of evolution consciousness awakened. But as polycellular organisms grow in size, and as their complexity increases with their size, a nervous system is formed to carry out ever more extensive and elaborate operations of self-control. Already some 400 million years ago, at a stage represented today by worms, our ancestors had formed a major ganglion in the forward tip of their elongated body. The segment which first meets and tries out the unknown world, into which the animal is advancing, thus acquired a controlling position. It henceforth will direct locomotion and also control growth and regeneration. A gradient is established thereby between the higher and lower functions within the organism. An animal pole is set up which uses the other parts of the body for its sustenance and as its tools. Within this active centre the animal's personhood is intensified in relation to a subservient body. So we find prefigured the cranial dominance which gives rise to the characteristic position of the mind in the body of man.

The groping movements by which worms explore the path in front of them are the precursors of the far more effective exploratory functions of visual, auditory and olfactory perception. The use of sense organs extends the animal's area of mental control into the surrounding space. But seeing is foreseeing and is hence also believing; perception involves judgment and the possibility of error. Therefore, as the personhood of our ancestors was enriched and expanded by the power of new senses, it was intensified still further in undertaking to control new hazards. The polarity of subject and object began to develop, and with it the fateful obligation to form expectations based on necessarily insufficient evidence.

The beginnings of such acts of judgment are shown in the capacity to learn from experience. Some observers have traced this faculty back to unicellular organisms, and it can certainly be found as far down as the level of worms. But the capacity for learning was greatly expanded by the advent of perception, which developed the rudiments of generalization, contriving and understanding. A whole firmament of self-set standards was prefigured here and soon the first faint thrills of intellectual joy appeared in the emotional life of the animal. And it became also liable to puzzlement and frustration.

But 500 million years of this growth and hardening of personhood still only lead up to the threshold of true mental life, which was to be achieved in little more than 500 centuries by man's sudden rise from mute beasthood. Teilhard de Chardin has called this ultimate evolutionary step, by which human knowledge was born, *noogenesis*.[1] It was achieved by men who, forming societies, invented language and created by it a lasting articulate framework of thought. Teilhard calls this framework the *noosphere*. We have seen that the child achieves responsible personhood by entering

[1] Teilhard de Chardin, *Le Phénomène Humain*, Paris, 1955, p. 200.

a traditional noosphere. Our race as a whole achieved such personhood by creating its own noosphere: the only noosphere in the world.

This was the second major rebellion against meaningless inanimate being. The first had consisted in the rise of self-centred individuals, pre-dominantly vegetative and quite unaware of the rationality of their per-formances. In these individuals the germ plasm lived on through many evolving generations until, at last, noogenesis created a new fabric of life *not* centred on individuals and transcending the natural death of indi-viduals. When man participates in this life his body ceases to be merely an instrument of self-indulgence and becomes a condition of his calling. The inarticulate mental capacities developed in our body by the process of evolution become then the tacit coefficients of articulate thought. By the forming and assimilation of an articulate framework these tacit powers kindle a multitude of new intellectual passions. They set in motion heuristic endeavours. They make us love human greatness and accept as our guides those who have achieved it. By accepting such teaching man testifies to the existence of grounds on which he can claim freedom.

While the first rise of living individuals overcame the meaninglessness of the universe by establishing in it centres of subjective interests, the rise of human thought in its turn overcame these subjective interests by its universal intent. The first revolution was incomplete, for a self-centred life ending in death has little meaning. The second revolution aspires to eternal meaning, but owing to the finitude of man's condition it too remains blatantly incomplete. Yet the precarious foothold gained by man in the realm of ideas lends sufficient meaning to his brief existence; the inherent stability of man seems to me adequately supported and certified by his submission to ideals which I believe to be universal.

This great spectacle, the spectacle of anthropogenesis, confronts us with a panorama of emergence; it offers massive examples of emergence in the gradual intensification of personal consciousness. At each successive stage of this epic process we see arising some novel operations not specifiable in terms of the preceding level; and the whole range of them is unspecifiable in terms of their inanimate particulars. For no events occurring according to the known laws of physics and chemistry can be conscious. Alchemists used to attribute conscious desires to the mating of acids and bases, but chemistry accounts for such processes without any such imputation. The 'action' of a reagent is no action, for it cannot fail; hydrochloric acid will never dissolve platinum by mistake. Nor can self-regulating machines operating in accordance with the known laws of physics and chemistry represent human beings. For such machines are insentient automata and men are not insentient automata. Some say that we merely speak in two different languages when referring to thoughts on the one hand and to neural processes on the other. But we speak in two languages because we are talking of two different things. We speak of the thoughts Shakespeare had while writing his plays and not of the thoughts of hydrochloric acid dissolving zinc, because men think and acids don't. It is obvious, therefore, that the rise of man can be accounted for only by other principles than those known today to physics and chemistry. If this be vitalism, then vitalism is mere common sense, which can be ignored only by a truculently bigotted mechanistic outlook.[1] And so long as we can form no idea of the way a material system may become a conscious, responsible person, it is an empty pretence to suggest that we have an explanation for the descent of man. Darwinism has diverted attention for a century from the descent of man by investigating the *conditions* of evolution and overlooking its *action*. Evolution can be understood only as a feat of emergence.

### 3. RANDOMNESS, AN EXAMPLE OF EMERGENCE

But emergence begins already in the inanimate domain, as can be seen from the relation of randomness to the particulars of the random system. Many years of fruitless endeavour have proved that it is impossible to define the probabilities derived from the random character of a system by the microscopic details of the system. This should encourage us to align randomness with other comprehensive features which are unspecifiable in terms of their particulars; and the analogy between these various cases will strengthen the concept of emergence as being that which they have in common.[2]

The shuffling of a pack of cards is a process of emergence. Card players think they know how to produce a well shuffled pack of cards, and writers on probability tend to agree that we may talk of such a pack of cards.[3] *But we can produce a well shuffled pack only if we do not know how we do it.* For if we knew the details of the process of shuffling, we would know the final arrangement of the cards, so that the pack would no longer be in a random state and no statistical statements could be made about the chances of pulling out a particular card from it. This holds generally. If I knew exactly the conditions of a throw of dice I could predict the result,

---

[1] Let me repeat that, contrary to a widespread opinion, the change from classical mechanics to quantum mechanics makes no difference to this argument. The behaviour of human beings whose particles were ruled by the equations of quantum mechanics would be completely predetermined by these, except for a certain range of random variations which would be strictly unaccountable. Since human judgment is anything but a strictly unaccountable random choice, a quantum mechanical automaton is no better a representation of intelligent behaviour than a mechanical automaton would be; and it offers no possibility either for the presence of human consciousness.

[2] This analysis of randomness is required here also, to show that randomness is in fact (as I have said before in Part One, ch. 3, p. 38) the ultimate, not further analysable, condition for the applicability of the calculus of probability. This view has been emphatically stated before by N. C. Campbell, *Physics, The Elements*, Cambridge, 1920, e.g. on p. 207: 'I urge that we must accept the conception of a random distribution as fundamental to all the study of chance and probability; we are prepared to accept the statement that some distribution is random as an ultimate statement and as one that requires no explanation. All chance events are to be explained in terms of random distributions and when we have so explained them there is nothing more to be said.'

[3] See e.g. I. J. Good, *Probability and the Weighing of Evidence*, London, 1950, p. 15.

but could no longer *guess* it. I could say nothing about the statistical properties of dice from a description which would tell me the result of future throws made by a machine.

I call this a case of emergence, for we can know the randomness of a system, yet cannot know it in terms of a more detailed knowledge of the system. Our knowledge of this emergent quality, randomness, is in fact destroyed by observing the particulars which determine the system below the emergent level. Moreover, randomness, as an emergent quality, offers a possibility for a new system of manipulations. In the case of a well-shuffled pack of cards or an unbiassed dice, these consist in estimating the chances of alternative events and betting accordingly on their outcome.

In science the most important random system is the molecular motion in a gas. For the better part of the past hundred years mathematical physicists have tried to specify the randomness of a gaseous molecular aggregate in terms of its mechanical particulars. But this is logically impossible. If we knew exactly the position and velocity of each molecule (within the limits of wave mechanics) we could only predict the behaviour of the molecules, but not the comprehensive features defined by randomness. Two comprehensive features of a gas which determine its condition are its temperature and pressure. The gas can be said to have a definite temperature and a definite pressure only if we assume that its molecules are in random motion; an assumption which is incompatible with our knowing the configuration of molecular motions in the gas.[1]

It could be objected that from the detailed knowledge of all the molecules in a gas we could calculate what a thermometer, or a gauge, would show at different places in the gas. We may be able to predict such readings. But the results would mean nothing unless they could be assumed to originate from a random condition of the gas.

For this we would have to fall back on a manner of stirring—like the shuffling by which cards are randomized. And if we trusted ourselves with the capacity of randomizing also several separate parts of a gas, we could also establish differences in temperature and pressure between them and predict that these differences would be equalized by a process of self randomization inherent in a system of particles in random motion. These processes would be irreversible, since it would be contrary to our assumption of randomness that a random aggregate should sort itself out—except

---

[1] Suppose the overall condition (temperature and the pressure of a gas is compatible with $n$ different microscopic states and the probabilities of these are $W_1\ W_2 \ldots W_n$ ($\Sigma\ W_n = 1$). The entropy ($S$) of the gas is then $S = -k \Sigma_n W_n\ ln\ W_n$ ($k = $ Boltzman's constant) and $S$ will always be a finite positive magnitude. We note further that if any $W$ is 0 or 1 the corresponding term drops out.

Suppose now that we know the molecular particulars of the gas; we know then in which microscopic state it is to be found. Consequently, the value of $W$ for this state is 1, while $W$ for all other states is 0. It follows that $S = 0$, i.e. the specification of its molecular particulars has wiped out the entropy of the gas. Since both the temperature and the pressure of a gas depend on its entropy, this result corroborates the statement made in the text.

by occasional fluctuations—into a less random state, unless compelled by inner forces or external intervention.

Randomization may be unsuccessful; a trace of order may always remain undestroyed.[1] In this sense randomness may be regarded as an achievement. In any case, as a comprehensive feature, randomness is subject to the logic of achievement. We can identify this logic here with the logic of emergence. The emergent form of existence is identified by our comprehensive judgment of it, which judgment accredits thereby, indirectly, a correlated context of properties, and of problems and manipulations, all of which presuppose the emergent form of existence and serve to elaborate its reality. This entire emergent system (consisting in the present case of randomness and probabilities, of averages, temperatures and pressures, of irreversible processes and thermal fluctuations, etc.) is unspecifiable in terms of its detailed particulars. But the particulars have a bearing on higher-level features. If the molecular motion in a gas is known to be random, we can evaluate from its particulars the temperature, pressure, entropy, etc. of the gas.

It is clear in this case also that unspecifiability is not simply ignorance. It has been frequently pointed out that you cannot identify a random system if you know nothing about it; I obviously cannot say that a sequence of numbers is random, if I do not know it to be random. But this holds even conversely. I can tell (from the nature of irrational numbers) that the sequence of digits in $\sqrt{27}$ is *not* random, though I may not know *anything else* about it; while on the other hand I may be familiar with the derivation of the number $\pi$, and yet affirm that the sequence of its digits is random. For statistical tests have shown that the first 2000 digits of $\pi$ follow no recognizable pattern,[2] except of course that of being derived by a computation of $\pi$, which is too cumbersome to be carried out mentally. In the case of $\pi$ we can also tell for once fairly well what it is that we must *know* and what we must *not know* in order to identify randomness. The principle by which $\pi$ is computed identifies the random number $\pi$, but any actual computation of $\pi$ would destroy the randomness.

Quantum mechanics does not affect the argument either in respect of a well shuffled pack of cards, or of the throws of dice, or of a gas conceived as an aggregate of molecules chasing around at random. Its wording would merely have to be changed by replacing the 'laws of mechanics' everywhere by the 'laws of quantum mechanics'. This will make no difference for packs of cards or for dice and little difference for gases other than hydrogen, for the laws of quantum mechanics coincide with those of mechanics for reasonably heavy particles. However, to be precise, the

[1] G. Spencer Brown suggests that randomization might always be incomplete and tries to explain Rhine's results on these grounds (G. Spencer Brown, *Nature*, 72 (1953), pp. 154, 594).

[2] Hilda Geiringer, 'On the Statistical Investigation of Transcendental Numbers', in *Studies in Mathematics and Mechanics presented to Richard von Mises*, Academic Press, New York, 1954, p. 310.

classical predictions of positions and velocities would have to be replaced by predictions of the probability distribution of positions and velocities.[1]

## 4. THE LOGIC OF EMERGENCE

We can now return to our proper subject matter, which is the contemplation of anthropogenesis. We have reached the point at which we must confront the unspecifiability of higher levels in terms of particulars belonging to lower levels, with the fact that the higher levels have in fact come into existence spontaneously from elements of these lower levels. How can the emergent have arisen from particulars that cannot constitute it? Does some new creative agent enter the emergent system at every new stage? If so, how can we account for the continuity of the process of anthropogenesis?

To answer these questions we must add some further questions to them. The rise of man culminates in the unfolding of the noosphere. Is this firmament of superior knowledge a last-minute improvisation of the anthropogenic process? Or were all the works of the human mind already inscribed invisibly in the configuration of primeval incandescent gases? Or must, alternatively, each new discovery of man be ascribed to a new divine intervention?

The first thing to observe here is that, strictly speaking, it is not the emerged higher form of being, but our knowledge of it, that is

---

[1] Niels Bohr in his Faraday Lecture (*J. Chem. Soc.*, 1932, Pt. I, pp. 349 ff.) expressed the view that the relation between the macroscopic and microscopic description of a gas is an instance of complementarity in the sense established in quantum mechanics between the position and velocity of an electron. This theory supports the unspecifiable character of randomness, but otherwise it is not acceptable. I want to show this here in some detail, for this argument will reveal my dissent from yet another widely held opinion of great importance. In quantum mechanics any attempt at specifying the position and velocity of an electron must be defined in terms of the electron's interaction with a definite measuring instrument. The result will depend on the instrument chosen and will again be a statement of probability. The more narrowly our measurement defines the position of a particle the more widely does it leave its velocity undetermined and the product of the two ranges is constant. The complementarity of these two ranges of knowledge differs however from the two alternative kinds of knowledge that we can have of a pack of cards, or of the sequence of digits in the number $\pi$. For the same pack of cards could be well shuffled for one man and perfectly stacked for another; and though one man can use the digits of the number $\pi$ as a random sequence, another may compute them with perfect assurance. This is not so for the probable positions and velocities of an electron. There is nothing present in this case that is hidden to one observer and known to another. In fact, the outcome of the observation does not depend here on the participation of the *observer*, but on the action of a *measuring instrument*, the result being the same for any observer. This contradicts on the one hand the view that the relation between the macroscopic and microscopic descriptions of an atomistic system is an instance of complementarity; and it shows also on the other hand that (contrary to a widespread opinion) the indeterminacy principle of quantum mechanics establishes no effect of the observer on the observed object. The supposed effect vanishes if we include the 'measuring instrument' in the 'observed object'. The latter becomes then 'the observed phenomenon' in the sense now accepted by Bohr's school of interpretation. (See L. Rosenfeld, 'The Strife about Complementarity' (*Science Progress*, No. 163, July, 1953, p. 395).

unspecifiable in terms of its lower level particulars. We cannot speak of emergence, therefore, except in conjunction with a corresponding progression from a lower to a higher *conceptual* level. And we realize then that conceptual progression may not always be existential, but that it becomes so by degrees.

For example: pour a handful of shot into a flat-bottomed saucepan, and you will find the grains forming a regular pattern. Crystals owe their symmetrical shapes to a similar principle: molecules of identical sizes and shapes tend to form regular aggregates in the same way as grains of shot in a saucepan. Is this the emergence of a new comprehensive feature? It is arguable that we could know the complete topography of the atoms in a crystal, without seeing that they form a regular pattern. There is, indeed, always a noticeable logical gap between a topography and a pattern derived from it, and to this extent no pattern is specifiable in terms of its topography. Yet since in the case of a crystal we can easily pass from the pattern to the topography and back again, the conception of such a pattern is in fact not destroyed by a knowledge of its topographic particulars. I would acknowledge, therefore, in this case two distinguishable conceptual levels but not two separate levels of existence.

We can even widen the conceptual gap between two levels, to the point where it precludes altogether the representation of the higher level in terms of the lower, without establishing a complete existential disjunction between the two. Consider the chemical aspects of matter. They are fully determined by atomic physics; yet no Laplacean Mind schooled in quantum mechanics could replace the science of chemistry. For chemistry answers questions regarding the interaction of more or less stable chemical substances, and these questions cannot be raised without experience of these substances and of the practical conditions in which they are to be handled. A Laplacean knowledge which merely predicts what will happen under *any given* conditions cannot tell us what conditions *should be given*; these conditions are determined by the technical skill and peculiar interests of chemists and hence cannot be worked out on paper. Therefore, while quantum mechanics can explain in principle all chemical reactions, it cannot replace, even in principle, our knowledge of chemistry. We may acknowledge this as an incipient separation of two forms of existence.

We have seen that two sharply separated levels of existence emerge by randomization. But even in this case, the emergent reality is comparatively poor in new features. No richly endowed new reality can be seen emerging in the inanimate domain. This happens for the first time in the emergence of a living being from inanimate constituents. I have described this process as a chance fluctuation which releases the action of certain self-sustaining operational principles. This results in the formation of two levels of existence: an upper level governed by physiology, and a subsidiary, lower level defined by physics and chemistry—the operations on the upper level being predicated on the emergence of an individual, whose interests they serve. In the course of anthropogenesis, individuality develops from beginnings of a purely vegetative character to successive stages of active, perceptive, and eventually responsible, personhood. This phylogenetic emergence is continuous—just as ontogenetic emergence clearly is. Hence the higher principles governing the emergent forms of evolution presumably gain control gradually of the evolving beings, in the same way as they gradually become more pronounced and predominant in the course of man's embryonic and infantile development. We shall say, in particular, that the rise of man includes a continuous intensification of individuality, similar to that which normally takes place in the formation of a human person from the parental zygote. No new creative agent, therefore, need be said to enter an emergent system at consecutive new stages of being. Novel forms of existence take control of the system by a process of *maturation*.

Admittedly, this conception still leaves open an unresolved conflict between continuity and essential progress. It presents us with an unwelcome alternative: either to regard the process of maturation itself as predetermined from the start, or else to assume that it results from the continuous intensification of an external creative agency. We shall have to reconsider the concept of maturation in order to reconcile these alternatives. The argument will fall into two parts, the first dealing with determinism *a fronte* by the universal target of a commitment, the other with determinism *a tergo* by the bodily mechanism of the person entering on a commitment.

(1) I shall recall for a start the ontogenetic emergence of human intelligence, as described by Piaget. The infant's understanding of its surroundings is self-centred. It goes on plunging irreversibly from one form of comprehension to another. Then, gradually, it develops a solid interpretative framework, each successive stage of which offers a possibility for increasingly elaborate logical operations. Irreversible comprehension is replaced by the steady deployment of discursive thought. The appetitive, motoric, perceptive child is transformed into an intelligent person, reasoning with universal intent. We have here a process of maturation closely analogous to the corresponding step of anthropogenetic emergence, leading from the self-centred individuality of the animal to the responsible personhood of thoughtful man: in fact, to the emergence of the noosphere.

This kind of emergence is known to us from inside. We experience intellectual growth in the process of education and, in more dramatic forms, in the creative acts of the mind. I may recall in particular the process of scientific discovery. This process is not specifiable in terms of strict rules, for it involves a modification of the existing interpretative framework. It crosses a heuristic gap and causes thereby a self-modification of the intelligence achieving discovery. In the absence of any formal procedure on which the discoverer could rely, he is guided by his intimations of a hidden knowledge. He senses the proximity of something unknown and strives passionately towards it. Where great originality is at work in science or, even more clearly, in artistic creation, the innovating mind sets itself new standards more satisfying to itself, and modifies itself by the process of innovation so as to become more satisfying to itself in the light of these self-set standards. Yet all the time the creative mind is searching for something believed to be real; which, being real, will—when discovered—be entitled to claim universal validity—something the knowledge of which mu.t indeed passionately insist on its own universal validity. Such are the acts by which man improves his own mind; such the steps by which our noosphere was brought into existence. For in the ontogenesis of the innovator we meet a step in the phylogenesis of the human mind.

Looking back on this process of emergence, it seems clear enough what has happened. The passionate urge to fulfil self-set standards will appear *completely determinate* if we too accept the same standards as real and valid; but it is also seen to be *quite indeterminate,* for it is achieved by a supreme intensification of uniquely personal intimations. Such is the logic of self-compulsion with universal intent. Action and submission are totally blended in a heuristic communion with reality; determinism and spontaneity mutually require each other when embodied in the universal and the personal poles of commitment. We have no difficulty in acknowledging this seemingly paradoxical situation every time we are confronted with human greatness. Wherever men have truly spoken in the name of truth, saying, Here I stand and cannot do otherwise, we instantly recognize both the power of impersonal truth and the greatness of a mind upholding it. We readily pay our respect to both poles of such a commitment.

Difficulties arise only when we look at the fragments of the commitment non-committally. If we ask whether Euclid's theorems existed before they were discovered, the answer is obviously No, in the same sense as we would say that Shakespeare's sonnets did not exist before he wrote them. But we cannot therefore say that the truth of geometry or the beauty of poetry came into existence at any particular place and time, for these constitute the universal pole of our appreciation, which cannot be observed non-committally like objects in space and time.

(2) Another difficulty arises, for similar reasons, at the personal pole of human greatness, if we regard the innovator as a material system, controlled by the laws of physics and chemistry. Such a vision traces Shakespeare's sonnets back to a pattern inscribed in the primordial incandescent gases in which our universe originated; it is the Laplacean idea of a universe determined from the start for all times. My answer to this view is to accredit once more my capacity for comprehending entities which are not specifiable in terms of their particulars—of particulars which are themselves usually comprehensive features and hence in their turn are unspecifiable in terms of their own particulars, and so on. Thus the ultimate Laplacean particulars turn out to be almost completely meaningless, and

certainly cannot be said to determine any significant feature of a universe enriched by emergent strata of being.

But I must elaborate this conclusion further if I am to take in from its point of view the whole panorama of emergence from its first beginnings. Admitting that no process governed by the laws of physics and chemistry as known today can be accompanied by consciousness, we may yet suppose that some enlarged laws of nature may make possible the realization of operational principles acting by consciousness. It would be unwarranted to retain then for structures operating on such principles the conception of automatic functioning derived from our *present* physics and chemistry. Since action and reaction usually arise together in nature, it would seem reasonable, on the contrary, that the new laws of nature, which would allow for the rise of consciousness in material processes, should also allow for the *reverse* action, that is, of conscious processes acting on their material substrate. Such laws of nature would not comprise psychology, which is a convivial study of mental operations, but their assumption would make it conceivable that material structures should offer *conditions for the occurrence of mental operations* and should *account for their occasional failure*. This assumption would enable us to envisage the rise from inanimate matter of sentient, motoric, perceptive individuals, and, at a higher stage, of thinking, responsible persons. And it would allow us also to bring this process of emergence into continuous alignment with the heuristic strivings of innovators.

Looking back in this light on a process of human ontogenesis, we can trace now the activities of the mature mind to ever further descending levels of sentient effort. I have identified these levels of action already before as the roots of the tacit component which participates decisively in all articulate thought. We have seen there, and later also in surveying the rising levels of biotic existence, that the outcome of these actions is always indeterminate; for they are commitments which have a bearing on reality to the very extent to which they are hazardous. They are irreversible processes of comprehension, guided only by vague maxims. Descending therefore from the person of a great man down to the level of the newborn infant and beyond that to the lowest animals, we find a continuous series of centres whose a-critical decisions account ultimately for every action of sentient individuality. Thus the personal pole of commitment retains its autonomy everywhere, exercising its calling within a material milieu which conditions but never fully determines its actions. Unopposed, the circumstances of a commitment would overwhelm and wipe out the impulse of commitment; but a centre actively committing itself resists and limits these circumstances to the point of turning them into instruments of its own operations.

## 5. Conception of a Generalized Field

We can now perceive more clearly the roots of the parallelism between comprehension and morphogenesis hinted at by Spemann in his Silliman Lectures.[1] Comprehension is an unformalizable process striving towards an unspecifiable achievement, and is accordingly attributed to the agency of a centre seeking satisfaction in the light of its own standards. For it cannot be defined without accrediting the intellectual satisfaction of the comprehending centre. The unspecifiability of a conscious act of comprehension implies the impossibility of accounting for it in terms of a fixed neurological mechanism; and the intellectual commitment involved in such an act excludes any representation of it in terms of a physico-chemical equilibration which cannot distinguish between success and failure. Comprehension and the somatic process which accompanies comprehension, represent therefore a kind of equilibration that can be defined only in terms of *intellectual rightness*. Morphogenesis, operating under the direction of a morphogenetic field, is a somatic process of the same kind, but following *morphological rightness* as its standard of achievement. Yet it may be described as equilibration, to distinguish it from the operation of a machine-like framework, and also to illustrate the inexhaustible resourcefulness shown by the morphogenetic process. Once it is recognized that this resourcefulness is mobilized in the service of an achievement which can be appreciated only in morphological terms, we find that this implies awarding to it success or failure, by standards which we ourselves set to the process as being appropriate to itself. The morphogenetic field (or its organizer, if there is one) is then defined as the agency of this success and as that which has failed if success is not achieved.

This situation can be described more precisely by a generalization of the field concept in a strictly biological sense, purified of any *arrière-pensée* of physico-chemical equilibration. All the operations of the 'tacit component' (whether self-centred or seeking universality, whether conscious or unconscious) will be subsumed under this field conception. All mental unease that seeks appeasement of itself will be regarded as a line of force in such a field. Just as mechanical forces are the gradients of a potential energy, so this field of forces would also be the gradient of a potentiality: a gradient arising from the proximity of a possible achievement. Our sense of approaching the unknown solution of a problem, and the urge to pursue it, are manifestly responses to a gradient of potential achievement; and when we identify a morphogenetic field, we see in it in fact a set of events co-ordinated by a common gradient of achievement. We may recall also that muscular co-ordination appears likewise unformalizable in terms of any fixed anatomical machinery, and that the functional stability and recuperation of the central nervous system after widespread injuries,

---

[1] See p. 338n. above.

as well as the search for lost memories, offer further instances of apparently unformalizable operations. These again are evidence of fields of forces derived from various gradients of achievement.

The conception of such a field is of course finalistic. It attributes to certain achievements—whether self-centred or aiming at universality—the power to promote their own realization. Scientists will not be prepared even to consider such a suggestion, unless they have completely accepted the fact that biotic achievements cannot—*logically cannot*—be ever represented in terms of physics and chemistry; and very few do realize this. Besides, a biologist may reject the assumption that living beings have peculiar faculties for achieving biotic success, on the grounds that this would impute to them magical powers which could explain anything— and hence explain nothing. But this objection would misunderstand the kind of finalism I suggest here. For though biotic achievements are said to be unspecifiable, we do claim the capacity for identifying and appraising them; nor is their scope unlimited or the range of their resourcefulness unbounded. A biology and a psychology formulated in terms of achievement can therefore be studied quite systematically; in fact these sciences *are* mainly pursued in these very terms—though heavily disguised—in our own days. Yet, even though a biologist might recognize this situation, he may prefer not to acknowledge it, for fear that biology would degenerate to mere speculation if it abandoned the ideal of becoming a science as objective as physics and chemistry. For my part, I do not share this apprehension, and would expect, on the contrary, that biology would gain greatly in scope and depth by addressing itself more candidly to the fundamental features of life. In any case, the non-specialist who wishes to find his way about the world can certainly not accept the prudential policies imposed by scientific objectivism.

Returning therefore to the outline of generalized biological fields, we see now that their operations comprise three stages of originality, of which phylogenetic emergence is the highest. (1) There is the originality of a resourcefulness manifested in achieving something clearly foreseeable. This kind of originality was illustrated in my last chapter, e.g. by the way mutilated rats run a maze known to them from their previous unmutilated condition. (2) The ontogenetic maturation by which infants develop the faculty of logical thinking may be classed in a higher category, as representing a series of achievements, each producing a new field by which the next higher achievement will be performed. Such emergence—defined as an ordering principle capable of producing operational principles which the system had not previously possessed—has been adequately illustrated by the process of ontogenetic maturation. (3) Phylogenetic emergence exceeds this degree of originality by producing operational principles that are altogether unprecedented, and this fully developed emergence we were able to approach so far only by forming a continuous transition from ontogenetic maturation to heuristic achievements. It is on this connection that we have now to rely further in applying to the process of anthropo-genesis a field theory based on the gradient of achievement.

Though this homology was suggested long ago by Samuel Butler and was elaborated since by Henri Bergson, it may still appear far-fetched. But this is partly due to an error in perspective. The highest forms of originality are far more closely akin to the lowest biotic performances than the external circumstances would indicate. It is true that creative human achievements rely on a far flung, highly articulate, cultural structure, but the creative act itself is performed by informal comprehensive powers—by powers which the man of genius shares with all men and which all men share with infants, who in their turn are about on a par in this matter with the animals. Remember how slight, indeed almost imperceptible, is the superiority of tacit powers which enables man to develop the tre-mendous gift of speech. Consider also that children under two years can learn lip reading better than most adults, even though the adult may be highly literate, while the child has to learn to speak *and* to lip read all at the same time. Originality is greatest in youth; it is indeed arguable that children would surpass adult genius if they could command the intellectual equipment and possess the emotional experience of maturity.

In any case, it is not so much the suggested explanatory framework, but the fact of phylogenetic emergence itself that is so astounding; and this fact, I believe, is indisputable. It is a process of maturation which differs in the most curious manner from that of ontogenesis; for it is a maturation *of the potentialities of ontogenesis*. The evolutionary process takes place in the germ plasm, but it manifests itself in the novel organism which the germ plasm potentially embodies. It occurs in one place and manifests itself in another. Hence if, contemplating the process of anthropogenesis, we are clearly convinced—as I am—that this is in fact so, we are *driven* to assume that the maturation of the germ plasm is *guided* by the potentialities that are open to it through its possible germination into new individuals. We are actually facing then the operations of a phylogenetic field guiding anthropogenic maturation along the gradients of phylogenetic achieve-ment—as clearly as the embryologist faces morphogenetic fields derived from the gradient of ontogenetic achievement. Nor can we fail then to note that at least in some cases we can experience such gradients internally. We know clearly the approach of a recollection for which we have been racking our memory, and will tend to compare this with an ontogenetic maturation which re-produces things already achieved before; while we know also how the search for entirely novel achievements is guided by intimations of their growing proximity, even as the possibility of unpre-cedented achievements guides the maturation of the germ plasm to ever higher evolutionary stages.

### 6. The Emergence of Machine-like Operations

My survey of anthropogenesis has stressed the rise of sentience and personhood, and it is this spectacle that has guided so far my observations on evolution. I have hardly mentioned the elaborate structural and functional innovations which have led to the formation of the higher animals, though the emergence of these ingenious instruments—precisely performing most delicate operations—has been the principal problem of evolutionary theories in the past. I have done this because I believe that the unformalizable regulative functions, linked to the animal's mental processes, are the predominant, comprehensive agency of animal life. The evolution of personhood clearly produces novel centres of being, and this fact must be fully envisaged before we approach the evolution of anatomic and physiological equipments which, viewed in themselves, might appear merely as new implements serving unchanged centres. And not unless we first recognize that evolution can give rise to ever new unformalizable operations only by acting, itself, as an unformalizable principle, shall we be prepared to acknowledge also that new machine-like operations can likewise emerge only in the same unformalizable manner. Even so, I shall not be able to carry out this argument here fully. However, having established, in my opinion, that the evolution of human personhood could only have been actuated by a maturation of the germ plasm, I feel that the question whether a similar maturation is also involved in the phylogenesis of machine-like biotic structures has been reduced to a side issue of my enquiry. Hence I shall merely outline here the relevant argument, quite briefly.

A machine-like function is characterized by its operational principle, which cannot be defined in terms of physics and chemistry, and consequently the rise of new operational principles in living things cannot be defined either in terms of physics and chemistry. In so far, therefore, as an organism sustains itself by functioning as a machine, it is the embodiment of an ordering principle that cannot be defined in terms of physics and chemistry. Random impacts can *release* the functions of an ordering principle and suitable physico-chemical conditions can *sustain* its continued operation; but the *action* which *generates* the embodiment of a novel ordering principle always lies in this principle itself.

I have said all this before. I would now have to defend against various possible objections my view that this argument applies to the cases in point—for example to the evolution of lungs and pulmonary breathing. The question would arise then whether the lungs of one individual at different ages, as well as of members of different species, particularly when descended from the same lung-breathing ancestry, are in fact to be accepted as widely different embodiments of the same structural and functional principles. And if this be admitted, I would have to discriminate between such a rational structure and function *actively used* by an animal, and the kind of *passive* advantage secured to some animals by a protective colouring. I

401

would then urge (as I have done before) that, since protective colouring is not an operational principle, it could be established by random mutation plus natural selection, by which the emergence of a new operational principle can be released but never established. I suppose that intermediate cases would have to be considered too, such as, e.g., the emergence of new habits by which an animal increases the advantage afforded to it by protective colouring. The distinction to be drawn here might prove difficult, in the same way as, for example, that between meaningless learning taking place by accident, and intelligent learning achieved by understanding. But once the latter form of emergence was fully established, it would be clear that it represented the achievement of a new way of life, induced in the germ plasm by a field based on the gradient of phylogenetic achievement.

I believe that this argument would show that all attempts at explaining the evolution of complex organs by chance variations in certain chemical bonds of the germ plasm must fail. But I must admit once more that I would not feel so certain of this, had I not before me the rise of human personhood, which manifestly demands the assumption of finalistic principles of evolution. I shall be satisfied, therefore, to rest my case for the acknowledgment of the principles in question on the argument dealing with the emergence of sentience and personhood.

## 7. FIRST CAUSES AND ULTIMATE ENDS

In the chapter on Probability I postulated that random impacts could produce biotic achievements only by releasing the operations of an ordering principle, and I suggested that the stability of open systems was a pointer towards the existence of ordering forces of this kind. The stability of living beings, and the even greater stability of the germ plasm carried by living beings—which can all be classed as open systems—added colour to this suggestion. I confronted this idea then with the spectacle of the ontogenesis and evolution of a human person, which I acknowledged as achievements of the highest order; and I appealed further to the evidence provided by various branches of biology (including psychology) which seem to cry out for the acknowledgment of a field as the agent of biotic performances. So, finally, I was led to the belief stated in the last two sections, that the pathways of biotic achievements have dynamic properties analogous to those of pathways along which the potential energy of a system decreases. An anthropogenesis induced by random mutations would then be essentially on a par with the ordering of a set of biassed dice under the impact of Brownian motion at low temperatures.

But again (as in the analysis of maturation) we must remember here the decisive fact that biotic achievements are those of an active centre. This completely transforms the picture at the upper levels where centres are called upon to exercise responsible choices—and continuity demands that we should take this active component into account likewise down to the lowest levels. The emergence of man and of the thoughts of man must, therefore, never be regarded as due to a passive shifting of matter and mind in a field of biotic achievement: it reflects the gradual rise of autonomous centres of decision.[1]

Of these decisions we know from experience those of the human mind, in addition to which we have also observations of the strain imposed by heuristic decisions on animals. Let me recall the characteristic features of such acts that are now to be reformulated in the light of the field conception.

I shall limit myself in the first place to acts of knowing which are the principal subjects of this book. Subjective knowing is classed as passive; only knowing that bears on reality is active, personal, and rightly to be called objective. It is for me, who use these terms confidently, to declare ultimately what knowing I believe to bear on objective reality, and this qualification is included in the foregoing definition of objective knowing. Let me now introduce the concept of a *heuristic field*. We assume that the gradient of a discovery, measured by the nearness of discovery prompts the mind towards it. This was implied already in the chapter on Intellectual Passions, but not yet explicitly stated there. The assumption of a heuristic field explains now how it is possible that we acquire knowledge and believe that we can hold it, though we can do this only on evidence which cannot justify these acts by any acceptable strict rules. It suggests that we may do so because an innate affinity for making contact with reality moves our thoughts—under the guidance of useful clues and plausible rules—to increase ever further our hold on reality.

Taken literally, however, this picture would be misleading, since it once more describes the movement of the mind as a passive event. The lines of force in a heuristic field should stand for *an access to an opportunity, and for the obligation and the resolve to make good this opportunity, in spite of its inherent uncertainties.* It is true that the assumption of such a field expresses more clearly than has been done before that our expectation to discover the truth is justified by our nature as living beings. It asserts the fact *that knowing belongs to the class of achievements that are comprised by all forms of living*, simply because every manifestation of life is a technical achievement, and is therefore—like the practice of technology—an

---

[1] It may be important to distinguish here between action and decision. The action of mechanical forces transforms potential into kinetic energy, and the action of biotic fields may be regarded as analogous to this. But mechanical effects can be produced without force, merely by selection, as in the case of a Maxwellian demon which can compress a gas indefinitely by effortlessly moving to and fro a frictionless, perfectly balanced shutter. This offers a possibility for conceiving the action of the mind on the body as exercising no force and transferring no energy of its own. Indeed, since it is the peculiar function of the mind to exercise discrimination, it may not even appear too far-fetched that the mind should exercise power over the body merely by sorting out the random impulses of the ambient thermal agitation. We may bear this possibility in mind whenever referring to autonomous centres of decision.

applied knowledge of nature.[1] But in order to express correctly this kinship of knowing and living, fields must be interpreted throughout biology
in accordance with their finalistic character, as fields of opportunity and
of a striving directed towards this opportunity. Biological fields normally
belong to a centre to which both the opportunities and the strivings are
attributable. Though these strivings are continuous with the conscious
strivings of higher animals, they are, of course, in general, neither conscious nor deliberate. By contrast to a field of forces operating on an inanimate system, a field of biological striving stands defined by the fact
that we attribute its operations to an active centre, and acknowledge these
operations as the successes or failures of this centre, while basing our
awards of success or failure on criteria which are largely unspecifiable. In
my description of anthropogenesis I have surveyed the gradual rise of field
centres to the rank of full personhood, and I have again spoken of this
rise when illustrating some aspects of emergence by the logical maturation of the mind from infancy to adulthood. At all levels of life it is these
centres which take the risks of living and believing. And it is still such
centres which, at the highest stage of development, actuate those men who
seek the truth and declare it to all comers—at all costs.

The point is reached here at which the observer's appraisal of biological
achievement turns into his submission to the leadership of superior minds.
This corresponds to the extrapolation of biology into ultra-biology, where
the appraisal of living beings merges into an acknowledgment of the ideals
transmitted by our intellectual heritage. This is the point at which the
theory of evolution finally bursts through the bounds of natural science
and becomes entirely an affirmation of man's ultimate aims. For the
emergent noosphere is wholly determined as that which we believe to be
true and right; it is the external pole of our commitments, the service of
which is our freedom. It defines a free society as a fellowship fostering
truth and respecting the right. It comprises everything in which we may
be totally mistaken.

Looking back from this point on the immensities of the past, we realize
that all that we see there, throughout the universe, is shaped by what we
now ultimately believe. We see primordial inanimate matter, the motions
of which are determined—whether mechanically or statistically—by intrinsic fields of forces. We see its particles settling down into orderly
configurations which our physical theories can trace back (however imperfectly) to the fundamental properties of inanimate matter. This universe
is still dead, but it already has the capacity of coming to life.

Can we see then all the works of the human mind invisibly inscribed
already in the configuration of primeval incandescent gases? No, we cannot;

[1] To this extent the suggestion that the faculty of acquiring knowledge originated in
the selective advantage offered by it appears tautologous; you obviously cannot explain
the origin of life by the survival of living beings. Admittedly, the cognitive faculty may
be developed further by its selective advantage; but since this does not explain how it
works, neither can it be said to explain its origin.

for the capacity of coming to life is due to the power of a field to consolidate centres of first causes. Each such centre bears a possibility of achievement which, however limited, uncertain, and unspecifiable in its outcome, characterizes this centre as an essentially new and autonomous prime mover. The centres of individual beings are short lived, but the centres of the phylogenetic fields of which individuals are offshoots go on operating through millions of years; indeed, some of these may endure for ever—we cannot tell. But we do know that the phylogenetic centres which formed our own primeval ancestry have now produced—by a deployment which, when compared with the long ages of life on earth, looks like a single sudden outburst—a life of the mind which claims to be guided by universal standards. By this act a prime cause emergent in time has directed itself at aims that are timeless.

So far as we know, the tiny fragments of the universe embodied in man are the only centres of thought and responsibility in the visible world. If that be so, the appearance of the human mind has been so far the ultimate stage in the awakening of the world; and all that has gone before, the strivings of a myriad centres that have taken the risks of living and believing, seem to have all been pursuing, along rival lines, the aim now achieved by us up to this point. They are all akin to us. For all these centres —those which led up to our own existence and the far more numerous others which produced different lines of which many are extinct—may be seen engaged in the same endeavour towards ultimate liberation. We may envisage then a cosmic field which called forth all these centres by offering them a short-lived, limited, hazardous opportunity for making some progress of their own towards an unthinkable consummation. And that is also, I believe, how a Christian is placed when worshipping God.

# INDEX

This index was compiled by Dr. Marjorie Grene.
The main entries offer systematic surveys of certain points which readers may find useful as summaries.

Abetti, G., 146n.
ab-reaction, 243.
abstract arts, 193–5.
accident, see *chance*.
accommodation (Piaget), 105n.
accrediting: essential to language, 80, 84; articulation, 91; perception, 96; speaker's judgment, 113; vision of reality, in discovery, 130; solution by heuristic craving, 130; intellectual passions, 142; scientists, 163; inference machine, 169, 259, cf. 261–4; mathematics, 188; articulate framework, 201–2; facts in society, 240–3; persons as knowers, 264, 343; philosophical reflection, 265; self-set standards, 267–8; validity of religious system, 283–4; facts in commitment, 304; machines, 329; myself as living being, 343; faculties of living beings, 347; reality relied on by living beings, 347; success of organisms, 363; commitments, part of biology, 373; emergent existence, 392; capacity for comprehension of unspecifiable entities, 396; personal centres, 398. See also *appraisal, belief, commitment, fiduciary mode, fiduciary programme, personal participation*.
accreditive: decisions in speech, 209; statements, 280–1.
accuracy, 135, 136, 138, 141. See also *precision*.
Ach, N., 129n.
achievement(s): inarticulate, 100; of science, 165; in use of tools, 175; of knowledge, 313, 347, 403; intellectual, & rightness, 321; logic of, 327–46, 381–2, 392, 399; physiology as study of, 334; of rightness in living processes in general, 340; of rightness in perception, 342–3; knowledge of, 347; biological, first level, 349, second level, 354–8; of ontogenetic success, 357, 400; types of biological, 357; evolutionary, 361, 382–90, 399–405, and adaptation, 385; human evolution, 385–390, 399–400, 404–5; acknowledged in learning experiments, 369; of living beings acknowledged in psychology,

370; of animals, 373; randomness as, 392; and comprehension, 398; gradient of, in anthropogenesis, 400.
a-critical: acts, 264, 280, 305, 312, 376; belief, 272; choices, 286; commitment to perception, 297; decisions, 397; processes of cultural transmission, 208; statements, 268. See also *uncritical*.
action: taught by technology, 176; and perception, 361–4; in learning, 364; of ordering principles, 401; vs. decision, 403n.
active: centre(s), in living individuals, 336, 344, 364, 388, 397, 402, 403, 404; principle, in perception, 96, in mathematics, 189; vs. passive, 63, 300, 312–13, 320, 345, 401. See also *commitment, personal, personhood, persons*.
Acton, H. B., 230n.
Adams, J. C., 30, 145, 182n.
adaptation (Piaget's 'accommodation'), 105.
adaptation in evolution, 385.
aesthetic: recognition, 351; rules, and ideal-blind medium, 334.
affiliation of child to society, 207–8.
affirmation(s): implies skill, 71; personal coefficient of, 81; in vision, 99; in appetite, 99; of intellectual passions, 134–5, 159; of mathematics, 131, 187–90; of scientific theory, 204; & shared convictions, 212; logic of, 249–68, 312, 336; critical, 264; & doubt, 272; vs. indwelling, 279; implies responsibility, 312; hierarchy of, 346; & superior knowledge, 375. See also *assertion*.
affirmative: content of emotions, 172–3; use of language, see *indicative*.
Africa; science in, 182; primitive beliefs in, 286–92.
Agassiz, L., 155n.
agnostic doubt, 272, 273–4; in law, 274, 277–279.
agreement and difference (Mill's canons), 167, 371.
alchemy, 354, 389.
Alexander, S., 182, 382.
algebra, 86, 185, 186.

Helmholtz, H. L. F. v., 319.
heuristic: achievements, & emergence, 399.
   acts, 76, 77, 172; in speaking, 105; in learning language, 106; in modifying frameworks, 106; in re-interpreting language,
   110; levels of, 123-4; in mathematical
   problem-solving, 125-31; paradigm of
   emergence, 382.
   commitments, 311-12, 316.
   decisions, in animals & men, 403.
   efficacy of doubt? 276-7.
   effort: in problem-solving, 365; in animals,
   367-8.
   feeling & assertion of fact, 254.
   field, 403.
   gap, see *logical gap.*
   impulse, degrees of, 366.
   maxims, & doubt, 277.
   passion, 142-5, 150, 159; of technologist,
   178; & Christian worship, 199, 280-1; in
   art, 200; degrees of, 305.
   powers, & objectivism, 371.
   process, in mathematics, 190.
   progress, see *problem-solving.*
   tension, see *incubation.*
   vision, 196, 280, 283, 285.
heuristics: mathematical, 124-31, 259;
   routine, 261; paradigm of commitment,
   301-2, 306; in framework of commitment, 310-12; & intellectual beauty, 320;
   equipotentiality &, 337n.; & ontogenesis, 339; evolution of, 389; & mental
   maturation, 395; action & submission
   in, 396; & emergence, 397. See also *discovery, heuristic, originality, problem-
   solving.*
Hicks, W. M., 12.
Hilgard, E. R., 26n., 71, 73n., 120n., 122n.,
   371n.
Himmler, H., 205, 232n.
Hiss, A., 241.
historical: context of scientific value, 183-4;
   prediction, in Marxism, 230-1; setting of
   commitment, 324.
historicism, 229.
history, 137, 138, 321; of science, 158, 164,
   170-1; & ritual, 211; in totalitarian
   society, 242, 243.
Hitchcock, A. S., 352n.
Hitler, A., 225, 226, 232.
Hoff, J. H. van't, 145, 155-6, 158, 160, 164,
   190, 275.
Hofmeister, W., 155n.
Hollo, J., 50, 51n.
Holst, E. v., 338n.
Holtfreter, J., 358n.
Homans, G. C., 211n.
Honzik, C. H., 74n.
Hook, S., 158n., 239n.
Hooker, J., 351.
Hoppe, F., 122n.
Horney, K., 288.
Hörstadius, S., 355n.
Horton, G. P., 120n.
Housman, A. E., 194n.
Huguenots, 53.
Hull, C. L., 320n., 369, 371.

*Humani Generis*, 153, 297.
Hume, D., 9, 137, 238, 270, 279, 284, 304.
Humphrey, G., 77n., 102.
Humphreys, L. G., 25.
Hungary, 52, 244.
Huxley, A., 197n.
Huxley, J., 383n.; ——, Hardy, A. C., &
   Ford, E. B., 35n., 384n.
Huxley, L., 351n.
hypnosis, 51-2, 108, 129, 157n., 167, 168,
   274-5, 320.
hypotheses: probability of, 24-7, 29-30;
   evidence for, 24, 29, 30; positivistic view
   of, 146, 170 (cf. 370); & regulative
   principles, 307.

I-it vs. I-thou, 346, 348.
idiom: theory =, 47; & interpretative framework, 105; of group, 112; & action, 112-
   113; of belief, 287; of Zande belief, 288; of
   objectivist belief, 288; of thought, 376, 380
Ignotus, Mr. and Mrs. P., 290n.
Illingworth, K. K., 13n.
illumination (problem-solving), 121, 123, 130,
   172.
imagination, 46, 186, 187, 334.
imitation, in learning, 206.
immanence, two-way (Marxism), 229, 230,
   231, 235.
'impersonal allegation', 256.
implements, see *tools.*
Impressionists, 164, 200, 319.
improbability: of past events, 35-6; of orderly
   patterns, 37; & open systems, 384.
inarticulate application of language, 81, 82,
   83, 86-7.
   confusion, 108.
   conviviality, 209-11.
   intelligence, 60, 62, 64, 69-76, 100, 132,
   194, 206, 335; in animals, 71-4, 120-2,
   132; in children, 74-5, 82; in wolf
   children, 296; evolution into articulate
   thought, 389; see also *comprehension,
   conception, insight, problem-solving, tacit
   component, understanding.*
interpretation of primitive terms & axioms,
   131.
knowledge: in animals & children, 90; vs.
   articulate knowledge, 103.
   See also *comprehension, conception, conceptual, ineffable, insight, intuition, tacit,
   understanding, unspecifiability.*
Incubation (problem-solving), 121-2, 126,
   129, 339.
indefinite regress: of precision, 251-2; of
   'true', 254-5; in objectivist theory of
   knowledge, 305.
indeterminacy: in use of language, 81, 86-7;
   in mathematics, 94; of conceptions, 104;
   of intensions, 116; of meaning, 150; of
   personal knowledge, 249; of knowledge,
   264, 336; & responsibility, 310; of commitment, 316; in biological achievement,
   397. See also *unspecifiability.*
indeterminate implications: of heuristic passion, 143; of knowledge, 311. See also
   *objectivity, reality, truth, unspecifiability.*

# Index

memory, 84–5, 127, 128, 399, 400.
Mendel, Mendelism, 43n., 158, 238.
mental derangement, see *obsessiveness*.
Mesmer, F. A., 51–2, 107–8, 157n., 274, 275n.
meta-theory, 258n., 344.
meteorites, 138, 274–5.
Michelson, Pease, & Pearson, 13n.
Michelson-Morley experiment, 9–13, 152n., 167.
Michelangelo, 284.
Michurin, I. V., 182.
Mill, J. S., 161, 167, 270, 371.
Miller, D. C., 12–13, 30, 167.
Milosz, Cz., 235.
mind: mechanistic conception of, 37, 261–4, 336, 382; epiphenomenal interpretation of, 158–9; of other persons, 263; ontology of, 264; knowledge of, 372; maturation of, 395–7; -body relation, 403n.
Minkowski, H., 15.
molar vs. molecular, 327. See also *appraisal, gestalt, order*.
Moniz, E., 368.
Montaigne, 297.
moral: consensus, 223; dynamism, see *dynamism, moral*; inversion, 231–5, partial, 233, spurious, 233–5; judgments, 214–16; neutrality of science, 153, 158; principles, 222–4, 334, 346; purpose, of Marxism, 231; reform of society, 222–3; standards, in society, 215–16.
morality, 133, 138, 180, 244, 380; & power, 142, 226–7; & science, 227–35.
Morand, P., 200n.
Morgan, C. Lloyd, 382.
morphogenesis, morphogenetic, 342, 356, 398; field, see *field, morphogenetic*; originality, 339; regulation, 338. See also *ontogenesis*.
morphological: concepts, 112; types, 383.
morphology, 352, 353, 357, 363, 373, 377. See also *taxonomy*.
mosaic principle, (ontogenesis) 338, 355, 356.
motion, Newtonian conception of, 10, 296; perpetual, 109, 249, 273, 332.
motoric learning, 71–2.
Mowrer, O. H., 71.
Murdoch, I., 102n., 113n.
music, 58, 193–5, 196, 199, 200, 319, 345.
Musil, R., 236.
mutation, 35, 159, 385, 402.
mysticism, 197–8; of Azande, 292.

Naef, A., 353n.
Nagel, E., 46.
Natural History, 353, 357, 360.
natural selection, 35, 40, 383–5, 402, 404n.
*Naturphilosophie*, 153–5.
Nazism, 298.
Needham, A. E., 355n.
Needham, J., 181n.
negative theology, 198–9.
neo-Darwinism, see *Darwinism, evolution*.
neo-Marxian theory of science, 238–9.
Neptune, 20, 30, 145, 181, 182n.
nervous system, 338, 340, 341, 369, 388, 398.

See also *mind, neural model, neurology*.
neural model of mind, 121, 158–9, 262–4, 340, 369, 382, 390, 398.
neurology, 121, 158–9, 262–4, 339, 344.
neurosis, experimental, see *experimenta neurosis*.
Nevill, W. E., 90n.
New Zealand, 182.
Newton, Newtonian, 5, 42, 104, 147, 148, 152, 164, 181, 277; gravitation, 5, 14, 20, 170; physics, 5, 8, 18–19, 20, 26, 36, 41, 63, 144, 148, 296, 306, 341, 390n., 392–3; space, 10, 11–12, 110, 114.
nihilism, 232, 234, 235–6, 268.
Nó, Lorente de, 340n.
nominalism, 113.
non-Euclidean geometry, see *geometry, non-Euclidean*.
noogenesis, 388, 389.
noosphere, 388–9, 393, 395, 396, 404.
Northrop, F. S. C., 358.
null hypothesis, 22–4, 26, 36, 260n.
numbers, 40–3, 144n., 164, 184, 186, 187, 192, 193, 194, 260–1, 392.

object: -creating science, 76 (see also *mathematical, mathematics*); -directed science, 76; theory, 258n.
objective: dynamo-, see *dynamo-objective*; vs. personal, 300.
objectivism, 15–17, 187, 214, 234, 239, 249, 253, 264, 265, 267, 268, 269, 275, 286, 288, 292, 304–6, 315, 323–4, 328, 350, 358, 371–3, 380, 381, 399. See also *Laplacean, mechanistic, objectivity, positivism*.
objectivity, 3–16; & contact with reality, vii, 5; & the personal, viii, 64, 113, 403; indeterminate implications of, 5, 43, 64, 104; of mechanistic world view, 8; & rationality in nature, 11, 15; vs. subjectivity, 15, 17, 48, 300; of measurement, 55; false ideal of, 136–7, 139–42, 144, 256; & art, 199; & Marxism, 228; in Soviet theory of science, 239; & doubt, 269–70; in psychology of learning, 370–373. See *objectivism, reality, theory*.
objects: opaque, 88–90; perception of, 96–7; & sense, data, 98–9; classes of, 114; of contemplation, 197; & machines, 329–330; vs. persons, 346; cf. 261–4.
obligation(s), 63, 65, 203, 324, 380, 386, 387, 403.
oblique use of words, 249–50.
observation: positivistic conception of, 9; sign-learning as, 76, 82; natural science =, 76, 328; & reading, 92; vs. contemplation, 98–9, 196–7; = affirmation, 99; vs. worship, 198, 279, 284; & personal judgment, 254; vs. understanding, 331, 346; in biological appraisal, 364; in psychology, 372; vs. indwelling, 378.
observing, 73, 76, 174–5, 184, 328.
obsessiveness, 363, 366, 374.
ontogenesis, 338, 355; & genic change, 383; human, 395, 397; & emergence, 399.

# Index

U.S.S.R., see *Soviet*.
Ullmann, S., 79n., 80n., 91n., 112n.
ultra-biology, 363, 377, 387, 404.
uncritical: aspect of tradition, 53; acceptance of scientific beliefs, 60; distinct from a-critical, 264.
undecidable sentences, 259, 260, 273.
understanding: deepened by discovery, 143; experiment clue to, 150; in mathematics, 184, 185, 186, 189–90; in music, 193; in abstract painting, 195; & contemplation, 195–6; & meaning, 250; of faith, through theology, 286; of machines, 331; of animal behaviour, 364; evolution of, 388. See also *appraisal, comprehension, conception, insight, intuition*.
Uniformity of Nature, 161–2.
uniqueness of man, 152, 285, 404–5.
universal: doubt, 294–8.
  intent: in appraisal of order, 37, 48; in commitment, 17, 32, 64–5, 301–3, 308, 309, 316, 324, 327, 343, 346, 396; in speech, 106, 265; of intellectual passions, 145, 150, 174; of parochial beliefs, 183, 203–4; in art appreciation, 201; of moral standards, 214, 215; of morality, & Marxism, 231; degrees of, 366; of induction, 370; of responsible commitments, 377; & superior knowledge, 379; in evolution, 389; development of, in child, 395.
  knowledge, Laplacean, 139–42.
  mathematics (Descartes), 8.
  pole of commitment, 313, 379, 396, 404.
  validity: of probability statements, 22; in mathematics, 189; of science, in Soviet theory, 238–9; in logic, 333; claim to, in discovery, 396.
universe, language = theory of, 80, 81, 94–5, 97, 112. See also *language*.
unknown: focus in problem-solving, 127–8; search for, 199, 395.
unmasking (Marxism), 229, 235, 238.
unspecifiability: of skill in science, 53, 55; of personal knowledge, 53, 62–3, 264, 343; of political maxims, 54; logical, 56, 63 (cf. 89–90); & heuristics, 77, 106; of 'consistency', 79; in denotation, 81; & ineffability, 88; of subsidiary knowledge, 88; of connotation, 112; of inventions, 337; in technology, 176; in learning language, 206; in confident use of language, 251; of mental control of machine, 262; of mind, 312; originality &, 336; in psychology & embryology, 342; of living shapes, 348–9; in taxonomy, 350, 351–2; double, of embryological knowledge, 357; in physiology vs. engineering, 359–360; in knowledge of animal behaviour, 364; of animal learning, 365; of biological achievements, 379, 399, 404; of randomness, 390; of emergent existence, 392; vs. ignorance, 392; of knowledge of emergence, 393–4; of patterns, 394; of entities in terms of particulars, 396; & comprehension, 398. See also *appraisal,*

*inarticulate, indeterminacy, indeterminate, tacit*.
Urey, H. C., 111.
utilitarianism, 180, 182, 192, 211, 232, 234, 239.

validation vs. verification, 121n., 201–2, 284, 321; cf. 15, 22, 42, 46–8, 170.
value, see *cultural value, scientific value*.
van der Waerden, B. L., 119n., 131n., 192n., 294n.
Vandel, A., 159.
Vavilov, N. I., 180.
vegetative commitments, 363, 377.
verbal: confusion, 108; error, 107.
verbalism, childish, see *childish*.
verification, 13, 20, 30, 64, 165, 167, 171, 172, 173, 202, 254, 284, 320, 321; in problem-solving, 121, 126.
Vesalius, 277.
visual perception, see *perception*.
vitalism, 358, 390.
vocabularies, 78–9, 80, 81; & appetite, 99; assured vs. marginal, 107; differences in, 112; rational, 114. See also *articulate, language, speech*.
Voltaire, 297.
Vossler, K., 102.

Waddington, C. H., 356n.
Wagner, R., 200.
Waismann, F., 95n., 113n., 307n.
Wakley, T., 52n.
Wald, A., 314n.
Wallas, G., 121.
Wallon, H., 56n.
Warburg, E., 136n.
Ward, S. W., 275n.
Wardlaw, C., 353n.
Wazyk, A., 244.
Weisgerber, J. L., 112n.
Weismann, A., 355.
Weiss, P., 338n., 340n., 355n., 356, 357.
Weissberg, A., 290n.
Weizsäcker, C. F. v., 154n.
Wertheimer, M., 125n.
Weyl, H., 7n., 16, 110n., 261n.
Wheeler, R. H., 314n.
White, E. W., 200n.
Whitehead, A. N., 28, 88n., 141.
Whittaker, E., 43n., 147n.
wholes: and meanings, 57–8, 63, 64; clues to (perception), 97–8; understanding of, 327, 344; & achievements, 381. See also *gestalt, molar, order*.
Wiame, Prigogine &, 384n.
Wigglesworth, V. B., 178n., 179n.
Williams, H., 52n., 275n.
Wilmott, A. J., 350n., 351.
Wislicenus, J. A., 155–6.
witchcraft, 93–4, 112–13, 168, 183, 287–94.
Wittgenstein, L., 87n., 113–14.
Wöhler, F., 137, 156–7, 275, 358.
Wolfe, B. D., 242n.
Wolynski, 147n.

427